Ceramic Coatings for
High-Temperature Applications

Ceramic Coatings for High-Temperature Applications

Editors

Amirhossein Pakseresht
Kamalan Kirubaharan Amirtharaj Mosas

Basel • Beijing • Wuhan • Barcelona • Belgrade • Novi Sad • Cluj • Manchester

Editors

Amirhossein Pakseresht
Coating Department, FunGlass
Alexander Dubcek University
of Trencin
Trencin
Slovakia

Kamalan Kirubaharan
Amirtharaj Mosas
Coatings Department, Funglass
Alexander Dubcek University
of Trencin
Trencin
Slovakia

Editorial Office
MDPI
St. Alban-Anlage 66
4052 Basel, Switzerland

This is a reprint of articles from the Special Issue published online in the open access journal *Ceramics* (ISSN 2571-6131) (available at: www.mdpi.com/journal/ceramics/special_issues/Cera_Coat).

For citation purposes, cite each article independently as indicated on the article page online and as indicated below:

Lastname, A.A.; Lastname, B.B. Article Title. *Journal Name* **Year**, *Volume Number*, Page Range.

ISBN 978-3-0365-9633-4 (Hbk)
ISBN 978-3-0365-9632-7 (PDF)
doi.org/10.3390/books978-3-0365-9632-7

© 2023 by the authors. Articles in this book are Open Access and distributed under the Creative Commons Attribution (CC BY) license. The book as a whole is distributed by MDPI under the terms and conditions of the Creative Commons Attribution-NonCommercial-NoDerivs (CC BY-NC-ND) license.

Contents

About the Editors . . . . **vii**

Dinesh Kumar Devarajan, Baskaran Rangasamy and Kamalan Kirubaharan Amirtharaj Mosas
State-of-the-Art Developments in Advanced Hard Ceramic Coatings Using PVD Techniques for High-Temperature Tribological Applications
Reprinted from: *Ceramics* **2023**, *6*, 301-329, doi:10.3390/ceramics6010019 . . . . **1**

Jing Liang, Marc Serra, Sandra Gordon, Jonathan Fernández de Ara, Eluxka Almandoz and Luis Llanes et al.
Comparative Study of Mechanical Performance of AlCrSiN Coating Deposited on WC-Co and cBN Hard Substrates
Reprinted from: *Ceramics* **2023**, *6*, 1238-1250, doi:10.3390/ceramics6020075 . . . . **30**

Vera Petrova, Siegfried Schmauder and Alexandros Georgiadis
Thermal Fracture of Functionally Graded Coatings with Systems of Cracks: Application of a Model Based on the Rule of Mixtures
Reprinted from: *Ceramics* **2023**, *6*, 255-264, doi:10.3390/ceramics6010015 . . . . **43**

Marina Fedorischeva, Mark Kalashnikov, Irina Bozhko, Tamara Dorofeeva and Victor Sergeev
Structural-Phase Change of Multilayer Ceramics Zr-Y-O/Si-Al-N under High Temperature
Reprinted from: *Ceramics* **2023**, *6*, 1227-1237, doi:10.3390/ceramics6020074 . . . . **53**

Francesca Curà, Raffaella Sesana, Luca Corsaro and Riccardo Mantoan
Characterization of Thermal Barrier Coatings Using an Active Thermography Approach
Reprinted from: *Ceramics* **2022**, *5*, 848-861, doi:10.3390/ceramics5040062 . . . . **64**

Ahmed Faramawy, Hamada Elsayed, Carlo Scian and Giovanni Mattei
Structural, Optical, Magnetic and Electrical Properties of Sputtered ZnO and ZnO:Fe Thin Films: The Role of Deposition Power
Reprinted from: *Ceramics* **2022**, *5*, 1128-1153, doi:10.3390/ceramics5040080 . . . . **78**

Florian Kerber, Magda Hollenbach, Marc Neumann, Tony Wetzig, Thomas Schemmel and Helge Jansen et al.
On the Statistics of Mechanical Failure in Flame-Sprayed Self-Supporting Components
Reprinted from: *Ceramics* **2023**, *6*, 1050-1066, doi:10.3390/ceramics6020062 . . . . **104**

Maria Berkes Maros and Shiraz Ahmed Siddiqui
Tribological Study of Simply and Duplex-Coated CrN-X42Cr13 Tribosystems under Dry Sliding Wear and Progressive Loading Scratching
Reprinted from: *Ceramics* **2022**, *5*, 1084-1101, doi:10.3390/ceramics5040077 . . . . **121**

Shiraz Ahmed Siddiqui and Maria Berkes Maros
Comparative Study on the Scratch and Wear Resistance of Diamond-like Carbon (DLC) Coatings Deposited on X42Cr13 Steel of Different Surface Conditions
Reprinted from: *Ceramics* **2022**, *5*, 1207-1224, doi:10.3390/ceramics5040086 . . . . **139**

Liubomyr Ropyak, Thaer Shihab, Andrii Velychkovych, Vitalii Bilinskyi, Volodymyr Malinin and Mykola Romaniv
Optimization of Plasma Electrolytic Oxidation Technological Parameters of Deformed Aluminum Alloy D16T in Flowing Electrolyte
Reprinted from: *Ceramics* **2023**, *6*, 146-167, doi:10.3390/ceramics6010010 . . . . **157**

Artemiy V. Aborkin, Dmitriy V. Bokaryov, Sergey A. Pankratov and Alexey I. Elkin
Increasing the Flow Stress during High-Temperature Deformation of Aluminum Matrix Composites Reinforced with TiC-Coated CNTs
Reprinted from: *Ceramics* **2023**, *6*, 231-240, doi:10.3390/ceramics6010013 179

Andrey Zayatzev, Albina Lukianova, Dmitry Demoretsky and Yulia Alexandrova
Tensile Adhesion Strength of Atmospheric Plasma Sprayed $MgAl_2O_4$, Al_2O_3 Coatings
Reprinted from: *Ceramics* **2022**, *5*, 1242-1254, doi:10.3390/ceramics5040088 189

A. V. Tyunkov, A. S. Klimov, K. P. Savkin, Y. G. Yushkov and D. B. Zolotukhin
Electron-Beam Deposition of Metal and Ceramic-Based Composite Coatings in the Fore-Vacuum Pressure Range
Reprinted from: *Ceramics* **2022**, *5*, 789-797, doi:10.3390/ceramics5040057 202

Aleksandr Klimov, Ilya Bakeev, Anna Dolgova, Efim Oks, Van Tu Tran and Aleksey Zenin
Electron-Beam Sintering of Al_2O_3-Cr-Based Composites Using a Forevacuum Electron Source
Reprinted from: *Ceramics* **2022**, *5*, 748-760, doi:10.3390/ceramics5040054 211

Yury Yushkov, Efim Oks, Andrey Kazakov, Andrey Tyunkov and Denis Zolotukhin
Electron-Beam Synthesis and Modification and Properties of Boron Coatings on Alloy Surfaces
Reprinted from: *Ceramics* **2022**, *5*, 706-720, doi:10.3390/ceramics5040051 224

Andrey O. Zhigachev, Ekaterina A. Agarkova, Danila V. Matveev and Sergey I. Bredikhin
CaO-SiO_2-B_2O_3 Glass as a Sealant for Solid Oxide Fuel Cells
Reprinted from: *Ceramics* **2022**, *5*, 642-654, doi:10.3390/ceramics5040047 239

About the Editors

Amirhossein Pakseresht

Amirhossein Pakseresht is an associate professor and head of coating department at FunGlass Centre for Functional and Surface Functionalized Glass, Alexander Dubček University of Trencin, Slovakia. He received his undergraduate study in materials science and engineering from Tehran University, which is the biggest and oldest university in Iran. Pakseresht completed his Ph.D. in materials science and engineering at the Materials and Energy Research Center (MERC). During his PhD, he also worked as a researcher in MERC and involved in so many industrial projects. After completion of his Ph.D. He joined the Tehran University as a postdoctoral research associate in the Department of material engineering. In 2019, he joined Amirkabir University (Tehran Polytechnic) as an adjunct research assistant professor and involved in so many projects about coating and composite. Generally speaking, his research interests lie in the area of plasma spraying, with a focus on splat morphology and new thermal barrier coatings. He also has collaborated actively in multidisciplinary materials science, particularly surface science, anti-corrosion coating and composite materials. Dr. Pakseresht authored two books titled *Production, Properties, and Application of High-Temperature Coatings* and *Handbook of Research on Tribology in Coating and Surface Treatment* in IGI Global, USA. He also authored a book titled *Physical Chemistry and Thermodynamic of Materials* (in Persian). He has published more than 65 journal papers, 5 patents and 5 international book chapters that were cited 2700 times (h-index 30) until 2023. He is a member of the editorial board and reviewer of more than 10 international scientific journals.

Kamalan Kirubaharan Amirtharaj Mosas

Kamalan Kirubaharan Amirtharaj Mosas is a researcher at Alexander Dubcek University in Trencin, Slovakia, developing superalloy coatings for high-temperature applications. His most recent findings are the design and development of multilayer diffusion barrier coatings based on Ni/YSZ to minimize corrosion and high-temperature oxidation in superalloys for nuclear vitrification components. He has a background in ceramic coatings for corrosion and high-temperature applications, ceramic matrix composites, and metal/ceramic thin films created using various physical vapor deposition (PVD) processes. He has prior experience with thermal evaporation, electron beam physical vapor deposition (EBPVD), pulsed laser deposition (PLD), and RF/DC magnetron sputtering systems. During the years 2017–2022, he obtained eight sponsored projects from the defence and aerospace sectors, as well as one institutional funding for product development and commercialization. He has 60 papers published with over 550+ citations and eight Indian patents. He has served as a guest editor for the *Ceramics*—MDPI and *Coatings*—MDPI journals and as a reviewer for a number of prestigious publications.

Review

State-of-the-Art Developments in Advanced Hard Ceramic Coatings Using PVD Techniques for High-Temperature Tribological Applications

Dinesh Kumar Devarajan [1,2,*], Baskaran Rangasamy [3] and Kamalan Kirubaharan Amirtharaj Mosas [4,*]

1. Centre for Nanoscience and Nanotechnology, Sathyabama Institute of Science and Technology, Jeppiaar Nagar 600119, Tamil Nadu, India
2. Sathyabama Centre for Advanced Studies, Sathyabama Institute of Science and Technology, Jeppiaar Nagar 600119, Tamil Nadu, India
3. Energy Storage Materials and Devices Lab, Department of Physics, School of Mathematics and Natural Sciences, The Copperbelt University, Riverside, Jambo Drive, P.O. Box 21692, Kitwe 10101, Zambia
4. Coating Department, FunGlass, Centre for Functional and Surface Functionalized Glass, Alexander Dubcek University of Trencin, 91150 Trencin, Slovakia
* Correspondence: ddinesh.tribology@gmail.com (D.K.D.); kamalan.mosas@tnuni.sk (K.K.A.M.)

Abstract: Hard and wear-resistant coatings created utilizing physical vapor deposition (PVD) techniques are extensively used in extreme tribological applications. The friction and wear behavior of coatings vary significantly with temperature, indicating that advanced coating concepts are essential for prolonged load-bearing applications. Many coating concepts have recently been explored in this area, including multicomponent, multilayer, gradient coatings; high entropy alloy (HEA) nitride; and functionally modified coatings. In this review, we highlighted the most significant findings from ongoing research to comprehend crucial coating properties and design aspects. To obtain enhanced tribological properties, the microstructure, composition, residual stress, hardness, and HT oxidation resistance are tuned through doping or addition of appropriate materials at an optimized level into the primary coatings. Such improvements are achieved by optimizing PVD process parameters such as input power, partial pressure, reactive gas flow rates, substrate bias, and temperature. The incorporation of ideal amounts of Si, Cr, Mo, W, Ag, and Cu into ternary and quaternary coatings, as well as unique multilayer designs, considerably increases the tribological performance of the coatings. Recent discoveries show that not only mechanical hardness and fracture toughness govern wear resistance, but also that oxidation at HT plays a significant role in the lubrication or wear failure of coatings. The tribo-induced metal oxides and/or Magnéli phases concentrated in the tribolayer are the key governing factors of friction and wear behavior at high temperatures. This review includes detailed insights into the advancements in wear resistance as well as various failure mechanisms associated with temperature changes.

Keywords: hard coatings; high temperature; HiPIMS; wear resistance; oxidation; tribolayer

1. Introduction and Need for Hard Coatings

Hard, protective coatings on moving mechanical assemblies are essential in modern industries to protect components operated under high-temperature (HT) and harsh environments to extend their lifetime and reduce wear-/corrosion-related losses. Tribological, hard coatings include metals, nitrides, carbides, oxides, and borides of transition metals [1,2]. Nitride-based hard coatings are frequently used as protective coating materials to protect from wear and corrosion and to extend the prolonged sustainability of high-speed cutting tools, aerospace components, and gas turbines owing to their superior properties, such as high hardness, corrosion resistance, high-temperature stability, low friction, and wear resistance [3,4]. More specifically, recent developments in Al-incorporated transition metal nitride (TMN) coatings have already demonstrated their role in the enhancement of mechanical and HT tribological properties. The thermal stability of TMN coatings is greatly influenced by the defect structure, interdiffusion, grain refinement, and solid solutions. These phenomena are scientifically significant since the resulting structure has a large impact on the mechanical and tribological properties of the coating [5]. Therefore, the development of advanced hard ceramic coatings can be achieved by tailoring the coating structure of single-layer coatings by adding suitable elements and fabricating novel coating architectures, such as multilayer and multicomponent nanocomposite coatings. Many key findings have indicated that tuning the microstructure of hard coatings has a considerable effect on friction and wear resistance properties [6–8]. Sabzi et al. [8] observed a significant reduction in friction (0.08) and wear for a Ni-W_2C (20 wt% W_2C) nanocomposite coating compared to a Ni coating (CoF: 0.72), owing to the dendritic microstructure with a uniform distribution of W_2C nanoparticles.

Generally, tribological coatings are classified into three major categories: soft coatings with hardness less than 10 GPa, hard coatings with hardness greater than 10 GPa, and superhard coatings with hardness greater than 40 GPa [9,10]. Ceramic coatings such as aluminum oxide, titanium nitride (TiN), and titanium carbide (TiC) have successfully been applied to cutting tools, providing enhanced protection against abrasive and diffusion wear at high temperatures, resulting in a more than ten-fold increase in lifetime. Ceramic coatings have higher mechanical properties than metal coatings because of the solid solution strengthening caused by gas atoms that create point defects in the metallic crystal lattice, preventing dislocation motion [11].

Ternary hard coatings, TiAlN and CrAlN, are commercially employed in a wide range of industrial applications due to their exceptional physical and chemical properties, which include high hardness, fracture toughness, and chemical stability [12]. Many studies have been conducted to overcome the higher friction behavior of these coatings by designing unique coating architectures and doping and alloying various elements without degrading the mechanical and thermal stability. Interestingly, Yang et al. [13] achieved enhanced tribological properties of AlCrN coatings fabricated on volcano-shaped textured surfaces. These unique structures substantially reduced the real contact of sliding surfaces as well as the wear particles trapped between the textured surfaces, resulting in less frictional force and wear. The sliding mechanisms were rather contradictory at high temperatures, where several fracture deformations and a lower influence of humidity, oxidation process, extreme wear, tribochemical alterations, softening, and third body influences were predominant. Such extreme conditions ultimately require coatings with superior withstanding ability at HT, extreme hardness, and a more precise ability to protect the component against severe wear and oxidations. Worldwide, numerous studies have been conducted on various hard ceramic coatings for high-temperature applications.

Previous research findings are useful for understanding these coatings and advancing the knowledge of utilizing them in various industrial applications. However, due to the huge volume of research, it may be difficult to identify the needed information regarding the selection and application of these coatings. Therefore, the current review focuses on collecting the most essential information required for researchers to understand the recent advancements in hard coatings to be employed in high-temperature applications. We provide a detailed overview of the most recent developments, especially in the last 5 years, regarding the design of hard coatings with the potential to be used in high-temperature applications.

Coating Contact/Failure Mechanisms at Elevated Temperatures

In recent years, numerous research reports have been made available on different hard coatings for elevated temperature tribological applications [14–18]. The tribological behavior of the coating under HT conditions is evidently different from that under room temperature conditions. Many research reports and a few potential review articles are available regarding the comparative analysis of RT and HT tribological properties of hard coatings; however, some results only display the RT tribological behavior [19,20]. It is highly necessary to understand the various factors of contact and/or failure mechanisms for coatings that interact with counterbodies at elevated temperatures for any potential application. Based on the literature, it is clearly demonstrated that different contact and/or failure mechanisms occur at the coating/counterbody contact interfaces, as illustrated in the flow diagram in Figure 1. The major wear mechanisms of coatings at elevated temperatures include wear particle adhesion on the interacting surface, abrasive wear, excessive oxidation, phase transformations, chipping, cutting, mechanical property weakening, and coating delamination from substrates [21–23].

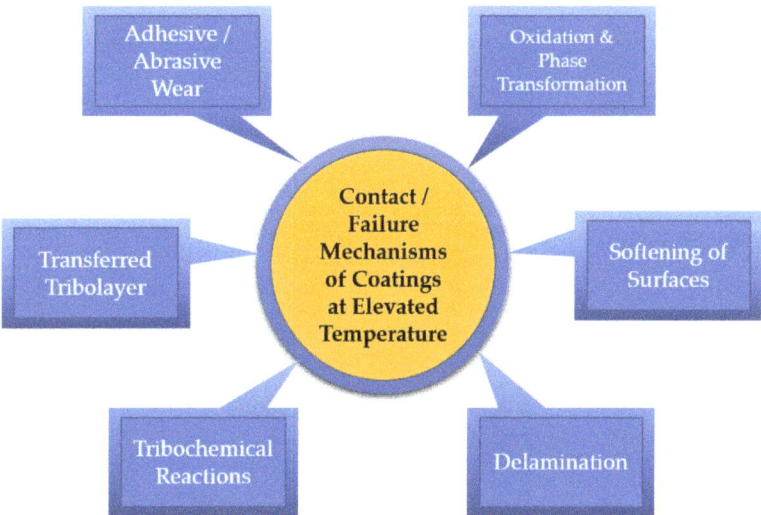

Figure 1. Various contact and failure mechanisms are involved during the tribological contact between the coating and counterbodies under high-temperature conditions.

2. Hard Refractory Ceramic Coatings for High-Temperature Applications

Nitrides and carbides based on transition metal nitrides have undergone extensive research and development over the last 20 years for use in high-temperature industrial applications. Titanium nitride (TiN) is the primary coating material used for HT applications because of its superior physiochemical and mechanical properties. However, the excessive oxidation of Ti at elevated temperatures (above 550 °C) leads to possible abrasive wear, and deterioration of mechanical hardness limits its applications [24–26].

To overcome this challenge, Al was incorporated into the TiN crystal lattice to create a stable TiAlN phase without changing the crystal structure (typically FCC). The operating temperature of the TiAlN coating was noticeably raised to 1000 °C with a significant increase in mechanical hardness. CrAlN is the other significant high-temperature protective coating material for industrial machining tools owing to its superior oxidation resistance, up to 900 °C. Al-based oxide layers that form on the coatings during HT exposure prevent further oxygen from diffusing inside the material, leading to high temperature stability [27–29]. Depending on the application requirements, the Al concentration was changed by up to 70% to extend the hardness and HT performance of the coatings [30,31]. However, the extremely high friction of these coatings at RT and HT settings severely restricts their use in a variety of industrial applications that demand low-friction properties.

3. Fabrication of Hard Ceramic Coatings

3.1. Fabrication Methods

Owing to the ease of operation and superior properties of the coatings, physical vapor deposition (PVD) techniques are the most popular for the deposition of hard ceramic coatings. The most preferable PVD methods for depositing hard coatings include magnetron sputtering, cathodic arc, pulsed laser deposition, and electron beam evaporation. The growth mechanisms of hard coatings are highly dependent on the deposition conditions, plasma kinetics, and physiochemical characteristics of the source elements. Figure 2 shows the many PVD techniques that have been employed over the last five years for fabricating hard coatings, as well as the common deposition parameters that affect the characteristics and functionality of the coatings. Among many PVD methods, magnetron sputtering methods are the most widely used techniques for fabricating hard coatings, owing to the ease of controlling the stoichiometry, microstructure, mechanical properties, and highly dense, defect-free, uniform structure of the coatings by controlling various processing parameters, such as ion energy, ion flux, sputtering power, substrate bias, gas flow rate, partial pressure, and substrate temperature [32].

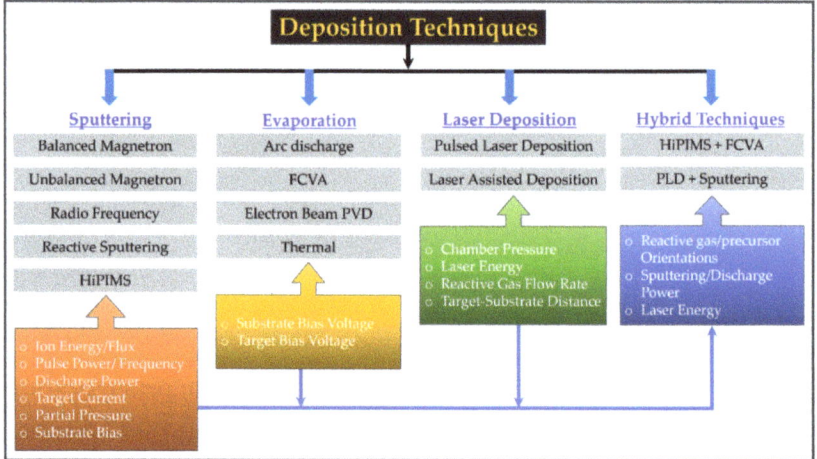

Figure 2. PVD methods to deposit hard ceramic coatings and the major process parameters influencing the coating properties.

Many types of advanced sputtering techniques, including balanced magnetron sputtering, radio frequency, reactive sputtering, and high-power impulse magnetron sputtering (HiPIMS), are frequently used for fabricating hard coatings. Owing to its ability to manufacture highly dense, stable, and hard coatings, HiPIMS is presently used as a flexible technology to deposit hard ceramic coatings for tribological applications [11,33–37]. Other potential techniques for fabricating materials for extreme environmental applications, such as filtered cathodic vacuum arc (FCVA) [38,39], pulsed laser deposition (PLD) [40,41], and their hybrid approaches with magnetron sputtering [42–44], have been used extensively in recent years. Figure 2 also includes the crucial parameters for tuning the microstructure and mechanical properties of the coatings produced by various PVD techniques.

The schematic representations of current coating techniques are shown in Figure 3, including filtered cathodic vacuum arc (FCVA), hybrid vacuum arc with RFMS, hybrid radio frequency magnetron sputtering (RFMS), HiPIMS, and ultrashort PLD. For the fabrication of ZrSiN coatings, Chang et al. [45] employed the co-sputtering approach of RFMS/HiPIMS (as depicted in Figure 3a). Owing to the combined ability of forming a denser coating morphology by HiPIMS and grain refinement by the addition of Si through RFMS, the residual stress of the coatings was reduced, and the resulting increase in hardness and elastic modulus were the most promising properties for use in extreme environmental applications. Pulsed laser deposition is another popular method for creating hard coatings (Figure 3b) [46,47]. Atoms from the source material are evaporated in this process using a high-energy, ultra-short, pulsed laser beam. As the atoms migrate toward the substrate, they produce a homogeneous layer. Furthermore, stoichiometry control from the source material to the coatings is superior using the PLD method compared to the other techniques, enabling the development of unique coatings with the desired compositions [48].

Figure 3c shows a schematic of the FCVA deposition system, where the substrates (anode) are mounted inside the vacuum chamber, while the target (cathode) is mounted at another end of the filter arch. To avoid larger particle and droplet formation in the coating, typically C-bend or S-bend arches are used in the FCVA system. Cao et al. [49] developed a multilayer coating of Ti-DLC on an Al alloy substrate using the FCVA method that exhibited ultralow friction (0.12), improved wear resistance (2.69 × 10^{-7} mm^3/Nm) at 300 °C, and decreased thermal conductivity, which could be beneficial for engine piston assemblies. However, hard coatings produced by cathodic arc methods frequently contain surface droplets that cause friction and wear during the initial period of sliding. Panjan et al. [50] revealed that post-polishing procedures for cathodic arc-deposited TiN coatings could improve the wear resistance and shorten the running-in duration. Another interesting fabrication method is the combined vacuum arc and magnetron sputtering hybrid process described by Hirata et al. [51], as shown in Figure 3d, for creating amorphous boron carbon nitride (a-BCN) films for tribological applications. Here, the BN target is mounted on the RFMS, while the graphite target is positioned on the cathodic arc gun to deposit the films. The use of a vacuum arc in the hybrid technique enhances the ionization rate and the kinetic energy of sputtered atoms/particles which bombard the substrate. Moreover, the diagonal placement of the arc gun in relation to the substrate allows for better control of droplet deposition in the coatings, a common problem associated with conventional vacuum arc processes.

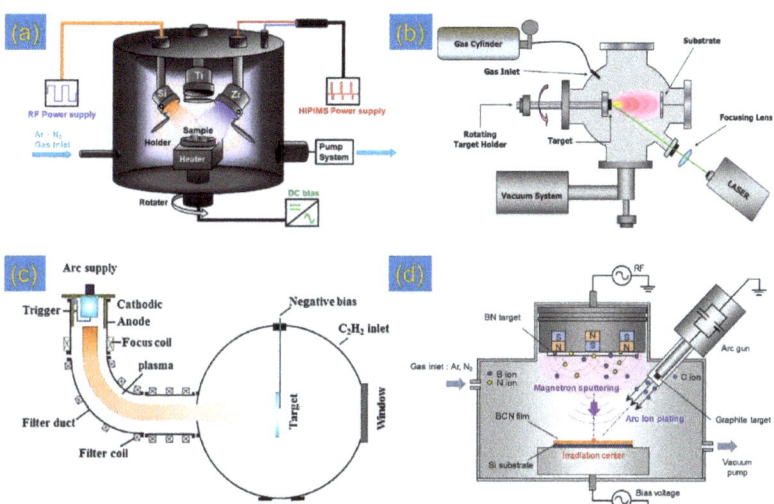

Figure 3. Schematic of the advanced PVD deposition methods used for the fabrication of cutting-edge hard coatings for tribological applications. (**a**) Co-sputtering method of RFMS and HiPIMS [45], (**b**) ultra-short PLD [46], (**c**) FCVA [49], and (**d**) arc-sputtering hybrid method [51]. (Reproduced with permission.)

3.2. Binary, Ternary, and Multicomponent Ceramic Coatings

Coating microstructures can be engineered to improve their properties by tailoring the deposition parameters of PVD techniques. Binary and ternary coatings have been most widely studied over the past few decades for a variety of industrial applications, including small-scale devices and large-scale mechanical components. In the last ten years, significant breakthroughs in coating concepts have been developed and investigated in the search for improved characteristics and long-lasting coatings for extreme environmental applications. Multicomponent, multilayer, nanocomposite, and functionally graded coatings have received much attention globally during the past five years [52–56]. Recent research trends indicate a notable interest in the development of multicomponent coatings by co-depositing or evaporating compound target materials in an atmosphere of reactive nitrogen (N_2). The introduction of elements into the crystalline lattice was developed to refine the grain growth and thereby enhance the mechanical strength and wear resistance in extreme sliding conditions [57,58]. For instance, Rodríguez et al. [59] observed a hardness enhancement from 32 GPa to 36 GPa for a small fraction (0.2 at%) of Hf doped into $Al_xTi_{1-x}N$ coatings to form c-$Al_{0.64}Ti_{0.36}Hf_{0.02}N$ coatings using the cathodic arc method. They suggested that the oxidation temperature of the coating was increased to 900 °C owing to the addition of Hf, which forms Hf-based oxynitrides and Al/Ti-based oxide layers during HT exposure. On the other hand, a study by Grigoriev et al. [60] showed that Ti-TiN-(Ti, Al, Nb, and Zr)N multicomponent coatings significantly enhanced the wear resistance and reduced the friction and performance of cutting tools with increasing the temperature from 700 to 900 °C.

On the other hand, the multilayered coatings showed remarkably improved properties, such as hardness, fracture toughness, and elastic modulus, compared to the single-layer coatings. It was revealed that a number of multilayer structures with varied compositions between sublayers improved the mechanical properties. The interface/superlattice structures between two alternate layers in this case impede the mobility of plastic deformation, acting as dislocation gliding barriers [61,62]. Xiao et al. [63], for example, achieved a maximum hardness of 38.3 GPa and an elastic modulus of 463.7 GPa, as well as the highest thermal stability (up to 1000 °C) and wear resistance at 800 °C, for the AlCrN/TiAlSiN multilayer compared to single-layer coatings. The multilayer structure, as well as the unique nanocomposite structure of the Si_3N_4 matrix around the TiAlN crystals from the TiAlSiN layers, were responsible for these properties. Similar enhanced mechanical and tribological properties of different multilayer architectures, such as CrAlN/TiSiN [64], AlCrSiN/VN [65], TiAlSiN/VSiN [66], and AlCrN/TiAlSiN [63], have been reported. The consolidated mechanical and HT tribological performances of multicomponent and multilayer coatings are described in Table 1.

Table 1. Recent developments in hard coatings for high-temperature tribological applications from publications in the last five years.

Coating	Deposition Method/Post Treatments	Thickness (μm)	Mechanical Properties (GPa)	Tribological Properties at HT	Major Outcomes	Ref.
Multicomponent and nanocomposite hard ceramic coatings fabricated using PVD techniques for HT tribological applications						
AlTiVCuN	HiPIMS	1.0–1.6	H: 34–41 GPa	μ: 0.5 k: 3.2×10^{-15} m^3/Nm (600 °C)	Cu rich coating involved more outwards diffusion of Cu to form the lubricious CuO at HT	[67]
W-DLC	Hybrid DCMS + HiPIMS	1.7	E: 200 GPa	μ: 0.1 k: 2×10^{-7} mm^3/Nm (150 °C)	As a result of very compact morphology of nanocomposite coatings, detachment of larger hard W–C particles is prevented, resulting in low-friction and higher wear resistance	[68]
Al$_2$O$_3$/ta-C	Lateral Arc with Central Sputtering and ALD for Al$_2$O$_3$ layer	(ta-C) and 200 nm (Al$_2$O$_3$ top layer)	–	μ: 0.1 (500 °C) Wear Volume: 1.4×10^{-3} mm^3 (500 °C)	The suppression of oxidation by a thin Al$_2$O$_3$ multifunctional layer improves the thermal stability and durability of the ta-C coating.	[69]
(Cr, V)N	Cathodic Arc ion-plated	4.5	H: 24 GPa	μ: 0.28–0.37 Wear Volume: $1.4–12.9 \times 10^{-5}$ mm^3/Nm (700–900 °C)	Because of the formation of V-O phases, Cr$_{0.58}$V$_{0.14}$N$_{0.28}$ coatings demonstrated superior tribological properties at 700–800 °C	[70]
AlCrON, and α-(Al,Cr)$_2$O$_3$	Cathodic Arc	4.0	H: 34.6 GPa E: 467 GPa (AlCrON) H: 26 GPa E: 446 GPa (α-(Al,Cr)$_2$O$_3$)	μ: 0.5 k: 150×10^{-17} m^3/Nm (AlCrON at 800 °C) μ: 0.25 k: 10×10^{-17} m^3/Nm (α-(Al,Cr)$_2$O$_3$ at 800 °C)	Superior tribological properties of the α-(Al,Cr)$_2$O$_3$ coating due to the nitrogen-free, stable alpha-alumina structure that inhibited HT oxidation and subsequent wear	[22]
Cr-V-Al-C	Hybrid Arc and Magnetron Sputtering	7.5	H: ~22.5 GPa E: ~280 GPa	μ: 0.5 k: No measurable wear (900 °C)	The formation of a combined Cr$_2$O$_3$ and Al$_2$O$_3$ tribolayer can be favored by optimizing the solid solution content of V in Cr$_2$AlC coating, leading to high hardness and HT tribological performance	[71]
MoCuVN	HiPIMS	2.1–2.4	H: 19.0–15.5 GPa E: 393–316 GPa (with increasing N$_2$ flow rate)	μ: 0.43–0.51 k: $3.1–13.5 \times 10^{-8}$ mm^3/Nm (400 °C)	At 400 °C, formation of mixed lubricious oxides of MoO$_3$/CuMoO$_4$ and V$_2$O$_5$ decreases the wear resistance compared to tests conducted at RT due to the loss of N and severe oxidation at HT	[72]
Multilayer hard ceramic coatings fabricated using PVD techniques for HT tribological applications						
CrAlN/TiSiN	Arc Ion Plating	6.8	2850 HV	μ: 0.6 k: 3.95×10^{-6} mm^3/Nm (300 °C)	Coatings are highly stable at high temperatures, while the adhesive wear of the ball on the coating surface forms the Fe$_2$O$_3$ tribolayer	[73]
TiAlSiN/VSiN	RF magnetron co-sputtering	1.0–1.2 (10–40 nm bilayer periods)	H: 29 GPa E: 260 GPa	μ: 0.28 k: 7.01×10^{-6} mm^3/Nm (700 °C)	Improved tribological properties due to improved mechanical properties and the formation of a self-lubricating V$_2$O$_5$ Magnéli phase at 700 °C	[74]
CrMoN/SiN$_x$	RF magnetron co-sputtering	1.2 (1 nm SiN$_x$ and 10–200 nm CrMoN)	H: 27 GPa E: 200 GPa (for 100 nm CrMoN/1 nm SiN$_x$)	μ: 0.22 k: 1.68×10^{-5} mm^3/Nm (600 °C)	The optimal thickness of bilayer periods and laminated architecture for stress dispersal and deflection results in improved performance	[66]
AlCrSiN/VN	Arc Ion Plating and Pulsed DC Sputtering	3.0 (4.6 nm modulation period)	H: 30.7 GPa (Multilayer) H: 28 GPa (AlCrSiN) H: 20.5 GPa (VN)	μ: 0.26 (800 °C) k: 2.6×10^{-15} m^3/Nm (600 °C) & 39.4×10^{-15} m^3/Nm (800 °C)	Outwards diffusion pf V and Al to form V$_2$O$_5$ and AlVO$_4$ phases results in low friction and increased wear resistance at HT	[65]
AlCrN/TiAlSiN	Cathodic Arc	2.2	H: 38 GPa E: 463 GPa	μ: 0.45 (800 °C) k: 2.5×10^{-6} mm^3/N m (800 °C)	Enhanced wear resistance is provided by the formation of a dense Al$_2$O$_3$ oxide lubricant layer on the wear tracks	[63]
TiN-AlTiN/nACo-CrN/AlCrN-AlCrOAlTiCrN	Cathodic Arc	3.6	H: 36–41 GPa	μ: 0.45 (800 °C)	The formation of stable protective oxides ((Al,Cr)$_2$O$_3$) increases the wear resistance at 800 °C compared to 600 °C	[75]

3.3. Significance of the Structural and Mechanical Properties of Ceramic Coatings

The tribological properties of hard coatings are primarily governed by several internal factors of coatings, including compositional, structural, microstructural, and mechanical properties, as well as other external factors, such as the sliding atmosphere, load, sliding speed, and humidity. The microstructure of the coating can be tailored to enhance the mechanical and HT properties depending on the concentration of added components in the primary phases of the coating. For the $Al_{60}Cr_{30}Si_{10}$ compound target, Fan et al. [76] observed a gradual change in Al and Si composition with a periodic change in sputtering power from 0.6 to 1.2, then to 2.0, and finally to 2.8 kW. They also observed the variation in crystallinity and microstructure of quaternary CrAlSiN coatings with the change in sputtering power from 0.6–1.2, then to 2.0, and finally to 2.8 kW. The coating samples deposited at 0.6–2.8 kW displayed poor wear resistance due to weaker crack resistance and higher surface roughness, whereas the highly dense and smooth surface coatings deposited at 0.6–2.0 kW exhibited improved wear resistance. To compare the microstructure and morphologies of CrN coatings, Ferreira et al. [77] developed two different coatings under varying process conditions of N_2 gas flow, partial pressure, and bias voltage. With a low N_2 gas flow and partial pressure with a higher bias voltage, a highly dense, void free, enhanced hardness microstructure coating with low friction (0.15) and wear loss was achieved. A second coating was deposited at higher N_2 and lower bias voltage, and the friction and wear loss were higher because of the presence of more voids inside the coating. Substrate rotational speed during the sputtering deposition of hard coatings also has a significant impact on the microstructure and mechanical properties of hard coatings. With increasing rotation speed, the dense microstructures are transferred to the coarse microstructure. The film growth direction and the mobility of adatoms during sputtering are greater at lower rotational speeds due to the almost perpendicular position of sputtered ions to the film growth surface, resulting in a dense microstructure. At higher speeds, the angle of growth of the coating structure predominantly varies, reducing the energy of sputtering ions [78]. Wang et al. [79] investigated the effect of reactive N_2 partial pressure on TiBN/TiAlSiN nano-multilayered coatings fabricated by the cathodic arc technique. The coating deposited at 2.0 Pa partial pressure had a defect-free, smoother surface and a dense microstructure, resulting in improved hardness (34 GPa), H/E, and H^3/E^{*2}; low friction (0.28); and wear resistance (4.3×10^{-7} mm^3/Nm). A higher partial pressure (3.0 Pa) caused target poisoning and a reduction in coating thickness.

Similarly, changing the bias voltage causes significant changes in the tribological properties of cathodic arc-deposited ternary and quaternary nitride coatings [80,81]. Cao et al. [82] observed a change in the preferred orientation from (200) to (111) for TiAlN coatings deposited using FCVA when the bias voltage was increased from 50 to 75 V. This is due to an increase in atomic mobility and lattice distortion, which results in a higher hardness (~30.3 GPa) and wear resistance (~4.4×10^{-5} mm^3/Nm). Akhter et al. [83] achieved a remarkable 85% increase in wear resistance for arc-produced TiNiN coatings at a 100 V bias voltage rather compared to the coating deposited at 0 V. The enhanced mechanical and tribological properties in this case were due to the very fine equiaxed structure and higher compressive residual stress. A similar improvement in tribological performance was observed for CrN/NbN multilayers deposited via HiPIMS with substrate biases ranging from −40 to −150 V [84]. The coating deposited at −65 V demonstrated reduced friction and enhanced wear resistance owing to the increased density of columnar grains. However, a higher substrate bias caused grain coarsening and increased defect concentration, which increased friction and wear. It is clear that the coating microstructure, thickness, composition, roughness, hardness, and toughness are clearly influenced by the applied substrate bias voltage. Therefore, the optimum bias voltage during PVD coating deposition plays a critical role in reducing internal stress and improving load-bearing ability. Interlayer thickness, on the other hand, has a significant impact on the residual stress and wear properties of hard coatings, according to Lin et al. [85]. They observed that increasing the Ti interlayer thickness of TiZrN coatings from 50 to 250 nm decreased the residual stress from −5.67 to −3.75 GPa, and the wear resistance increased by 16%.

4. Recent Progress in Advanced Coatings for Application in Harsh Environments

Many researchers around the world have designed and developed several hard coatings in the last five years to improve load-bearing ability under extreme environmental conditions. The majority of researchers are concerned with tailoring the microstructure by adding appropriate elements to achieve adaptive coatings with multifunctional properties. Many researchers are interested in multicomponent coatings and novel multilayer structures, as previously discussed. Table 1 shows the recently developed advanced coatings deposited using PVD methods, as well as their mechanical and tribological properties, particularly under harsh operating conditions.

4.1. Recent Advancements in Coating Design Aspects

The improved design of hard coatings shows beneficial effects on the mechanical and tribological properties of coatings through controlling the microstructure, varying the composition, altering the nanocomposite design, and combining hard/soft multilayer coatings. Figure 4 depicts the various coating architectures that have been created using PVD methods for use in HT environments. For the selection of hard coatings to extend the performance and life of the base components, there are several primary concerns. Considering the extreme mechanical hardness of the coating, the influence of internal stress would limit the coating's adhesion properties to the substrate, resulting in delamination during highly stressed sliding operations. Therefore, special coating architectures are highly inevitable, with specific properties at different zones for HT tribological applications. Figure 4 depicts the coating concept with desirable properties at different zones, such as higher adhesion strength between the coating and substrate (zone 1), a mechanical layer with high fracture toughness and wear resistance in the middle (zones 2 and 3), and active surface layers (zone 4) that should provide lubricity and superior oxidation resistance properties.

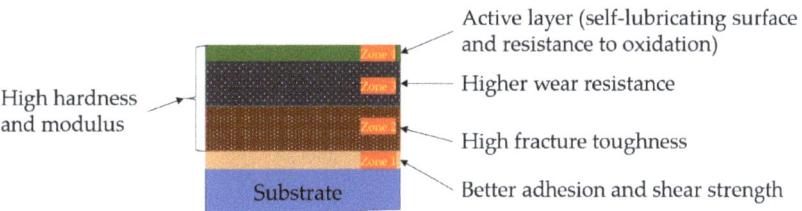

Figure 4. Schematic representation of the advanced coatings with important properties at different zones for high-temperature tribological coatings.

Many researchers have reported different coating architectures using PVD techniques in the last five years, taking into account the above coating requirements for extreme environmental applications. Figure 5 depicts the most recent coating architectures studied for HT tribological applications from the literature. Haung et al. [86] designed gradient composite TiAlSiN coatings and achieved improved tribological performance as well as superhardness (42 GPa) and superior adhesion strength (85 N). Figure 5a depicts the three-layer microstructure of this coating, where TiN and AlTiN provide mechanical stability, while the formation of SiO_2 surface oxides from the top TiAlSiN layer improves lubrication and protection efficiency. In another study, multicomponent gradient (Ti, Al, Si, Cr, Mo, S, O, and N) coatings (Figure 5b), fabricated using UBMS, demonstrated improved tribological performance due to the presence of many nanocrystalline phases, which effectively impeded the deformation during sliding [87]. Bondarev et al. [88] developed TiSiN/TiN(Ag) multilayer coatings using DCMS with a total thickness of 2.3 μm and a bilayer thickness of 40 nm for tribological applications. Chang et al. [89] fabricated novel multilayer AlTiN/CrN/ZrN coatings using a cathodic arc, as shown by the cross-sectional TEM micrograph, for tribological applications. They observed that fabricating an AlTiN/CrN/ZrN multilayer greatly reduced residual stress, resulting in superior wear resistance (4.21 × 10^{-7} mm^3/Nm).

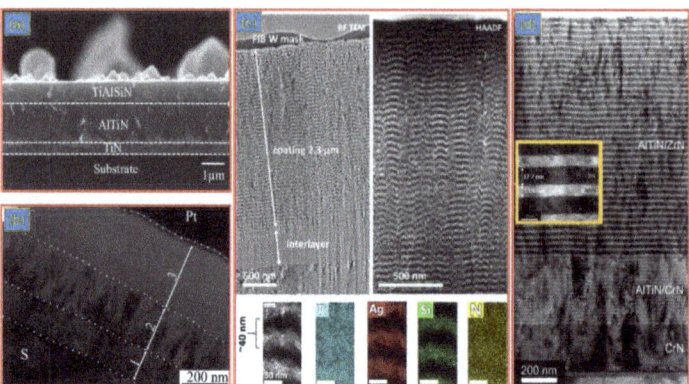

Figure 5. Cross-sectional microstructure of various coating architectures in the recent literature. (**a**) FESEM micrograph of a three-layer TiN/AlTiN/TiAlSiN coating [86], (**b**) bright field HRTEM image of a gradient layered structure (Ti, Al, Si, Cr, Mo, S, O, and N) coating [87], (**c**) bright field TEM, HAADF STEM, and EDS elemental distribution of TiSiN/TiN(Ag) multilayer coatings [88], and (**d**) TEM micrograph of novel nanoscale multilayered CrN/[AlTiN/CrN]$_n$/[AlTiN/ZrN]$_n$ coatings [89]. (Reproduced with permission.)

As a result, in recent studies, the main consideration to improve the tribological performance of hard ceramic coatings is mainly controlled by tailoring the coating microstructure by introducing additional elements and/or fabricating new coating architectures. The selected additive materials should possess solubility in the primary coating to some extent, be able to form strong protective oxides on the coating surface during HT sliding conditions, control the inwards diffusion of oxides, and promote self-lubricating effects. In the case of multilayer coatings, the interface strengthening mechanism with periodic changes between sublayers exhibits enhanced resistance to plastic deformation and wear even at high temperatures. Several metals, nonmetals, and soft metals have been considered for this purpose in recent research, and the advancements in these nanocomposite and multicomponent coatings for HT applications are described in the following sections.

4.2. Role of Dopants (Mo, Cr, W, Si, and C) in Hard Ceramic Coatings

As a result of their superior mechanical stability over a wide operating temperature range, nanocomposite coatings have recently received much attention for high-temperature tribological applications. However, there is still a problem associated with the lack of high-temperature lubricity of nanocomposite coatings compared to at room temperature, resulting in poor wear lifetime. To address this issue, transition metal-based dopants such as Ti, Mo, V, and W, as well as nonmetallic elements such as Si, C, and B, have been used to improve the tribological performance at elevated temperatures. The formation of the Magnéli phase of transition metal oxides with layered crystal structures during exposure of these metals to high-temperatures provides an improved load-bearing ability at high temperatures. These layered structures slip more easily during sliding motion, providing enhanced lubrication due to their attenuated Van der Waals forces between each layer's crystal. For example, Tao et al. [90] discovered that adding Mo to magnetron-sputtered CrAlSiN coatings reduced the friction with increasing Mo concentration. They observed that the stable MoO$_3$ species, observed in the tribolayer in ESCA analysis, which provides the self-lubrication properties of nanocomposite CrAlMoSiN coatings at 600 °C, is more stable than the CrAlSiN coatings.

Another study found that adding Mo and C to the TiN crystalline lattice significantly improved the mechanical and tribological properties. In this case, Mo addition promoted hardness and toughness by increasing the crystallite charge density as well as forming Magnéli phases, and C addition promoted the formation of amorphous carbon (a-C) phases at grain boundaries, resulting in friction reduction and wear resistance [91]. Similarly, Cr-rich additions to TiAlN-coated tools provided a greater cutting performance above 600 °C due to the formation of a protective Cr-O tribolayer, which is the strongest tribolayer, compared to Al-O and Ti-O gained from TiAlN coatings at higher cutting speeds [92]. Similar behavior of WO_3 triboinduced oxides at high temperatures for the addition of W in magnetron-sputtered HfN coatings with diverse additive compositions revealed minimal friction and wear resistance at high temperatures [93]. However, excessive amounts of these transition metals are prone to oxidative damage at high temperatures.

As previously stated, oxidation of coatings plays a critical role in the formation of tribolayers on interacting surfaces in the HT environment. Metal oxides, depending on their nature, may be prone to abrasive and third body wear, as well as some protective oxides that are stable under extreme temperature conditions. For example, TiO_2 formation from TiAlN coatings under HT (typically above 600 °C) provides lubricity, but these oxides have poor wear resistance. In contrast, the formation of Al- and Cr-based oxides in CrAlN coatings provides superior wear resistance at extremely high temperatures (900 °C) but with relatively high friction [94,95]. In this regard, adding a small fraction of Si to TiAlN coatings may prevent Ti from diffusing out of the coating, thereby improving oxidation and wear resistance. The formation of a stable Si_3N_4 structure in the TiAlSiN coating, depending on the Si concentration, significantly improves the mechanical and tribological properties. However, excessive Si addition will cause the coatings to become brittle, resulting in increased friction and wear.

Drnovšek et al. [96] investigated the correlation between HT mechanical and tribological properties for the addition of Si in CrAlN coatings using the magnetron sputtering method. The mechanical and tribological properties of the CrAlSiN coatings were superior to those of CrAlN coatings from RT to HT (up to 700 °C). However, the mechanical properties decreased with increasing temperature during indentation (H: 37 GPa to 36 GPa for CrAlSiN and 31 GPa to 24 GPa for CrAlN). Figure 6a depicts the comparison of the observed friction and wear rates for both coatings tested from RT to 700 °C. The friction increases in both coatings at intermediate temperatures up to 500 °C due to the formation of wear debris and third body wear. Comparatively, wear predominates in the CrAlN coatings, as shown in Figure 6b, where abrasive wear of oxides and three body wear predominate. However, wear resistance was superior for the CrAlSiN coatings, as shown in Figure 6c, due to the enhanced resistance to oxidation and a critical H/E* ratio of 0.08–0.085 at HT, resulting in improved wear resistance. As a result, they suggested that optimizing the H/E* ratio of the coatings could be the critical governing parameter for the wear resistance characteristics.

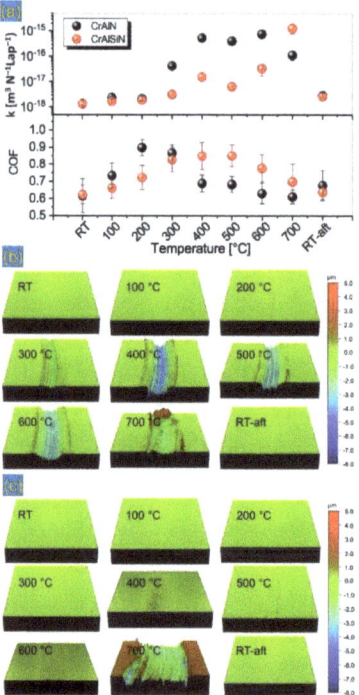

Figure 6. (**a**) Friction curves of CrAlN and CrAlSiN coatings tested under inert atmosphere at elevated temperature (RT to 700 °C). 3D optical profiler images of the wear tracks of (**b**) CrAlN and (**c**) CrAlSiN coatings after testing from RT to 700 °C [96]. (Reproduced with permission.)

Cai et al. [97] investigated the effect of B addition to AlCrN coatings using the arc ion plating method for high-temperature tribological applications. For this, they fabricated a coating with Cr and CrN layers on the substrate, followed by an AlCrN layer, then by the incorporation of B in the top layer with different B contents of 1.5 at%, 2.9 at%, and 4.8 at% to obtain the AlCrBN top layer. They found that a solid solution of B in the AlCrN columnar structure at 1.5 at% and a small fraction of a-BNx, as seen in the TEM microstructure and schematic (Figure 7a,d), resulted in superior hardness (38.3 GPa) and modulus (622 GPa). When the B content exceeded 2.9 at%, composite (Al, Cr, and B)N nanograins and a-BNx composite structures were formed and crystallinity decreased, as shown in Figure 7b,c,e,f. This was followed by a decrease in the mechanical properties to 26 GPa (H) and 389 GPa (E*). The HT tribological results showed that the B content of 4.8 at% had superior wear resistance at 800 °C due to the formation of continuous tribofilms on the wear tracks. Figure 7g depicts the different tribological mechanisms of AlCrBN (4.8 at%) coatings with respect to temperature change based on the wear track micrograph and tribochemical analysis. Continuous Cr_2O_3 tribofilms were observed at 400 °C, while at 600 °C, continuous Al_2O_3 and Cr_2O_3 with a small fraction of discontinuous $B_2O_3/B(OH)_3$ tribofilms were obtained, and at 800 °C, a stable and continuous $B_2O_3/B(OH)_3$ tribofilm was formed as a protective layer of the coating from wear. Korneev et al. [98] also observed improved mechanical, HT tribological, and oxidation resistance with C and N doping, when they fabricated TaZrSiBCN hard coatings through a sputtering technique.

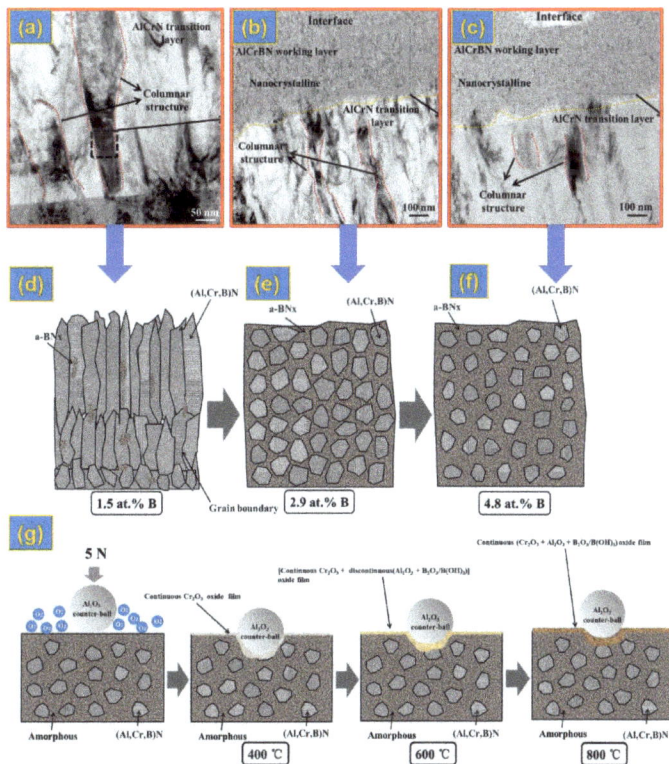

Figure 7. TEM microstructure and corresponding schematic representation of the AlCrBN coatings with B contents of (**a**,**d**) 1.5 at%, (**b**,**e**) 2.9 at%, and (**c**,**f**) 4.8 at%. (**g**) Schematic of the wear mechanism of the AlCrBN coating with 4.8 at% B at different temperatures (RT—800 °C) [97]. (Reproduced with permission).

4.3. Soft/noble Metal (Ag and Cu)-Doped Hard Coatings for High-Temperature Applications

Another promising coating design concept for elevated-temperature tribological applications is the incorporation of soft/noble metals into hard ceramic coatings. In recent years, metals such as Ag, Au, and Cu have been incorporated as a second phase as well as a lubricant phase into hard nitride coatings to form a nanocomposite structure (nc-MeN/metal) to improve high-temperature tribological properties. These materials not only have improved fracture toughness and hardness due to grain refinement but also have an excellent lubrication effect due to out-diffusion of soft metals at high temperatures [99]. The morphologies and microstructures of soft metals (Ag and Ag-Cu) doped into hard coatings for tribological applications are depicted in Figure 8a–c. The microstructure of the coatings substantially varies depending on the dopant concentration, beginning with the formation of a solid solution in the primary crystal lattice and progressing to nanosized grains at lower concentrations. The microstructure of the coating is altered or amorphized with higher concentrations of soft metals (Ag and Cu), and larger particles are then out diffused onto the coating's surface. Figure 8d illustrates the change in the microstructure with varying Ag content from 0 to 25.3 at%; the columnar microstructure disappears, and the amorphous structure predominates, which accelerates wear at high temperatures [100].

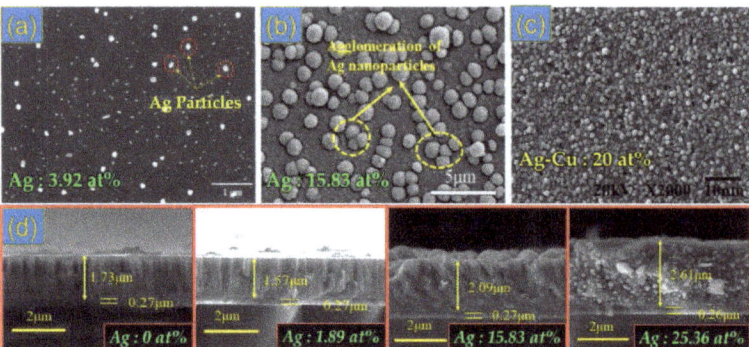

Figure 8. Microstructure of soft metal doping/addition to hard ceramic coatings. SEM morphologies of (**a**) CrMoSiCN coatings with Ag doping (3.92 at%) [101], (**b**) nanocomposite NbN-Ag coatings (Ag content of 15.83 at%) [102], and (**c**) Ag-Cu incorporated TiAlN coatings (Ag-Cu contents of 20 at%) [102]. (**d**) Growth morphologies of NbN-Ag (Ag content 0, 1.89, 15.83, and 25.36 at%) [100]. (Reproduced with permission.)

Recently, Rajput et al. [103] conducted a comparative tribological performance analysis of CrAlN coatings with Ag additions ranging from 2.4 at% to 15.6 at%. They observed that adding Ag up to 8.6 at% increased the mechanical hardness from 18 GPa to 23 GPa due to grain refinement, but it tended to decrease (to 14.4 GPa) with further Ag addition due to the enrichment of softer phases. The friction and wear resistance characteristics did not change significantly at room temperature, but there was a more than two-fold reduction in low-friction values for coatings with higher Ag concentrations (9–16 at%) at elevated temperature (600 °C), as shown in Figure 9a,b. Such enhancement in lubricity is anticipated from the Ag- and $AgCrO_2$-rich tribolayers on the wear tracks of these samples, as represented in the FESEM and EDX analysis (Figure 9c–g). The self-lubricating properties are caused by the outwards diffusion of Ag at high temperatures. However, the wear resistance and mechanical properties of higher Ag content (12 and 16 at%) samples were drastically reduced due to the excessive presence of soft phases, as seen in the wear micrographs and EDS results. They found that the optimal addition of Ag (9 at%) shows low friction and improved wear resistance at HT (600 °C) due to the formation of Ag and $AgCrO_2$ (as seen in Figure 7e).

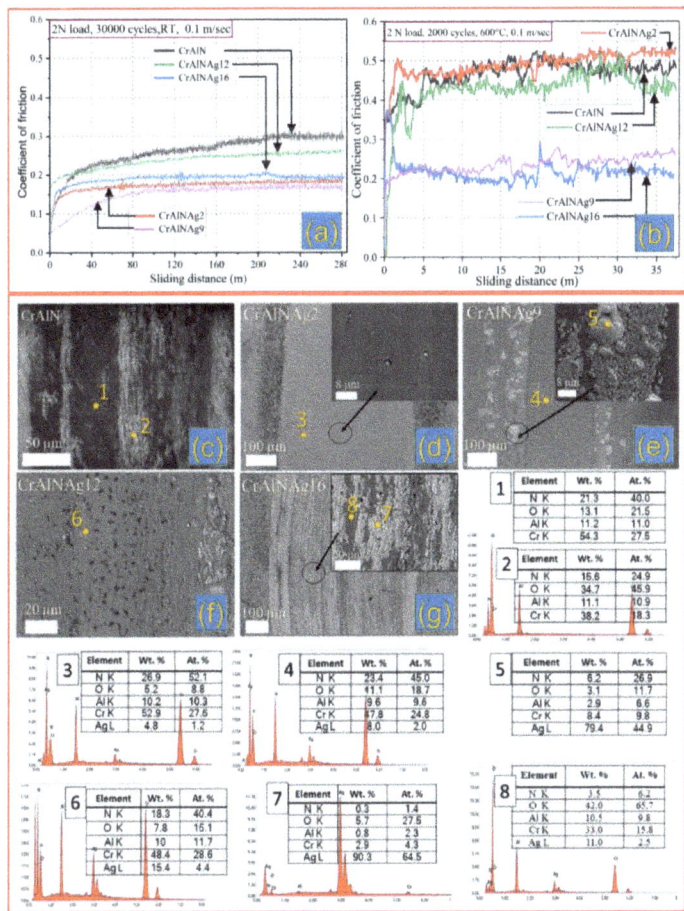

Figure 9. Friction, wear behavior, and tribochemical analysis of CrAlN and CrAlN(Ag) coatings. Variation in the friction coefficient of CrAlN and CrAlN(Ag) coatings tested at (**a**) RT and (**b**) 600 °C, and (**c–g**) wear micrographs of the bare CrAlN and CrAlN(Ag) with different Ag content coatings tested at 600 °C. The EDX spectra (1–8) represent the chemical information of wear tracks for the coatings tested at 600 °C [103]. (Reproduced with permission).

Similarly, various studies have revealed that the optimal concentration of soft/noble metals in primary ceramic hard coatings has a beneficial effect on elevated temperature applications [104–107]. Similar observations on Cu incorporated into AlTiVN coatings using HIPIMS demonstrated that the friction (0.45) and wear resistance (10−16 m^3/N·m) were significantly enhanced at the optimum Cu concentration of 10.7 at% at high temperature (600 °C), as reported by Mei et al. [99]. The friction-induced oxide species, AlVO$_4$, was primarily transferred to the lubricious Al$_2$O$_3$, V$_2$O$_5$, and CuO oxides with the addition of Cu, improving the friction and wear characteristics at elevated temperatures. On the other hand, the addition of soft metals not only improved the tribological performance of hard coatings under HT but also reduced the internal stress of the primary coatings, providing further enhancement of the desired properties. Experimental results of Ren et al. [100] on Ag incorporation into NbN coatings (Figure 10) by sputtering show a significant reduction in residual stress, which improves the coating adhesion strength, as illustrated in Figure 10a,b. The change in stress with the addition of Ag from 0–25.3 at% has a direct influence on the friction and wear behavior of NbN-Ag coatings at different temperatures, as represented in Figure 10c,d. In HT sliding conditions, the soft metals promote the oxides, and the soft phases of these metal-rich tribolayers have self-lubricating effects. Table 2 shows a recent literature review on the properties of soft metals with hard coatings specifically for HT applications.

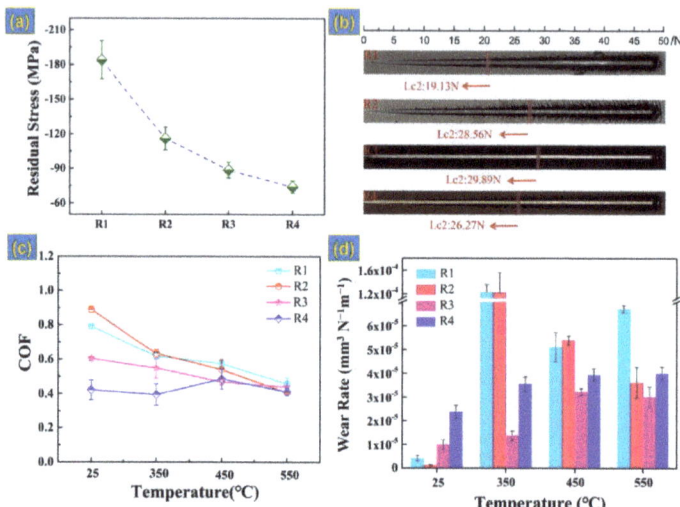

Figure 10. (**a**) Change in residual stress with the addition of Ag to NbN nanocomposite coatings at different concentrations of 0 at% (R1), 1.89 at% (R2), 15.83 at% (R3), and 25.36 at% (R4) and their corresponding (**b**) adhesion strength, (**c**) coefficient of friction, and (**d**) wear rates [100]. (Reproduced with permission).

Table 2. Recent results on Ag- and Cu-doped coatings for high-temperature tribological applications from publications in the last five years.

Coating and Dopants	Deposition Method/Post Treatments	Thickness (μm)	Mechanical Properties (GPa)	Tribological Properties at HT	Major Outcomes	Ref.
MoN–Ag (Ag: 0, 2.2, 7.9, 17.3 at%)	DC/RF magnetron sputtering	3.6–4.4	H: 14.4 GPa E: 232 GPa (for 2.2 at% of Ag)	μ: 0.27 (700 °C) k: 2.52×10^{-6} mm^3/Nm (for 2.2 at% of Ag)	The formation of lubricating oxides (MoO$_3$, Ag$_2$MoO$_4$, and Ag$_2$Mo$_4$O$_{13}$) reduce the friction coefficient, but wear resistance decreases above 300 °C.	[108]
TiSiN(Ag) (Ag: 0–17 at%)	HiPIMS	2.2–2.8	H: 20 GPa E: 218 GPa (for 6 at% of Ag)	μ: 0.5 k: no wear (600 °C)	At 600 °C, the tribolayer consists of superficial Ag and the adhesive material from the ball counterpart to form a stable protective layer, reducing friction and causing no noticeable wear.	[104]
NbN–Ag (Ag: 0, 2.62, 15.83, 25.36 at%)	UBMS	1.5–3.0	H: 14 GPa E: 261 GPa (for 2.62 at% of Ag)	μ: 0.4 k: 3.24×10^{-5} mm^3/Nm (for 15.83 at% (Ag) at 550 °C)	The reduction in friction and the wear rate at 550 °C for the Ag (15.83 at%) sample is due to the formation of tribo-induced compacted glaze tribolayer, which is primarily composed of Nb$_2$O$_5$ and AgNbO$_3$.	[100]
Al-Ti-V-Cu-N (Cu: 6.2, 8.0, 10.2, 10.7 and 11.7 at%)	HiPIMS	1.1–1.5	H: 35.2 GPa (for 6.2 at% of Cu)	μ: 0.45 (600 °C) k: 10^{-16} m^3/Nm (600 °C)	The formation of predominant V$_2$O$_5$ and CuO lubricating oxide phases on worn surfaces results in low friction and wear at 600 °C.	[99]
Mo(Cu)N (Cu: 0, 5.5, 7.5, 17.8 and 24.3 at%)	Magnetron Sputtering	1.5	H: 25 GPa E: 359 GPa (for 5.5 at% of Cu, i.e., 9.2% of Cu/(Cu and Mo))	μ: 0.4 k: 3.5×10^{-5} mm^3/N m (800 °C)	The formation of strong oxide phases, CuMoO$_4$ and MoO$_3$, at 600 °C results in low friction and wear resistance, whereas CuO predominated at 200 °C.	[106]
TiAlN (Ag, Cu) (Ag and Cu: 0, 11, 16, 17 and 20 at%)	Magnetron Sputtering	2.0	H: 15.2 GPa E: 216 GPa (for 11 at% of Ag and Cu) H: 6.7 GPa E: 140 GPa (for 20 at% of Ag and Cu)	μ: 0.25 k: 7.7×10^{-5} mm^3/N m (for 17 at% of Ag and Cu)	Friction and wear reduction were due to the solid lubrication effect of out-diffused Au-Cu nanoparticles up to 17 at% in TiAlN coatings	[102]

4.4. Functionally Modified Coatings

As previously discussed, the addition of Si to the CrAlN and TiAlN ternary ceramic coatings improves their high-temperature stability and wear and corrosion resistance properties. In the most promising structures of CrAlSiN and TiAlSiN coatings, a small amount of Si is used to substitute Cr/Ti atoms, causing lattice distortion due to the different atomic sizes. Additionally, the amorphous Si$_3$N$_4$ matrix structure surrounding the crystalline phases regulates the grain growth, resulting in the superhard properties of these coatings. However, these solid solution and amorphous matrix structures increase the residual stress and brittle behavior of the coatings, leading to delamination of the coatings from substrates. To address these issues, coating structures are modified to finetune the microstructure to functionalize the coatings from the surface towards the substrate using gradient composition coatings. The elimination of sharp interfaces in the gradient coatings is extremely beneficial for reducing the internal stress of the coatings and improving the adhesion strength and wear resistance. According to Lu et al. [109], the adhesion strength and cutting properties of magnetron-sputtered TiAlSiN coatings were superior due to the newly fabricated out-of-plane gradient distribution of Si in the coatings. In this gradient structure, the Si content was higher on the TiAlSiN coating surface and decreases towards the substrate, effectively reducing the internal coating stress.

It can be clearly presumed that oxidation and tribolayer formation are the most crucial factors in determining the mechanical and tribological properties of hard coatings. Almost all of the literature shows that abrasive wear of sliding surfaces occurs during the initial stage and that oxidation of wear particles and coating/sliding counterbodies occurs in the majority of the samples tested under HT conditions. Depending on the temperature and nature of oxidative elements present in the coatings, the metal oxide tribolayer is the critical governing factor of friction and wear behavior. Therefore, before studying tribological experiments, some studies have performed pre-oxidation treatments of hard coatings at elevated temperatures in an air atmosphere. In these treatments, the oxygen species will diffuse on the coatings and form oxidized surface layers to a certain depth.

For example, Lim et al. [110] conducted a comparative analysis of AlSiTiN, CrAlTiN, and CrAlSiTiN coatings with oxidized coatings annealed at 800 and 900 °C in an air atmosphere. Interestingly, the oxidized samples had lower friction and wear rates due to the formation of lubricious SiO_2 layers on the surface, which provided better protection against friction and wear. The machining performance of oxidized coated tools was improved due to the high presence of protective SiO_2 against flank wear. Surface oxides with improved hardness and low friction are formed on the sliding interfaces of Cr-, Al-, Ti-, and Si-based hard coatings during post-deposition annealing processes. As a result, coatings with surface oxide layers can easily form smooth and dense adhesive layers, potentially reducing friction and wear [111]. Other remarkable functional coating concepts from recent studies to enhance tribological performance have been achieved by designing compositionally gradient coatings [109,112–114], novel multilayer coatings [73,75,89,115, 116], and nanocomposite coatings [60].

4.5. High Entropy Alloy-Based Nitride Coatings

High entropy alloys (HEAs) are the most recently developed multimetallic alloys, in which five or more metallic elements are mixed with equal atomic ratios for the structural components in extremely high temperature and corrosive environments. The alloying elements are mixed to form a single-phase alloy based on the requirements. For instance, Chen et al. [117] reported the study of AlCrNiTiV amorphous coatings fabricated using the FCVA technique and their oxidation at 400–800 °C. The coatings annealed at 600 °C demonstrated higher hardness (22.6 GPa) and lower friction coefficients (0.22) (Figure 11a) and wear rates (1.1×10^{-5} mm^3/Nm) in the tribological test performed at 600 °C. The 800 °C annealed samples have extremely high oxide contents, such as Al_2O_3, $NiCr_2O_4$, and Cr_2O_3, which protect the coatings from further oxidation; however, the sudden decrease in mechanical properties results in poor tribological performance, as shown in Figure 11a. On the other hand, nitrides of these HEAs have recently been developed and their superior hardness and high-temperature stability have been demonstrated, making them suitable for harsh environmental applications. Recently, Li et al. [118] investigated the effect of nitrogen content on the mechanical and tribological properties of $(MoSiTiVZr)N_x$, using a confocal magnetron sputtering technique with varying N_2 flow rates. Remarkably, a superhardness of 45.6 GPa, an elastic modulus 408 GPa, and low friction (~0.3) (Figure 11b) were attained for the N concentration of approximately 53.7 at%. The wear micrographs also show that this coating has wear resistance due to its enhanced resistance to plastic deformation with almost 50% N content present in the coating.

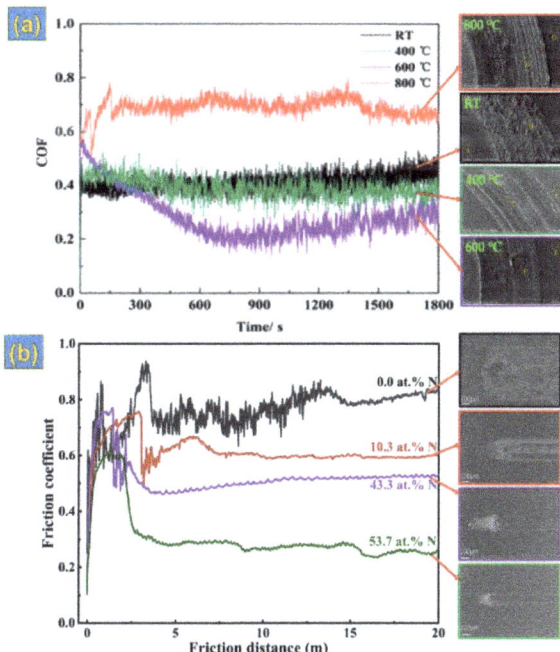

Figure 11. Tribological properties of HEA-based nitride coatings: (**a**) friction curves of AlCrNi-TiV amorphous HEA coatings annealed at different temperatures (400–800 °C) tested from RT to 800 °C [117] and (**b**) (MoSiTiVZr)N_x coatings deposited using DC sputtering with varying N_2 contents of 0, 10.3, 43.3, and 53.7 at% [118]. (Corresponding wear morphologies of the HEA coatings with respect to change in temperature and N content). (Reproduced with permission).

Similar enhancements in wear as well as corrosion resistance properties were observed for other HEA nitride coatings, such as (AlCrMoSiTi)N_x deposited using FVCA [119], TiZrNbTaFeN deposited using HiPIMS [120], (AlCrTiZrHf)N deposited using reactive magnetron sputtering [121], and AlCrMoZrTi/(AlCrMoZrTi)N multilayer coatings deposited using RF magnetron sputtering techniques [122]. Similar to other ceramic coatings, the residual stress plays a critical role in the mechanical and tribological properties of HEA nitride coatings, and many reports have demonstrated the optimization of substrate bias to reduce the residual stress. Lo et al. [123] investigated the effect of substrate bias on the fabrication of (AlCrNbSiTiMo)N using RF magnetron sputtering; the results revealed that the coating deposited with a (−100 V) substrate bias exhibited a higher hardness of 34.5 GPa and the lowest wear rate of approximately 1.2×10^{-6} mm^3/Nm at 700 °C. Furthermore, the presence of the MoO_3 Magnéli phase with the addition of Mo in these HEA nitride coatings also benefits lubrication under HT sliding conditions. The higher the substrate bias, the denser the structure and smoother the coatings, which improves the wear resistance of HEA nitride coatings [124].

5. Failure Mechanisms of Hard Ceramic Coatings Tested under HT Sliding Conditions

The friction and wear behavior of hard coatings are primarily related to the physical, chemical, and mechanical properties at RT and elevated temperatures. Surface roughness and composition also play important roles in friction, wear debris formation, and coating deformations during initial sliding contacts. As shown in Figure 12a, asperities from both contacting surfaces interact with each other, determining the initial frictions [125]. When the friction energy is exceeded during continuous sliding motions, wear of one or both of the surfaces occurs, followed by a change in friction behavior. Depending on the nature of the wear particles, the friction decreases when the wear particles are lubricated, and friction increases when the wear particles have difficulty producing scratches. Coatings are plastically deformed as a result of continuous sliding interactions, as locally generated frictional heat decreases the mechanical properties and accelerates the loose wear particles, resulting in coating material loss. Therefore, one of the most important criteria for harsh environmental wear protection is the selection of ceramic coatings with smoother surfaces.

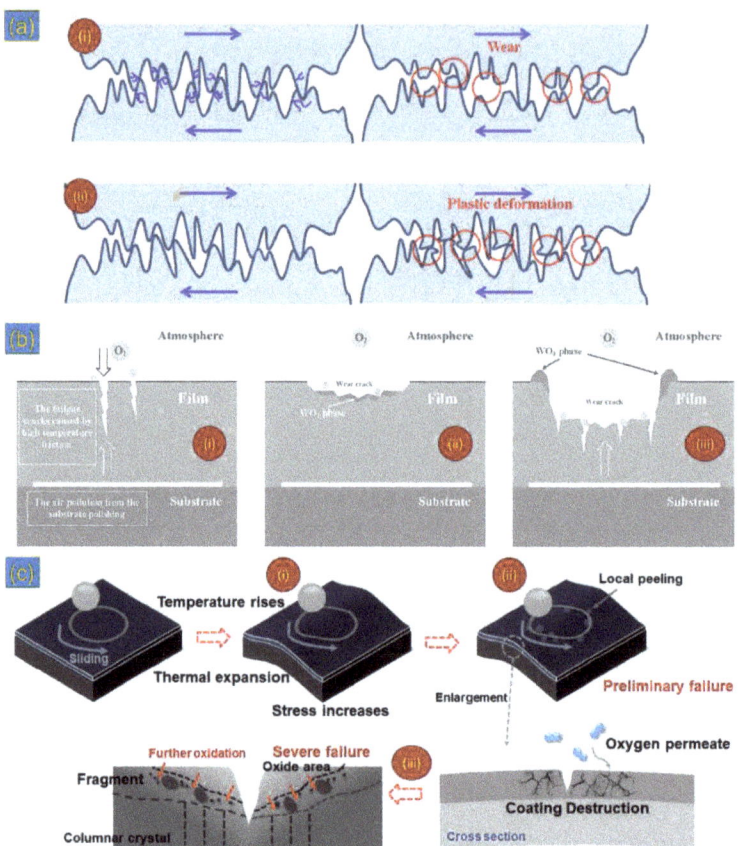

Figure 12. Different wear mechanisms of coatings under elevated temperature conditions. (a) Schematic of wear and deformation mechanisms [125], (b) cracking and oxidation of $Hf_{1-x}W_xN$ coatings during high-temperature tribological experiments [93], (c) wear mechanism of TiSiN coatings tested at elevated temperature [126]. (Reproduced with permission).

Oxidation is the most unavoidable factor in determining the friction and wear mechanism of the coatings, especially at elevated temperatures. The formation of cracks during tribological interactions promotes oxidation at high temperatures. To represent the oxidation behavior, Yu et al. [93] illustrated a schematic of the oxidation process of $Hf_{1-x}W_xN$ coatings with different W contents, as shown in Figure 12b. The crack initiated and developed at lower W (x = 0.37) content coatings at 600 °C, due to the lower fracture toughness. Therefore, the atmospheric oxides enter through the crack regions and predominantly cause oxidation across the coating. Higher W additions (x = 0.73) promote higher fracture toughness as well as WO_3 lubricating oxides, resulting in a lower friction coefficient and improved wear resistance at elevated temperatures (up to 600 °C). With the addition of

W (x = 1.0), the coating became severely oxidized and the surface became more brittle, resulting in oxidative wear and deformation.

On the other hand, oxidation of coatings at high temperatures is more common in columnar structured coatings, which degrades the coating composition and causes total coating wear failure. Figure 12c depicts a schematic representation of the wear life failure mechanism of columnar coatings. Generally, three steps of failure mechanisms are involved for such columnar coatings: (i) First, thermal expansion of the coating and substrate occurs with increasing temperature (stage 1, Figure 12c(i)). (ii) Second, higher contact stress at HT conditions damage the surface and create cracks and fragments in the coating, allowing oxygen to easily penetrate and form unstable and shapeless oxides (stage 2, Figure 12c(ii)). (iii) Finally, oxygen atoms easily propagate through the columnar grains, and the wear continues with further sliding, resulting in total wear loss of the coating (stage 3, Figure 12c(iii)) [126]. Therefore, highly dense microstructures would be greatly advantageous for HT tribological applications; such coatings can be designed using novel coating structures such as nanocomposite, multicomponent, and multilayer coatings.

Tribochemical Layer Formation and Failure Mechanisms of Coatings under HT Conditions

Most of the research articles used in this present review demonstrate that the friction and wear behavior of hard coatings under RT to elevated temperatures is highly dependent on the nature of the tribolayer. Figure 13 depicts the tribological mechanism and formation of tribolayers in various nanocomposite coatings tested at RT and elevated temperatures. Figure 13a represents the variation in tribolayer formations of CrMoSiCN/Ag nanocomposite coatings with respect to Ag content (0.83, 1.64, and 2.51 at%). At lower Ag concentrations, the tribolayer, composed primarily of SiO_2, transfers wear debris from Si_3N_4 balls, as well as the oxide tribochemical species Cr_2O_3 and MoO_3 from the coating surface (Figure 13a(i)). With increasing Ag, the tribolayer consists of mild SiO_2 from the ball, Cr_2O_3 and MoO_3 oxide species from the coating, and a small fraction of Ag nanoparticles that are subjected to abrasive wear (Figure 13a(ii)). At higher Ag concentrations, a tribolayer composed of enriched Ag lubricating layers with Cr_2O_3, MoO_3, and Ag_2MoO_4 species, as well as transfer on the ball surface, results in combined tribochemical and abrasive wear behavior, as illustrated in Figure 13a(iii) [101].

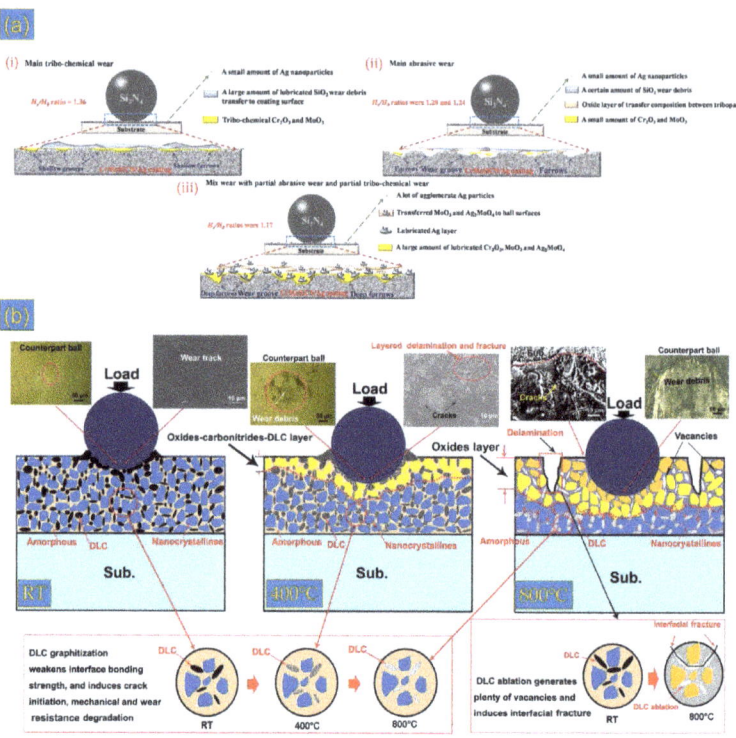

Figure 13. Schematic of different tribological mechanisms of nanocomposite coatings and the role of the tribolayer at RT and HT in an air atmosphere. (**a**) RT tribological behavior of nanocomposite Cr-MoSiCN/Ag coatings with different Ag contents: (1) 0.83 at%, (2) 1.64 at%, and (3) 2.51 at% [103] and (**b**) wear mechanisms of TiAlSiCN coatings at RT and elevated temperatures (400 and 800 °C) [127]. (Reproduced with permission).

Changes in the wear behavior of nanocomposite coatings and their tribological mechanisms have been discussed by Guo et al. [127] for TiAlSiCN coatings tested at RT–800 °C. The coatings are composed of nanocrystalline (Ti,Al)(C,N), diamond-like carbon (DLC), and amorphous Si_3N_4/SiC phases. Under RT conditions, the segregation of DLC at the sliding interfaces resulted in low friction and wear resistance; however, at 400 °C, the segregated DLC species started to graphitize, and carbonitride formation in the tribolayer began to deteriorate the coating fracture toughness. Extremely high graphitization and oxidative damage of coatings tested at 800 °C resulted in the severe interfacial fracture of coatings and further increased oxidation, resulting in reduced wear resistance. It can be demonstrated that carbon-rich coatings are beneficial under RT conditions, but excessive graphitization degrades the mechanical properties. Wang et al. [70] reported a similar trend in oxide forms on (Cr, V)N coatings at intermediate temperatures. Up to 600 °C, the coatings showed low friction and enhanced wear resistance properties, at 700–800 °C, the wear rate increased progressively with increased oxidation, and at 900 °C, coating breakdown due to high oxidative wear was observed.

6. Conclusions

In the search for hard, wear-resistant coatings for HT tribological applications, numerous research articles are published every day by researchers from all over the world. Many advanced PVD techniques have been used to create cutting-edge coating architectures with improved microstructures, grain refinement, hardness, and toughness, and their HT wear and oxidation properties for load-bearing applications have been tested. This review examined the most recent published research articles on state-of-the-art coating architectures and their HT tribological properties. HiPIMS, FVCA, and their hybrid deposition techniques have been most recently used for the deposition of hard coatings through optimization of power, substrate bias, temperature, and partial pressure in their respective methods. Reactive as well as co-deposition methods are promising routes for these PVD methods for fabricating novel multicomponent, nanocomposite, multilayer, and gradient coatings in current research. Highlights of the innovations in recent findings are summarized below with future research aspects.

- Appropriate addition of Si and B to the TiAlN and CrAlN coatings promotes the formation of a lubricating layer consisting of SiO_2 and $B_2O_3/B(OH)_3$, which provides lower friction and wear resistance at 800–900 °C. The a-Si_3N_4 and a-BN_x matrices around the ceramic nanocrystallites strengthen the coatings due to grain refinement. A similar effect was observed for Mo and V addition due to the formation of Mo-O and V-O Magnéli phase oxides in the tribolayer at HT (>700 °C);
- Multilayer coatings of binary, ternary, and quaternary nitride layers with nanoscale bilayer thickness exhibit remarkably high hardness (>30 GPa) and wear resistance under HT conditions. The combined protective surface oxide formation and multilayer structure restrict crack propagation, and further oxidation results in enhanced wear resistance;
- Gradient coatings with Si-rich surface layers of hard coatings demonstrate improved lubrication and ceramic nitride mechanical strength towards the substrate. Pre-oxidation of nitride coatings also favors lubricity; however, excessive oxidation deteriorates the mechanical properties;
- Appropriate soft metal (Ag and Cu) additions exhibit interesting low friction and wear resistance behavior at intermediate temperatures (up to 500 °C) due to the formation of an out-diffused Ag- and Cu-rich tribolayer;
- At very high temperatures, various coating failure mechanisms are related to the coating microstructure, compactness, and resistance to HT deformations, as well as excessive oxidation.

Author Contributions: Conceptualization, D.K.D. and K.K.A.M.; methodology, D.K.D. and B.R.; validation, B.R. and K.K.A.M.; resources, D.K.D.; data curation, B.R.; writing—original draft preparation, D.K.D. and B.R.; writing—review and editing, K.K.A.M.; supervision, D.K.D.; funding acquisition, D.K.D. All authors have read and agreed to the published version of the manuscript.

Funding: D.K.D. acknowledges the Department of Science and Technology-Science and Engineering Research Board (DST-SERB), Government of India, for a sponsored project under the Core Research Grant Scheme (Sanction order number: CRG/2018/001448). This work was also created in the frame of the project Centre for Functional and Surface Functionalized Glass (CEGLASS), ITMS code is 313011R453, operational program Research and innovation, co-funded from the European Regional Development Fund (K.K.A.M.).

Institutional Review Board Statement: Not applicable.

Informed Consent Statement: Not applicable.

Data Availability Statement: Not applicable.

Acknowledgments: The authors thank the management at Sathyabama Institute of Science and Technology, Chennai.

Conflicts of Interest: The authors declare no conflict of interest.

References

1. Fox-Rabinovich, G.S.; Yamamoto, K.; Beake, B.D.; Gershman, I.S.; Kovalev, A.I.; Veldhuis, S.C.; Aguirre, M.H.; Dosbaeva, G.; Endrino, J.L. Hierarchical Adaptive Nanostructured PVD Coatings for Extreme Tribological Applications: The Quest for Nonequilibrium States and Emergent Behavior. *Sci. Technol. Adv. Mater.* **2012**, *13*, 043001. [CrossRef] [PubMed]

2. Kuo, C.-C.; Lin, Y.-T.; Chan, A.; Chang, J.-T. High Temperature Wear Behavior of Titanium Nitride Coating Deposited Using High Power Impulse Magnetron Sputtering. *Coatings* 2019, *9*, 555. [CrossRef]
3. Ul-Hamid, A. Deposition, Microstructure and Nanoindentation of Multilayer Zr Nitride and Carbonitride Nanostructured Coatings. *Sci. Rep.* 2022, *12*, 5591. [CrossRef] [PubMed]
4. Aissani, L.; Alhussein, A.; Zia, A.W.; Mamba, G.; Rtimi, S. Magnetron Sputtering of Transition Metal Nitride Thin Films for Environmental Remediation. *Coatings* 2022, *12*, 1746. [CrossRef]
5. Kirnbauer, A.; Kretschmer, A.; Koller, C.M.; Wojcik, T.; Paneta, V.; Hans, M.; Schneider, J.M.; Polcik, P.; Mayrhofer, P.H. Mechanical Properties and Thermal Stability of Reactively Sputtered Multi-Principal-Metal Hf-Ta-Ti-V-Zr Nitrides. *Surf. Coat. Technol.* 2020, *389*, 125674. [CrossRef]
6. Mersagh Dezfuli, S.; Sabzi, M. Deposition of Self-Healing Thin Films by the Sol–Gel Method: A Review of Layer-Deposition Mechanisms and Activation of Self-Healing Mechanisms. *Appl. Phys. A Mater. Sci. Process.* 2019, *125*, 557. [CrossRef]
7. Mersagh Dezfuli, S.; Sabzi, M. Deposition of Ceramic Nanocomposite Coatings by Electroplating Process: A Review of Layer-Deposition Mechanisms and Effective Parameters on the Formation of the Coating. *Ceram. Int.* 2019, *45*, 21835–21842. [CrossRef]
8. Sabzi, M.; Dezfuli, S.M.; Far, S.M. Deposition of Ni-Tungsten Carbide Nanocomposite Coating by TIG welding: Characterization and Control of Microstructure and Wear/Corrosion Responses. *Ceram. Int.* 2018, *44*, 22816–22829. [CrossRef]
9. Rapuc, A.; Simonovic, K.; Huminiuc, T.; Cavaleiro, A.; Polcar, T. Nanotribological Investigation of Sliding Properties of Transition Metal Dichalcogenide Thin Film Coatings. *ACS Appl. Mater. Interfaces* 2020, *12*, 54191–54202. [CrossRef]
10. Kumar, D.D.; Kumar, N.; Kalaiselvam, S.; Dash, S.; Jayavel, R. Wear Resistant Super-Hard Multilayer Transition Metal-Nitride Coatings. *Surf. Interfaces* 2017, *7*, 74–82. [CrossRef]
11. Ghailane, A.; Makha, M.; Larhlimi, H.; Alami, J. Design of Hard Coatings Deposited by HiPIMS and DcMS. *Mater. Lett.* 2020, *280*, 128540. [CrossRef]
12. Singh, A.; Ghosh, S.; Aravindan, S. Investigation of Oxidation Behaviour of AlCrN and AlTiN Coatings Deposited by Arc Enhanced HIPIMS Technique. *Appl. Surf. Sci.* 2020, *508*, 144812. [CrossRef]
13. Yang, J.; Fu, H.; He, Y.; Gu, Z.; Fu, Y.; Ji, J.; Zhang, Y.; Zhou, Y. Investigation on Friction and Wear Performance of Volcano-Shaped Textured PVD Coating. *Surf. Coat. Technol.* 2022, *431*, 128044. [CrossRef]
14. Khetan, V.; Valle, N.; Duday, D.; Michotte, C.; Mitterer, C.; Delplancke-Ogletree, M.P.; Choquet, P. Temperature-Dependent Wear Mechanisms for Magnetron-Sputtered AlTiTaN Hard Coatings. *ACS Appl. Mater. Interfaces* 2014, *6*, 15403–15411. [CrossRef] [PubMed]
15. Jang, Y.J.; Kim, J.-I.; Kim, W.-s.; Kim, D.H.; Kim, J. Thermal Stability of Si/SiC/Ta-C Composite Coatings and Improvement of Tribological Properties through High-Temperature Annealing. *Sci. Rep.* 2022, *12*, 3536. [CrossRef] [PubMed]
16. Zhang, C.; Cao, H.; Han, D.; Qiao, S.; Guo, Y. Influence of a TiAlN Coating on the Mechanical Properties of a Heat Resistant Steel at Room Temperature and 650 C. *J. Wuhan Univ. Technol. Mater. Sci. Ed.* 2013, *28*, 1029–1033. [CrossRef]
17. Sampath Kumar, T.; Balasivanandha Prabu, S.; Manivasagam, G.; Padmanabhan, K.A. Comparison of TiAlN, AlCrN, and AlCrN/TiAlN Coatings for Cutting-Tool Applications. *Int. J. Miner. Metall. Mater.* 2014, *21*, 796–805. [CrossRef]
18. Khetan, V.; Valle, N.; Duday, D.; Michotte, C.; Delplancke-Ogletree, M.P.; Choquet, P. Influence of Temperature on Oxidation Mechanisms of Fiber-Textured AlTiTaN Coatings. *ACS Appl. Mater. Interfaces* 2014, *6*, 4115–4125. [CrossRef]
19. Ul-Hamid, A. Microstructure, Properties and Applications of Zr-Carbide, Zr-Nitride and Zr-Carbonitride Coatings: A Review. *Mater. Adv.* 2020, *1*, 1012–1037. [CrossRef]
20. Al-Asadi, M.M.; Al-Tameemi, H.A. A Review of Tribological Properties and Deposition Methods for Selected Hard Protective Coatings. *Tribol. Int.* 2022, *176*, 107919. [CrossRef]
21. Courbon, C.; Fallqvist, M.; Hardell, J.; M'Saoubi, R.; Prakash, B. Adhesion Tendency of PVD TiAlN Coatings at Elevated Temperatures during Reciprocating Sliding against Carbon Steel. *Wear* 2015, *330–331*, 209–222. [CrossRef]
22. Nohava, J.; Dessarzin, P.; Karvankova, P.; Morstein, M. Characterization of Tribological Behavior and Wear Mechanisms of Novel Oxynitride PVD Coatings Designed for Applications at High Temperatures. *Tribol. Int.* 2015, *81*, 231–239. [CrossRef]
23. Luo, Y.; Ning, C.; Dong, Y.; Xiao, C.; Wang, X.; Peng, H.; Cai, Z. Impact Abrasive Wear Resistance of CrN and CrAlN Coatings. *Coatings* 2022, *12*, 427. [CrossRef]
24. Devia, D.M.; Restrepo-Parra, E.; Arango, P.J. Comparative Study of Titanium Carbide and Nitride Coatings Grown by Cathodic Vacuum Arc Technique. *Appl. Surf. Sci.* 2011, *258*, 1164–1174. [CrossRef]
25. Chen, L.; Paulitsch, J.; Du, Y.; Mayrhofer, P.H. Thermal Stability and Oxidation Resistance of Ti-Al-N Coatings. *Surf. Coat. Technol.* 2012, *206*, 2954–2960. [CrossRef]
26. Beake, B.D.; Smith, J.F.; Gray, A.; Fox-Rabinovich, G.S.; Veldhuis, S.C.; Endrino, J.L. Investigating the Correlation between Nano-Impact Fracture Resistance and Hardness/Modulus Ratio from Nanoindentation at 25–500 °C and the Fracture Resistance and Lifetime of Cutting Tools with Ti1−xAlxN (x = 0.5 and 0.67) PVD Coatings in Milling Operations. *Surf. Coat. Technol.* 2007, *201*, 4585–4593. [CrossRef]
27. Peng, J.; Su, D.; Wang, C. Combined Effect of Aluminum Content and Layer Structure on the Oxidation Performance of Ti1−xAlxN Based Coatings. *J. Mater. Sci. Technol.* 2014, *30*, 803–807. [CrossRef]
28. He, Q.; DePaiva, J.M.; Kohlscheen, J.; Beake, B.D.; Veldhuis, S.C. Study of Wear Performance and Tribological Characterization of AlTiN PVD Coatings with Different Al/Ti Ratios during Ultra-High Speed Turning of Stainless Steel 304. *Int. J. Refract. Met. Hard Mater.* 2021, *96*, 105488. [CrossRef]

29. Hu, C.; Xu, Y.X.; Chen, L.; Pei, F.; Zhang, L.J.; Du, Y. Structural, Mechanical and Thermal Properties of CrAlNbN Coatings. *Surf. Coat. Technol.* **2018**, *349*, 894–900. [CrossRef]
30. He, Q.; DePaiva, J.M.; Kohlscheen, J.; Veldhuis, S.C. Analysis of the Performance of PVD AlTiN Coating with Five Different Al/Ti Ratios during the High-Speed Turning of Stainless Steel 304 under Dry and Wet Cooling Conditions. *Wear* **2022**, *492–493*, 204213. [CrossRef]
31. Mayrhofer, P.H.; Willmann, H.; Reiter, A.E. Structure and Phase Evolution of Cr-Al-N Coatings during Annealing. *Surf. Coat. Technol.* **2008**, *202*, 4935–4938. [CrossRef]
32. Lim, K.S.; Kim, Y.S.; Hong, S.H.; Song, G.; Kim, K.B. Influence of N_2 Gas Flow Ratio and Working Pressure on Amorphous Mo–Si–N Coating during Magnetron Sputtering. *Coatings* **2020**, *10*, 34. [CrossRef]
33. Chang, C.-L.; Lin, C.-Y.; Yang, F.-C.; Tang, J.-F. The Effect of Match between High Power Impulse and Bias Voltage: TiN Coating Deposited by High Power Impulse Magnetron Sputtering. *Coatings* **2021**, *11*, 822. [CrossRef]
34. Sánchez-López, J.C.; Dominguez-Meister, S.; Rojas, T.C.; Colasuonno, M.; Bazzan, M.; Patelli, A. Tribological Properties of TiC/a-C:H Nanocomposite Coatings Prepared via HiPIMS. *Appl. Surf. Sci.* **2018**, *440*, 458–466. [CrossRef]
35. Anders, A. Tutorial: Reactive High Power Impulse Magnetron Sputtering (R-HiPIMS). *J. Appl. Phys.* **2017**, *121*, 171101. [CrossRef]
36. Reinhard, C.; Ehiasarian, A.P.; Hovsepian, P.E. CrN/NbN Superlattice Structured Coatings with Enhanced Corrosion Resistance Achieved by High Power Impulse Magnetron Sputtering Interface Pre-Treatment. *Thin Solid Films* **2007**, *515*, 3685–3692. [CrossRef]
37. Lakhonchai, A.; Chingsungnoen, A.; Poolcharuansin, P.; Pasaja, N.; Bunnak, P.; Suwanno, M. Comparison of the Structural and Optical Properties of Amorphous Silicon Thin Films Prepared by Direct Current, Bipolar Pulse, and High-Power Impulse Magnetron Sputtering Methods. *Thin Solid Films* **2022**, *747*, 139140. [CrossRef]
38. Grigoriev, S.; Vereschaka, A.; Uglov, V.; Milovich, F.; Cherenda, N.; Andreev, N.; Migranov, M.; Seleznev, A. Influence of Tribological Properties of Zr-ZrN-(Zr,Cr,Al)N and Zr-ZrN-(Zr,Mo,Al)N Multilayer Nanostructured Coatings on the Cutting Properties of Coated Tools during Dry Turning of Inconel 718 Alloy. *Wear* **2023**, *512–513*, 204521. [CrossRef]
39. Vereschaka, A.; Grigoriev, S.; Tabakov, V.; Migranov, M.; Sitnikov, N.; Milovich, F.; Andreev, N. Influence of the Nanostructure of Ti-TiN-(Ti,Al,Cr)N Multilayer Composite Coating on Tribological Properties and Cutting Tool Life. *Tribol. Int.* **2020**, *150*, 106388. [CrossRef]
40. Gayathri, S.; Kumar, N.; Krishnan, R.; Ravindran, T.R.; Dash, S.; Tyagi, A.K.; Sridharan, M. Influence of Cr Content on the Micro-Structural and Tribological Properties of PLD Grown Nanocomposite DLC-Cr Thin Films. *Mater. Chem. Phys.* **2015**, *167*, 194–200. [CrossRef]
41. Bonse, J.; Kirner, S.V.; Koter, R.; Pentzien, S.; Spaltmann, D.; Krüger, J. Femtosecond Laser-Induced Periodic Surface Structures on Titanium Nitride Coatings for Tribological Applications. *Appl. Surf. Sci.* **2017**, *418*, 572–579. [CrossRef]
42. Geng, D.; Li, H.; Chen, Z.; Xu, Y.X.; Wang, Q. Microstructure, Oxidation Behavior and Tribological Properties of AlCrN/Cu Coatings Deposited by a Hybrid PVD Technique. *J. Mater. Sci. Technol.* **2022**, *100*, 150–160. [CrossRef]
43. Alhafian, M.R.; Chemin, J.B.; Fleming, Y.; Bourgeois, L.; Penoy, M.; Useldinger, R.; Soldera, F.; Mücklich, F.; Choquet, P. Comparison on the Structural, Mechanical and Tribological Properties of TiAlN Coatings Deposited by HiPIMS and Cathodic Arc Evaporation. *Surf. Coat. Technol.* **2021**, *423*, 127529. [CrossRef]
44. de Castilho, B.C.N.M.; Rodrigues, A.M.; Avila, P.R.T.; Apolinario, R.C.; de Souza Nossa, T.; Walczak, M.; Fernandes, J.V.; Menezes, R.R.; de Araújo Neves, G.; Pinto, H.C. Hybrid Magnetron Sputtering of Ceramic Superlattices for Application in a next Generation of Combustion Engines. *Sci. Rep.* **2022**, *12*, 2342. [CrossRef]
45. Chang, L.-C.; Zheng, Y.-Z.; Chen, Y.-I. Mechanical Properties of Zr–Si–N Films Fabricated through HiPIMS/RFMS Co-Sputtering. *Coatings* **2018**, *8*, 263. [CrossRef]
46. de Bonis, A.; Teghil, R. Ultra-Short Pulsed Laser Deposition of Oxides, Borides and Carbides of Transition Elements. *Coatings* **2020**, *10*, 501. [CrossRef]
47. Gayathri, S.; Kumar, N.; Krishnan, R.; Ravindran, T.R.; Dash, S.; Tyagi, A.K.; Raj, B.; Sridharan, M. Tribological Properties of Pulsed Laser Deposited DLC/TM (TM=Cr, Ag, Ti and Ni) Multilayers. *Tribol. Int.* **2012**, *53*, 87–97. [CrossRef]
48. Groenen, R.; Smit, J.; Orsel, K.; Vailionis, A.; Bastiaens, B.; Huijben, M.; Boller, K.; Rijnders, G.; Koster, G. Research Update: Stoichiometry Controlled Oxide Thin Film Growth by Pulsed Laser Deposition. *APL Mater.* **2015**, *3*, 070701. [CrossRef]
49. Cao, H.; Liu, F.; Li, H.; Qi, F.; Ouyang, X.; Zhao, N.; Liao, B. High Temperature Tribological Performance and Thermal Conductivity of Thick Ti/Ti-DLC Multilayer Coatings with the Application Potential for Al Alloy Pistons. *Diam. Relat. Mater.* **2021**, *117*, 108466. [CrossRef]
50. Panjan, P.; Drnovšek, A.; Terek, P.; Miletić, A.; Čekada, M.; Panjan, M. Comparative Study of Tribological Behavior of TiN Hard Coatings Deposited by Various PVD Deposition Techniques. *Coatings* **2022**, *12*, 294. [CrossRef]
51. Hirata, Y.; Takeuchi, R.; Taniguchi, H.; Kawagoe, M.; Iwamoto, Y.; Yoshizato, M.; Akasaka, H.; Ohtake, N. Structural and Mechanical Properties of A-BCN Films Prepared by an Arc-Sputtering Hybrid Process. *Materials* **2021**, *14*, 719. [CrossRef] [PubMed]
52. Sawicki, J.; Paczkowski, T. Electrochemical Machining of Curvilinear Surfaces of Revolution: Analysis, Modelling, and Process Control. *Materials* **2022**, *15*, 7751. [CrossRef] [PubMed]
53. Li, Y.; Wang, Z.W.; Zhang, Z.H.; Shao, M.H.; Lu, J.P.; Yan, J.W.; Zhang, L.; He, Y.Y. Microstructure and Tribological Properties of Multilayered ZrCrW(C)N Coatings Fabricated by Cathodic Vacuum-Arc Deposition. *Ceram. Int.* **2022**, *48*, 36655–36669. [CrossRef]

54. Shugurov, A.R.; Kuzminov, E.D. Mechanical and Tribological Properties of Ti-Al-Ta-N/TiAl and Ti-Al-Ta-N/Ta Multilayer Coatings Deposited by DC Magnetron Sputtering. *Surf. Coat. Technol.* **2022**, *441*, 128582. [CrossRef]
55. Cao, H.; Momand, J.; Syari'ati, A.; Wen, F.; Rudolf, P.; Xiao, P.; de Hosson, J.T.M.; Pei, Y. Temperature-Adaptive Ultralubricity of a WS2/a-C Nanocomposite Coating: Performance from Room Temperature up to 500 °C. *ACS Appl. Mater. Interfaces* **2021**, *13*, 28843–28854. [CrossRef] [PubMed]
56. Vuchkov, T.; Leviandhika, V.; Cavaleiro, A. On the Tribological Performance of Magnetron Sputtered W-S-C Coatings with Conventional and Graded Composition. *Surf. Coat. Technol.* **2022**, *449*, 128929. [CrossRef]
57. Cai, Q.; Li, S.; Pu, J.; Cai, Z.; Lu, X.; Cui, Q.; Wang, L. Effect of Multicomponent Doping on the Structure and Tribological Properties of VN-Based Coatings. *J. Alloys Compd.* **2019**, *806*, 566–574. [CrossRef]
58. Grigoriev, S.; Vereschaka, A.; Milovich, F.; Sitnikov, N.; Andreev, N.; Bublikov, J.; Kutina, N. Investigation of the Properties of the Cr,Mo-(Cr,Mo,Zr,Nb)N-(Cr,Mo,Zr,Nb,Al)N Multilayer Composite Multicomponent Coating with Nanostructured Wear-Resistant Layer. *Wear* **2021**, *468–469*, 203597. [CrossRef]
59. Mondragón-Rodríguez, G.C.; Hernández-Mendoza, J.L.; Gómez-Ovalle, A.E.; González-Carmona, J.M.; Ortega-Portilla, C.; Camacho, N.; Hurtado-Macías, A.; Espinosa-Arbeláez, D.G.; Alvarado-Orozco, J.M. High-Temperature Tribology of Hf Doped c-Al0.64Ti0.36N Cathodic Arc PVD Coatings Deposited on M2 Tool Steel. *Surf. Coat. Technol.* **2021**, *422*, 127516. [CrossRef]
60. Grigoriev, S.; Vereschaka, A.; Milovich, F.; Migranov, M.; Andreev, N.; Bublikov, J.; Sitnikov, N.; Oganyan, G. Investigation of the Tribological Properties of Ti-TiN-(Ti,Al,Nb,Zr)N Composite Coating and Its Efficiency in Increasing Wear Resistance of Metal Cutting Tools. *Tribol. Int.* **2021**, *164*, 107236. [CrossRef]
61. Guan, X.; Wang, Y.; Xue, Q. Effects of Constituent Layers and Interfaces on the Mechanical and Tribological Properties of Metal (Cr, Zr)/Ceramic (CrN, ZrN) Multilayer Systems. *Appl. Surf. Sci.* **2020**, *502*, 144305. [CrossRef]
62. Frank, F.; Kainz, C.; Tkadletz, M.; Czettl, C.; Pohler, M.; Schalk, N. Microstructural and Micro-Mechanical Investigation of Cathodic Arc Evaporated ZrN/TiN Multilayer Coatings with Varying Bilayer Thickness. *Surf. Coat. Technol.* **2022**, *432*, 128070. [CrossRef]
63. Xiao, B.; Liu, J.; Liu, F.; Zhong, X.; Xiao, X.; Zhang, T.F.; Wang, Q. Effects of Microstructure Evolution on the Oxidation Behavior and High-Temperature Tribological Properties of AlCrN/TiAlSiN Multilayer Coatings. *Ceram. Int.* **2018**, *44*, 23150–23161. [CrossRef]
64. Miletić, A.; Panjan, P.; Čekada, M.; Kovačević, L.; Terek, P.; Kovač, J.; Dražič, G.; Škorić, B. Nanolayer CrAlN/TiSiN Coating Designed for Tribological Applications. *Ceram. Int.* **2021**, *47*, 2022–2033. [CrossRef]
65. Wang, R.; Li, H.Q.; Li, R.S.; Mei, H.J.; Zou, C.W.; Zhang, T.F.; Wang, Q.M.; Kim, K.H. Thermostability, Oxidation, and High-Temperature Tribological Properties of Nano-Multilayered AlCrSiN/VN Coatings. *Ceram. Int.* **2022**, *48*, 11915–11923. [CrossRef]
66. Wang, T.C.; Hsu, S.Y.; Lai, Y.T.; Tsai, X.; Duh, J.G. Microstructure and High-Temperature Tribological Characteristics of Self-Lubricating TiAlSiN/VSiN Multilayer Nitride Coatings. *Mater. Chem. Phys.* **2023**, *295*, 127149. [CrossRef]
67. Mei, H.; Ding, J.C.; Yan, K.; Peng, W.; Zhao, C.; Luo, Q.; Gong, W.; Ren, F.; Wang, Q. Effects of V and Cu Codoping on the Tribological Properties and Oxidation Behavior of AlTiN Coatings. *Ceram. Int.* **2022**, *48*, 22317–22327. [CrossRef]
68. Evaristo, M.; Fernandes, F.; Cavaleiro, A. Room and High Temperature Tribological Behaviour of W-DLC Coatings Produced by DCMS and Hybrid DCMS-HiPIMS Configuration. *Coatings* **2020**, *10*, 319. [CrossRef]
69. Alamgir, A.; Bogatov, A.; Jõgiaas, T.; Viljus, M.; Raadik, T.; Kübarsepp, J.; Sergejev, F.; Lümkemann, A.; Kluson, J.; Podgursky, V. High-Temperature Oxidation Resistance and Tribological Properties of Al$_2$O$_3$/Ta-C Coating. *Coatings* **2022**, *12*, 547. [CrossRef]
70. Wang, C.; Xu, B.; Wang, Z.; Li, H.; Wang, L.; Chen, R.; Wang, A.; Ke, P. Tribological Mechanism of (Cr, V)N Coating in the Temperature Range of 500–900 °C. *Tribol. Int.* **2021**, *159*, 106952. [CrossRef]
71. Wang, Z.; Wang, C.; Zhang, Y.; Wang, A.; Ke, P. M-Site Solid Solution of Vanadium Enables the Promising Mechanical and High-Temperature Tribological Properties of Cr$_2$AlC Coating. *Mater. Des.* **2022**, *222*, 111060. [CrossRef]
72. Mei, H.; Wang, R.; Zhong, X.; Dai, W.; Wang, Q. Influence of Nitrogen Partial Pressure on Microstructure and Tribological Properties of Mo-Cu-V-N Composite Coatings with High Cu Content. *Coatings* **2018**, *8*, 24. [CrossRef]
73. Song, Y.; Wang, L.; Shang, L.; Zhang, G.; Li, C. Temperature-Dependent Tribological Behavior of CrAlN/TiSiN Tool Coating Sliding against 7A09 Al Alloy and GCr15 Bearing Steel. *Tribol. Int.* **2023**, *177*, 107942. [CrossRef]
74. Yeh-Liu, L.K.; Hsu, S.Y.; Chen, P.Y.; Lee, J.W.; Duh, J.G. Improvement of CrMoN/SiNx Coatings on Mechanical and High Temperature Tribological Properties through Biomimetic Laminated Structure Design. *Surf. Coat. Technol.* **2020**, *393*, 125754. [CrossRef]
75. Alamgir, A.; Yashin, M.; Bogatov, A.; Viljus, M.; Traksmaa, R.; Sondor, J.; Lümkemann, A.; Sergejev, F.; Podgursky, V. High-Temperature Tribological Performance of Hard Multilayer TiN-AlTiN/NACo-CrN/AlCrN-AlCrOAlTiCrN Coating Deposited on WC-Co Substrate. *Coatings* **2020**, *10*, 909. [CrossRef]
76. Fan, Q.; Zhang, S.; Lin, J.; Cao, F.; Liu, Y.; Xue, R.; Wang, T. Microstructure, Mechanical and Tribological Properties of Gradient CrAlSiN Coatings Deposited by Magnetron Sputtering and Arc Ion Plating Technology. *Thin Solid Films* **2022**, *760*, 139490. [CrossRef]
77. Ferreira, R.; Carvalho, Ó.; Sobral, L.; Carvalho, S.; Silva, F. Influence of Morphology and Microstructure on the Tribological Behavior of Arc Deposited CrN Coatings for the Automotive Industry. *Surf. Coat. Technol.* **2020**, *397*, 126047. [CrossRef]
78. Yan, M.; Wang, C.; Sui, X.; Liu, J.; Lu, Y.; Hao, J.; Liu, W. Effect of Substrate Rotational Speed during Deposition on the Microstructure, Mechanical and Tribological Properties of a-C: Ta Coatings. *Ceram. Int.* **2022**. [CrossRef]

79. Wang, Z.; He, Z.; Chen, F.; Tian, C.; Valiev, U.v.; Zou, C.; Fu, D. Effects of N2 Partial Pressure on Microstructure and Mechanical Properties of Cathodic Arc Deposited TiBN/TiAlSiN Nano-Multilayered Coatings. *Mater. Today Commun.* **2022**, *31*, 103436. [CrossRef]
80. Chen, J.; Guo, Q.; Li, J.; Yang, Z.; Guo, Y.; Yang, W.; Xu, D.; Yang, B. Microstructure and Tribological Properties of CrAlTiN Coating Deposited via Multi-Arc Ion Plating. *Mater. Today Commun.* **2022**, *30*, 103136. [CrossRef]
81. Akhter, R.; Zhou, Z.; Xie, Z.; Munroe, P. Influence of Substrate Bias on the Scratch, Wear and Indentation Response of TiSiN Nanocomposite Coatings. *Surf. Coat. Technol.* **2021**, *425*, 127687. [CrossRef]
82. Cao, H.S.; Liu, F.J.; hao, L.I.; Luo, W.Z.; Qi, F.G.; Lu, L.W.; Nie, Z.H.; Ouyang, X.P. Effect of Bias Voltage on Microstructure, Mechanical and Tribological Properties of TiAlN Coatings. *Trans. Nonferrous Met. Soc. China* **2022**, *32*, 3596–3609. [CrossRef]
83. Akhter, R.; Bendavid, A.; Munroe, P. Tailoring the Scratch Adhesion Strength and Wear Performance of TiNiN Nanocomposite Coatings by Optimising Substrate Bias Voltage during Cathodic Arc Evaporation. *Surf. Coat. Technol.* **2022**, *445*, 128707. [CrossRef]
84. Biswas, B.; Purandare, Y.; Khan, I.; Hovsepian, P.E. Effect of Substrate Bias Voltage on Defect Generation and Their Influence on Corrosion and Tribological Properties of HIPIMS Deposited CrN/NbN Coatings. *Surf. Coat. Technol.* **2018**, *344*, 383–393. [CrossRef]
85. Lin, Y.-W.; Chih, P.-C.; Huang, J.-H. Effect of Ti Interlayer Thickness on Mechanical Properties and Wear Resistance of TiZrN Coatings on AISI D2 Steel. *Surf. Coat. Technol.* **2020**, *394*, 125690. [CrossRef]
86. Huang, B.; Zhou, Q.; An, Q.; Zhang, E.G.; Chen, Q.; Liang, D.D.; Du, H.M.; Li, Z.M. Tribological Performance of the Gradient Composite TiAlSiN Coating with Various Friction Pairs. *Surf. Coat. Technol.* **2022**, *429*, 127945. [CrossRef]
87. Ovchinnikov, S.; Kalashnikov, M. Structure and Tribological Properties of Gradient-Layered Coatings (Ti, Al, Si, Cr, Mo, S) O, N. *Surf. Coat. Technol.* **2021**, *408*, 126807. [CrossRef]
88. Bondarev, A.; Al-Rjoub, A.; Yaqub, T.B.; Polcar, T.; Fernandes, F. TEM Study of the Oxidation Resistance and Diffusion Processes in a Multilayered TiSiN/TiN(Ag) Coating Designed for Tribological Applications. *Appl. Surf. Sci.* **2023**, *609*, 155319. [CrossRef]
89. Chang, Y.Y.; Chang, B.Y.; Chen, C.S. Effect of CrN Addition on the Mechanical and Tribological Performances of Multilayered AlTiN/CrN/ZrN Hard Coatings. *Surf. Coat. Technol.* **2022**, *433*, 128107. [CrossRef]
90. Tao, H.; Tsai, M.T.; Chen, H.W.; Huang, J.C.; Duh, J.G. Improving High-Temperature Tribological Characteristics on Nanocomposite CrAlSiN Coating by Mo Doping. *Surf. Coat. Technol.* **2018**, *349*, 752–756. [CrossRef]
91. Yi, B.; Zhou, S.; Qiu, Z.; Zeng, D.C. The Influences of Pulsed Bias Duty Cycle on Tribological Properties of Solid Lubricating TiMoCN Coatings. *Vacuum* **2020**, *180*, 109552. [CrossRef]
92. Fernandes, F.; Danek, M.; Polcar, T.; Cavaleiro, A. Tribological and Cutting Performance of TiAlCrN Films with Different Cr Contents Deposited with Multilayered Structure. *Tribol. Int.* **2018**, *119*, 345–353. [CrossRef]
93. Yu, W.; Li, H.; Li, J.; Liu, Z.; Huang, J.; Kong, J.; Wu, Q.; Shi, Y.; Zhang, G.; Xiong, D. Balance between Oxidation and Tribological Behaviors at Elevated Temperatures of Hf1−xWxN Films by Optimizing W Content. *Vacuum* **2022**, *207*, 111673. [CrossRef]
94. Chim, Y.C.; Ding, X.Z.; Zeng, X.T.; Zhang, S. Oxidation Resistance of TiN, CrN, TiAlN and CrAlN Coatings Deposited by Lateral Rotating Cathode Arc. *Thin Solid Films* **2009**, *517*, 4845–4849. [CrossRef]
95. Sánchez-López, J.C.; Contreras, A.; Domínguez-Meister, S.; García-Luis, A.; Brizuela, M. Tribological Behaviour at High Temperature of Hard CrAlN Coatings Doped with y or Zr. *Thin Solid Films* **2014**, *550*, 413–420. [CrossRef]
96. Drnovšek, A.; Rebelo de Figueiredo, M.; Vo, H.; Xia, A.; Vachhani, S.J.; Kolozsvári, S.; Hosemann, P.; Franz, R. Correlating High Temperature Mechanical and Tribological Properties of CrAlN and CrAlSiN Hard Coatings. *Surf. Coat. Technol.* **2019**, *372*, 361–368. [CrossRef]
97. Cai, F.; Wang, J.; Zhou, Q.; Xue, H.; Zheng, J.; Wang, Q.; Kim, K.H. Microstructure Evolution and High-Temperature Tribological Behavior of AlCrBN Coatings with Various B Contents. *Surf. Coat. Technol.* **2022**, *430*, 127994. [CrossRef]
98. Kiryukhantsev-Korneev, P.V.; Sytchenko, A.D.; Vorotilo, S.A.; Klechkovskaya, V.V.; Lopatin, V.Y.; Levashov, E.A. Structure, Oxidation Resistance, Mechanical, and Tribological Properties of N- and C-Doped Ta-Zr-Si-B Hard Protective Coatings Obtained by Reactive D.C. Magnetron Sputtering of TaZrSiB Ceramic Cathode. *Coatings* **2020**, *10*, 946. [CrossRef]
99. Mei, H.; Ding, J.C.; Zhao, Z.; Li, Q.; Song, J.; Li, Y.; Gong, W.; Ren, F.; Wang, Q. Effect of Cu Content on High-Temperature Tribological Properties and Oxidation Behavior of Al-Ti-V-Cu-N Coatings Deposited by HIPIMS. *Surf. Coat. Technol.* **2022**, *434*, 128130. [CrossRef]
100. Ren, Y.; Jia, J.; Cao, X.; Zhang, G.; Ding, Q. Effect of Ag Contents on the Microstructure and Tribological Behaviors of NbN–Ag Coatings at Elevated Temperatures. *Vacuum* **2022**, *204*, 111330. [CrossRef]
101. Zhang, M.; Zhou, F.; Fu, Y.; Wang, Q.; Zhou, Z. Influence of Ag Target Current on the Structure and Tribological Properties of CrMoSiCN/Ag Coatings in Air and Water. *Tribol. Int.* **2021**, *160*, 107059. [CrossRef]
102. Perea, D.; Bejarano, G. Development and Characterization of TiAlN (Ag, Cu) Nanocomposite Coatings Deposited by DC Magnetron Sputtering for Tribological Applications. *Surf. Coat. Technol.* **2020**, *381*, 125095. [CrossRef]
103. Rajput, S.S.; Gangopadhyay, S.; Yaqub, T.B.; Cavaleiro, A.; Fernandes, F. Room and High Temperature Tribological Performance of CrAlN(Ag) Coatings: The Influence of Ag Additions. *Surf. Coat. Technol.* **2022**, *450*, 129011. [CrossRef]
104. Cavaleiro, D.; Veeregowda, D.; Cavaleiro, A.; Carvalho, S.; Fernandes, F. High Temperature Tribological Behaviour of TiSiN(Ag) Films Deposited by HiPIMS in DOMS Mode. *Surf. Coat. Technol.* **2020**, *399*, 126176. [CrossRef]

105. Fenker, M.; Balzer, M.; Kellner, S.; Polcar, T.; Richter, A.; Schmidl, F.; Vitu, T. Formation of Solid Lubricants during High Temperature Tribology of Silver-Doped Molybdenum Nitride Coatings Deposited by DcMS and HIPIMS. *Coatings* **2021**, *11*, 1415. [CrossRef]
106. Liu, C.; Ju, H.; Xu, J.; Yu, L.; Zhao, Z.; Geng, Y.; Zhao, Y. Influence of Copper on the Compositions, Microstructure and Room and Elevated Temperature Tribological Properties of the Molybdenum Nitride Film. *Surf. Coat. Technol.* **2020**, *395*, 125811. [CrossRef]
107. Fu, Y.; Li, H.; Chen, J.; Guo, H.; Wang, X. Microstructure, Mechanical and Tribological Properties of Arc Ion Plating NbN-Based Nanocomposite Films. *Nanomaterials* **2022**, *12*, 3909. [CrossRef]
108. Xu, X.; Sun, J.; Su, F.; Li, Z.; Chen, Y.; Xu, Z. Microstructure and Tribological Performance of Adaptive MoN–Ag Nanocomposite Coatings with Various Ag Contents. *Wear* **2022**, *488–489*, 204170. [CrossRef]
109. Lü, W.; Li, G.; Zhou, Y.; Liu, S.; Wang, K.; Wang, Q. Effect of High Hardness and Adhesion of Gradient TiAlSiN Coating on Cutting Performance of Titanium Alloy. *J. Alloys Compd.* **2020**, *820*, 153137. [CrossRef]
110. Lim, H.P.; Jiang, Z.-T.; Gan, J.H.M.; Nayan, N.; Chee, F.P.; Soon, C.F.; Hassan, N.; Liew, W.Y.H. A Systematic Investigation of the Tribological Behaviour of Oxides Formed on AlSiTiN, CrAlTiN, and CrAlSiTiN Coatings. *Wear* **2022**, *512–513*, 204552. [CrossRef]
111. Huang, B.; Zhang, E.G.; Du, H.M.; Chen, Q.; Liang, D.D.; An, Q.; Zhou, Q. Effects of Annealing Temperature on the Microstructure, Mechanical and Tribological Properties of CrAlTiN Coatings. *Surf. Coat. Technol.* **2022**, *449*, 128887. [CrossRef]
112. Tillmann, W.; Grisales, D.; Marin Tovar, C.; Contreras, E.; Apel, D.; Nienhaus, A.; Stangier, D.; Lopes Dias, N.F. Tribological Behaviour of Low Carbon-Containing TiAlCN Coatings Deposited by Hybrid (DCMS/HiPIMS) Technique. *Tribol. Int.* **2020**, *151*, 106528. [CrossRef]
113. Zhao, Y.; Xu, F.; Xu, J.; Li, D.; Sun, S.; Gao, C.; Zhao, W.; Lang, W.; Liu, J.; Zuo, D. Effect of the Bias-Graded Increment on the Tribological and Electrochemical Corrosion Properties of DLC Films. *Diam. Relat. Mater.* **2022**, *130*, 109421. [CrossRef]
114. Abedi, M.; Abdollah-zadeh, A.; Vicenzo, A.; Bestetti, M.; Movassagh-Alanagh, F.; Damerchi, E. A Comparative Study of the Mechanical and Tribological Properties of PECVD Single Layer and Compositionally Graded TiSiCN Coatings. *Ceram. Int.* **2019**, *45*, 21200–21207. [CrossRef]
115. Chang, Y.Y.; Huang, J.W. Nanostructured AlTiSiN/CrVN/ZrN Coatings Synthesized by Cathodic Arc Deposition-Mechanical Properties and Cutting Performance. *Surf. Coat. Technol.* **2022**, *442*, 128424. [CrossRef]
116. Carabillò, A.; Sordetti, F.; Querini, M.; Magnan, M.; Azzolini, O.; Fedrizzi, L.; Lanzutti, A. Tribological Optimization of Titanium-Based PVD Multilayer Hard Coatings Deposited on Steels Used for Cold Rolling Applications. *Mater. Today Commun.* **2023**, *34*, 105043. [CrossRef]
117. Chen, S.N.; Yan, W.Q.; Liao, B.; Wu, X.Y.; Chen, L.; Ouyang, X.; Ouyang, X.P. Effect of Temperature on the Tribocorrosion and High-Temperature Tribological Behaviour of Strong Amorphization AlCrNiTiV High Entropy Alloy Film in a Multifactor Environment. *Ceram. Int.* **2022**, *49*, 6880–6890. [CrossRef]
118. Li, J.; Chen, Y.; Zhao, Y.; Shi, X.; Wang, S.; Zhang, S. Super-Hard (MoSiTiVZr)Nx High-Entropy Nitride Coatings. *J. Alloys Compd.* **2022**, *926*, 166807. [CrossRef]
119. Zhao, Y.; Chen, S.; Chen, Y.; Wu, S.; Xie, W.; Yan, W.; Wang, S.; Liao, B.; Zhang, S. Super-Hard and Anti-Corrosion (AlCrMoSiTi)Nx High Entropy Nitride Coatings by Multi-Arc Cathodic Vacuum Magnetic Filtration Deposition. *Vacuum* **2022**, *195*, 110685. [CrossRef]
120. Bachani, S.K.; Wang, C.J.; Lou, B.S.; Chang, L.C.; Lee, J.W. Fabrication of TiZrNbTaFeN High-Entropy Alloys Coatings by HiPIMS: Effect of Nitrogen Flow Rate on the Microstructural Development, Mechanical and Tribological Performance, Electrical Properties and Corrosion Characteristics. *J. Alloys Compd.* **2021**, *873*, 159605. [CrossRef]
121. Cui, P.; Li, W.; Liu, P.; Zhang, K.; Ma, F.; Chen, X.; Feng, R.; Liaw, P.K. Effects of Nitrogen Content on Microstructures and Mechanical Properties of (AlCrTiZrHf)N High-Entropy Alloy Nitride Films. *J. Alloys Compd.* **2020**, *834*, 155063. [CrossRef]
122. Ren, B.; Zhao, R.F.; Zhang, G.P.; Liu, Z.X.; Cai, B.; Jiang, A.Y. Microstructure and Properties of the AlCrMoZrTi/(AlCrMoZrTi)N Multilayer High-Entropy Nitride Ceramics Films Deposited by Reactive RF Sputtering. *Ceram. Int.* **2022**, *48*, 16901–16911. [CrossRef]
123. Lo, W.L.; Hsu, S.Y.; Lin, Y.C.; Tsai, S.Y.; Lai, Y.T.; Duh, J.G. Improvement of High Entropy Alloy Nitride Coatings (AlCrNbSiTiMo)N on Mechanical and High Temperature Tribological Properties by Tuning Substrate Bias. *Surf. Coat. Technol.* **2020**, *401*, 126247. [CrossRef]
124. Zhang, X.; Pelenovich, V.; Liu, Y.; Ke, X.; Zhang, J.; Yang, B.; Ma, G.; Li, M.; Wang, X. Effect of Bias Voltages on Microstructure and Properties of (TiVCrNbSiTaBY)N High Entropy Alloy Nitride Coatings Deposited by RF Magnetron Sputtering. *Vacuum* **2022**, *195*, 110710. [CrossRef]
125. Wang, Q.; Jin, X.; Zhou, F. Comparison of Mechanical and Tribological Properties of CrBN Coatings Modified by Ni or Cu Incorporation. *Friction* **2022**, *10*, 516–529. [CrossRef]

126. Zhu, Y.; Dong, M.; Li, J.; Wang, L. Wear Failure Mechanism of TiSiN Coating at Elevated Temperatures. *Appl. Surf. Sci.* **2019**, *487*, 349–355. [CrossRef]
127. Guo, F.; Li, K.; Huang, X.; Xie, Z.; Gong, F. Understanding the Wear Failure Mechanism of TiAlSiCN Nanocomposite Coating at Evaluated Temperatures. *Tribol. Int.* **2021**, *154*, 106716. [CrossRef]

Disclaimer/Publisher's Note: The statements, opinions and data contained in all publications are solely those of the individual author(s) and contributor(s) and not of MDPI and/or the editor(s). MDPI and/or the editor(s) disclaim responsibility for any injury to people or property resulting from any ideas, methods, instructions or products referred to in the content.

Article

Comparative Study of Mechanical Performance of AlCrSiN Coating Deposited on WC-Co and cBN Hard Substrates

Jing Liang [1,*], Marc Serra [1], Sandra Gordon [1], Jonathan Fernández de Ara [2], Eluxka Almandoz [2,3], Luis Llanes [1,4] and Emilio Jimenez-Piqué [1,4,*]

1. CIEFMA—Department of Materials Science and Engineering, EEBE, Universitat Politècnica de Catalunya-BarcelonaTECH., 08019 Barcelona, Spain; marc.serra.fanals@upc.edu (M.S.); sandra.gordon@upc.edu (S.G.); luis.miguel.llanes@upc.edu (L.L.)
2. Centre of Advanced Surface Engineering, AIN, 31191 Cordovilla, Spain; jfernandez@ain.es (J.F.d.A.); ealmandoz@ain.es (E.A.)
3. Science Department, Universidad Pública de Navarra (UPNA), Campus de Arrosadía, 31006 Pamplona, Spain
4. Barcelona Research Center in Multiscale Science and Engineering, Universitat Politècnica de Catalunya-BarcelonaTECH., Avda. Eduard Maristany 16, 08019 Barcelona, Spain
* Correspondence: jing.liang@upc.edu (J.L.); emilio.jimenez@upc.edu (E.J.-P.)

Abstract: The objective of this study is to explore and compare the mechanical response of AlCrSiN coatings deposited on two different substrates, namely, WC-Co and cBN. Nano-indentation was used to measure the hardness and elastic modulus of the coatings, and micro-indentation was used for observing the contact damage under Hertzian contact with monotonic and cyclic (fatigue) loads. Microscratch and contact damage tests were also used to evaluate the strength of adhesion between the AlCrSiN coatings and the two substrates under progressive and constant loads, respectively. The surface damages induced via different mechanical tests were observed using scanning electron microscopy (SEM). A focused ion beam (FIB) was used to produce a cross-section of the coating–substrate system in order to further detect the mode and extent of failure that was induced. The results show that the AlCrSiN coating deposited on the WC-Co substrate performed better in regard to adhesion strength and contact damage response than the same coating deposited on the cBN substrate; this is attributed to the lower plasticity of the cBN substrate as well as its less powerful adhesion to the coating.

Keywords: AlCrSiN; quaternary coating; WC-Co; cBN; nano-indentation; microscratch; adhesion strength

Citation: Liang, J.; Serra, M.; Gordon, S.; Fernández de Ara, J.; Almandoz, E.; Llanes, L.; Jimenez-Piqué, E. Comparative Study of Mechanical Performance of AlCrSiN Coating Deposited on WC-Co and cBN Hard Substrates. *Ceramics* **2023**, *6*, 1238–1250. https://doi.org/10.3390/ceramics6020075

Academic Editors: Amirhossein Pakseresht and Kamalan Kirubaharan Amirtharaj Mosas

Received: 1 May 2023
Revised: 5 June 2023
Accepted: 5 June 2023
Published: 9 June 2023

Copyright: © 2023 by the authors. Licensee MDPI, Basel, Switzerland. This article is an open access article distributed under the terms and conditions of the Creative Commons Attribution (CC BY) license (https://creativecommons.org/licenses/by/4.0/).

1. Introduction

Hard protective coatings are frequently used in the tool industry to improve wear and corrosion resistance [1,2]. Based on traditional binary metal nitride ceramic coatings (such as CrN and TiN), aluminum (Al) and silicon (Si) are introduced to form quaternary coating systems (AlCrSiN and AlTiSiN) with enhanced mechanical properties, thermal stability, and wear resistance [3,4]. By adding Al and Si, an oxide-rich layer is formed on top, contributing to improved oxidation resistance and thermal stability of the coating [5,6]. In addition, the amorphous Si_3N_4 phase formed in the grain boundary inhibits the neighboring grains from sliding, which results in better hardness and thermal stability [7–9].

These coatings are normally deposited via physical vapor deposition (PVD) [10,11]. The crystalline phase coatings consist of a face-centered cubic lattice with high atomic density formed via crystallization of a metastable amorphous phase after thermal treatment, producing a coating with a columnar structure [12–14].

Although the mechanical properties of these coatings have been studied before [15], information about mechanical behavior when the coatings are deposited on different substrates is scarce. Performance of tool materials is governed by both the coating and the substrate, which may result in very different performances even for the same coating if the substrate is different. The two most used hard substrates in the coating industry are hardmetals (also known as cemented carbides) and cubic boron nitride.

Hardmetals are composed of a hard WC phase bonded with a metal, usually cobalt (WC-Co), presenting an outstanding combination of mechanical properties including hardness, fracture toughness, and wear resistance [16,17]. Cubic boron nitride (cBN) is a superhard material consisting of cBN particles bonded with a ceramic, with hardness only second to diamond, with an extremely high hardness and wear resistance together with exceptional thermal and chemical properties [18–24].

There are some examples of studies exploring the effects of different substrates on quaternary coatings, but they focus mainly on metallic substrates. For example, Gao Y. et al. focused on how the structure of AlCrSiN coating protected high-speed steel (HSS) [25], and K. Tuchid et al. studied thermal oxidation on HastelloyX substrate [26]. These papers studied just one substrate, so it is interesting to study the effect of different substrates in the mechanical responses of TiAlSiN coatings under contact loads. Similar studies on other types of coatings exist. Sveen et al. [27] studied the scratch adhesion of TIAlN on three different substrates (high-speed steel, cBN, and cemented carbides), but no evidence exists on quaternary AlCrSiN coatings.

Therefore, the present study focuses on comparing the mechanical responses of AlCrSiN quaternary coatings deposited on two different hard substrates, WC-Co and cBN, with the objective of understanding the mechanical performances of the coated materials.

2. Experimental Procedure

2.1. Sample Preparation

AlCrSiN coatings were deposited on two different commercially available hard substrates of cemented carbide (WC-Co) and cubic boron nitride (cBN), making two different coating–substrate systems: AlCrSiN/WC-Co and AlCrSiN/cBN. The WC-Co material consisted in WC grains with 0.9 ± 0.4 μm grain size and 10% of cobalt content, while the cBN material had a low cBN content with an average grain size of 1.5 microns and a TiN ceramic binder.

Cathodic arc evaporation was used to deposit the AlCrSiN coatings. The process was carried out using the industrial equipment Platit p80, in a vacuum chamber with an argon atmosphere at 0.8–2 Pa and a negative bias voltage of −65 V. Cr, and Al+Si cathodes were used as the material source. A description of the deposition process of the AlCrSiN coatings with a Cr adhesion layer can be found in [28]. The composition of the coatings is presented in Table 1.

Table 1. Element concentrations of AlCrSiN coating deposited on WC-Co substrate.

Element	Al	Cr	Si	N
Concentration (at.%)	41	13	8	38

The thicknesses of coatings were measured using the Calowear test, and calculations for the coating thicknesses were performed using Equation (1).

$$t = \frac{(R+r)(R-r)}{d} \quad (1)$$

where t is the thickness of coating, and d is the diameter of the hard steel sphere. R is the outer edge radius of the ring depression, and r is the inner edge radius of the depression, both of which were measured using electric scanning microscopy (Phenom XL Desktop SEM, Phenom-World, Eindhoven, The Netherlands).

2.2. Characterization of Structure and Composition

The crystallographic phase of the coating was characterized using glancing incident angle X-ray diffraction (GIXRD) with Cu X-ray tube radiation (D8 advance, Bruker, Billerica,

MA, USA), and the test voltage and current went up to 40 kV and 40 mA. The incident angle was fixed at 1°. The scan speed was 10 s per step, and a 0.01° scan size was performed from 20°–70°.

2.3. Nano-Indentation and Micro-Indentation

The hardness and elastic modulus were measured with a Nanoindenter XP, Oak Ridge, TN, USA, with a continuous stiffness measurement. All surfaces of the coatings were slightly polished using colloidal silica under the force of 1 N and cleaned using acetone to lessen the effect caused by the roughness of the coatings during measurement [29]. Nano-indentation tests were performed with a Berkovich diamond tip calibrated against a fused silica standard. An array of 25 imprints was made for each sample with a constant strain rate of $0.05\ \text{s}^{-1}$. All indentations were performed with up to 2000 nm maximum penetration depth. The calculations of the hardness (H) and elastic modulus (E) were obtained via the Oliver–Pharr method, and the Poisson ratio was assumed to be $\nu = 0.25$ [12]. The hardness was measured at 10% penetration depth, and the elastic modulus was estimated by extrapolating the results to null thickness.

The spherical contact response was assessed by means of spherical micro-indentation with both monotonic and cyclic loads. Hertzian tests were carried out using a servo hydraulic test machine (Instron 8500, Instron, Norwood, MA, USA) using a cemented carbide sphere with a radius of 5 mm. Monotonic loads were conducted following a trapezoidal waveform, at a loading rate of 500 N/s and holding 10 s at maximum load, under seven equally spaced loads: 2000 N, 3000 N, 4000 N, 5000 N, 6000 N, 7000 N, and 8000 N. The same loading waveform was applied for the cyclic loading with a frequency of 3 Hz and 10^3 circles. At least three indentations were made at each load and loading mode.

2.4. Adhesion Test

Adhesion strength between coating and substrate was evaluated with scratch and contact damage tests. Scratch tests were conducted using a Revetest Scratch tester (CSM Revetest, Buchs, Switzerland) with a progressive load from 0 N to 90 N at a constant loading rate of 10 N/min with a Rockwell C tip radius, and 120° apex angle, and length of 5 mm. Contact damage tests were performed using the same tip geometry [30]: a Rockwell C indenter was pressed against the surface of samples to generate deformation and cracking. Nine loads were used for the two coated systems, 9.8 N, 49 N, 98 N, 196 N, 294 N, 392 N, 490 N, and 613 N, with the intention of producing different types of damage. Failure and damage produced via scratch and contact damage tests were observed using scanning electron microscopy (SEM).

2.5. Microscopy

A Phenom XL SEM and a Zeiss Neon 40 SEM-FIB coupled with an energy dispersive X-ray detector (EDX) were used to observe the samples. The cross-section using FIB for the coated systems was performed by using gallium ions accelerated up to 30 kV with a decreasing ion current down to 500 pA. A protective layer of platinum was deposited on the area of interest to avoid the waterfall effect.

3. Results and Discussion

3.1. Coating Composition and Coating Thickness

To confirm crystallographic structure and composition, XRD patterns were created and are presented in Figure 1. The phase detected for the coatings on both substrates was (Cr,Al)N solid solution, which was the main phase of the AlCrSiN coating. The (111), (200), and (220) of peaks were shifted about 0.2 degrees due to the lattice expansion produced by the substitution of Cr atoms with Al atoms [31,32]. The WC-Co substrate was also detected, as shown in Figure 1a. In Figure 1b, indexing of the cBN substrate is also presented. The binder phase TiN of the cBN substrate was clearly identified as a cubic phase. The diffraction peak at 42.6 degrees was assigned to the face-centered cubic cBN [33].

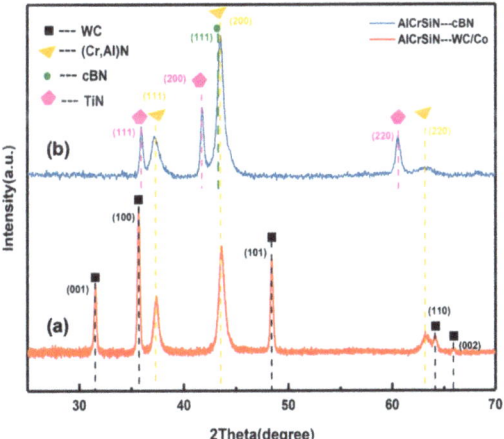

Figure 1. XRD patterns of AlCrSiN coatings deposited on (**a**) WC-Co and (**b**) cBN substrates. The coatings present a (Cr,Al)N crystallographic phase.

The deposited AlCrSiN coatings were uniform and dense both on WC-Co and cBN substrates, as seen in Figure 2a,b. Based on the EDX composition maps, elements of chromium (Cr), aluminum (Al), and nitrogen (N) were detected in the coatings. Elements of wolfram (W), cobalt (Co), titanium (Ti), boron (B), and nitrogen (N) were detected on the substrates. The results are consistent with the compositions of coatings and substrates.

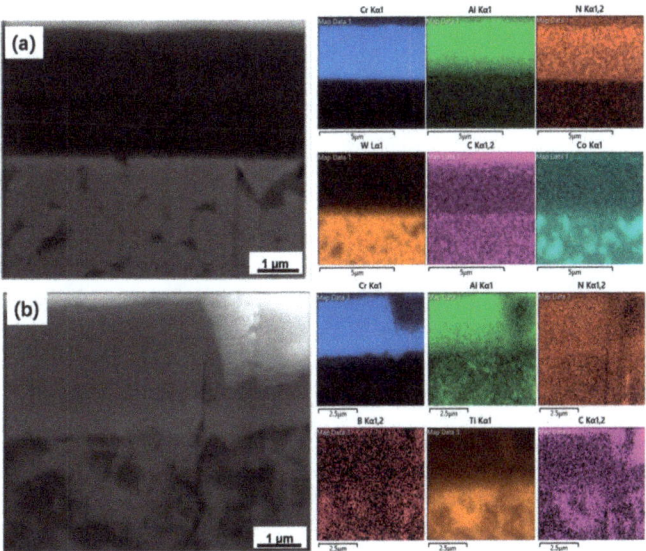

Figure 2. SEM micrograph of the cross-section and the EDX composition maps for corresponding areas. (**a**) AlCrSiN coating deposited on WC-Co substrate, (**b**) AlCrSiN coating deposited on cBN substrate together with the platinum layer to avoid the waterfall effect during milling procedure, and the crack in the coating and substrate was caused by the scratch test.

The measured thicknesses of the coatings are presented Table 2, showing similar values for both substrates. Consequently, the geometry is similar, and the results of the mechanical testing can be compared for both substrates.

Table 2. Thickness of coating–substrate systems.

Coating	AlCrSiN	
Substrate	WC-Co	cBN
Thickness (μm)	3.6 ± 0.4	3.5 ± 0.3

3.2. Mechanical Properties of the Coating–Substrate Systems

The results of nano-indentation are presented in Table 3, and the images of nano-indentation imprints are presented in Figure 3. According to Table 3, the AlCrSiN coatings exhibited similar values when deposited on different substrates, WC-Co and cBN, which means the mechanical properties of the coatings were not altered by the substrates. Some circumferential cracks, shown in Figure 3c, appeared at the inner edge of the nano-indentation and indicated that the cracking resistance of the AlCrSiN coating deposited on the cBN substrate may be lower than the sample with the WC-Co substrate, probably due to the higher stiffness of the substrate.

Table 3. Mechanical properties of coating–substrate systems.

	Material	Hardness (GPa)	Elastic Modulus (GPa)	H/E	H^3/E^2
Coating	AlCrSiN on WC-Co	40 ± 5	553 ± 60	0.073	0.211
	AlCrSiN on cBN	39 ± 4	508 ± 52	0.076	0.225
Substrate	WC-Co	31 ± 3	574 ± 54	0.054	0.091
	cBN	36 ± 5	510 ± 90	0.071	0.179

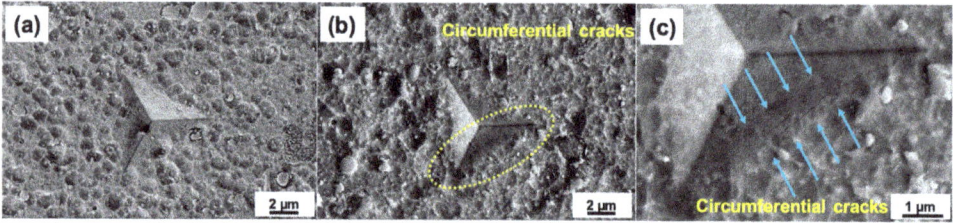

Figure 3. SEM images of nano-indentation, (**a**) AlCrSiN coating deposited on WC-Co substrate, (**b**) AlCrSiN coating deposited on cBN substrate, (**c**) magnification of the circumferential cracks highlighted with dotted circle on the image (**b**).

3.3. Adhesion Tests

Figure 4 presents the scratch tracks of the two coating–substrate systems obtained via optical microscopy together with the magnification of failure events via SEM. It is seen how plastic deformation, microcracking, and delamination occurred as the load was increased. First, cracking of the coating (indicated as L_{C1}) was present in both samples. For the WC-Co substrate, stick–slip deformation was induced via compressive stress and appeared at the contour. As the load increased, transverse cracks, which were induced by tensile stress, appeared until spallation of the coating occurred; this load is defined as the second critical load (L_{C2}), implying failure of the interface between coating and substrate [34]. Interfacial failure occurred much earlier in the cBN substrate (seen in Figure 4b,c) with clear spallation of the coating [35,36]. Fracture features were also different, with more substrate exposure in the case of cBN.

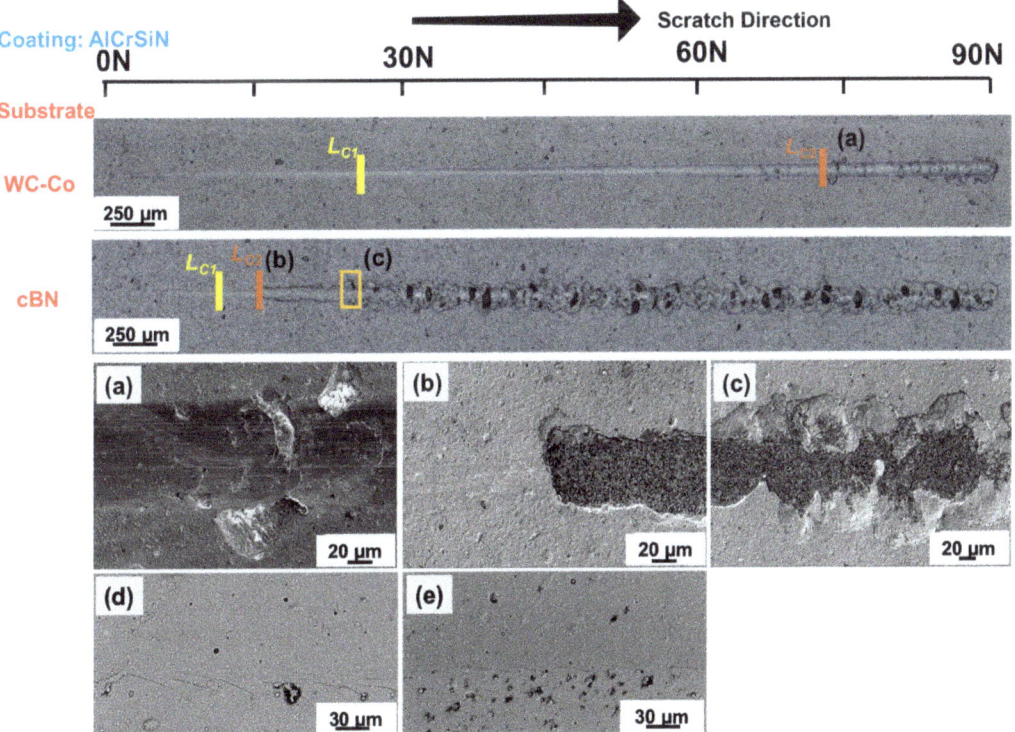

Figure 4. The optical profile and SEM images of failure after scratch tests of AlCrSiN coatings deposited on WC-Co and cBN, two different substrates. The second critical loads (L_{C2}) are magnified in the images of (**a**,**b**), respectively. (**c**) shows the scenario in which cBN substrate was totally exposed after receiving more of the second critical load. The first critical loads (L_{C1}) are magnified in the images of (**d**,**e**), respectively.

The scratch crack propagation resistance (CPRs) was calculated to rationalize the scratch resistance of the coatings in the coating–substrate systems, calculated using Equation (2) [37,38].

$$CPRs = L_{c1}(L_{c2} - L_{c1}) \qquad (2)$$

where L_{C1} is the first critical load and the start of lateral crack, and L_{C2} is the second critical load and the start of the coating delamination or spallation. The results are presented in Figure 4.

The critical stress σ_c was calculated using Equation (3).

$$\sigma_c = \left(\frac{2L_{c2}}{\pi d_c^2}\right)\left[\frac{(4+\nu_f)3\pi\mu}{8} - 1 + 2\nu_f\right] \qquad (3)$$

where d_c is the track width at L_{C2}, μ is the friction coefficient calculated using the friction force, and ν_f is the Poisson ratio of the coatings [39]. The surface energy of the known interfacial crack (adhesion energy) is then defined using Equation (4) [39–42].

$$G_c = \frac{\pi d_c^2}{2E} \qquad (4)$$

where t and E are the thickness and elastic modulus of coatings, respectively. The experimental values of t and E are presented in Tables 1 and 2, respectively.

The results of CPR and adhesion energy are presented in Table 4. The AlCrSiN coating deposited on the WC-Co substrate presents higher CPR values, which means better scratch resistance. The adhesion energy between the AlCrSiN coating and WC-Co substrate was higher than for the cBN substrate, which illustrates better adhesion strength between the coating and WC-Co substrate. Calculation of adhesion energy followed the method from the previous study by S. J. Bull et al., and the values of G_c were similar to previous research on similar coatings such as CrAlN and AlCrSiN [34,40,43,44].

Table 4. Scratch crack propagation resistance and adhesion energy of both coated samples.

Coating	Substrate	CPR (N^2)	G_C (J/m^2)
AlCrSiN	WC-Co	1244 ± 87	357 ± 36
	cBN	48 ± 7	210 ± 20

Figure 5 presents the SEM images after the contact damage tests at a series of normal loads of 98 N, 196 N, and 294 N. For the sample of AlCrSiN coating deposited on the WC-Co substrate, the coating does not show any cracking. At loads of 196 N and higher, radial cracking and delamination can be observed. In the case of the cBN substrate, delamination is clearly present even for the smaller loads. As the load increases, more spalling of the coating is observed, similar to previous research on similar materials [34,45,46].

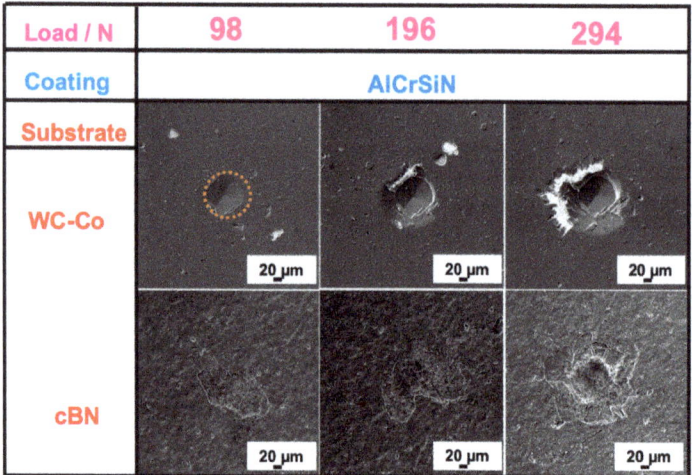

Figure 5. SEM images of the contact damage tests of AlCrSiN coatings at 98 N, 196 N, and 294 N with a Rockwell C tip for both substrates. Contact area of the coating on WC-Co for 98 N is indicated.

In order to further inspect the mechanisms of this delamination, lower loads of 9.8 N and 29.4 N were conducted on the cBN substrate sample, and the surface was observed using SEM, as presented in Figure 6. Under a lower normal load, delamination of the AlCrSiN coating appeared on the surface. Cohesive failure of the coating was evident with normal load of 9.8 N, as seen in Figure 6(a_2), where it can be seen that the coating had fractured, without exposing the substrate. However, when the normal load was increased to 29.4 N, the coating was totally delaminated from the cBN substrate, as seen in Figure 6(b_2). This conclusion is xposure of the cBN substrate indicates that adhesion failure between the AlCrSiN coating and cBN substrate was produced at that load. In any case, this illustrates worse mechanical performance as compared to the WC-Co substrate.

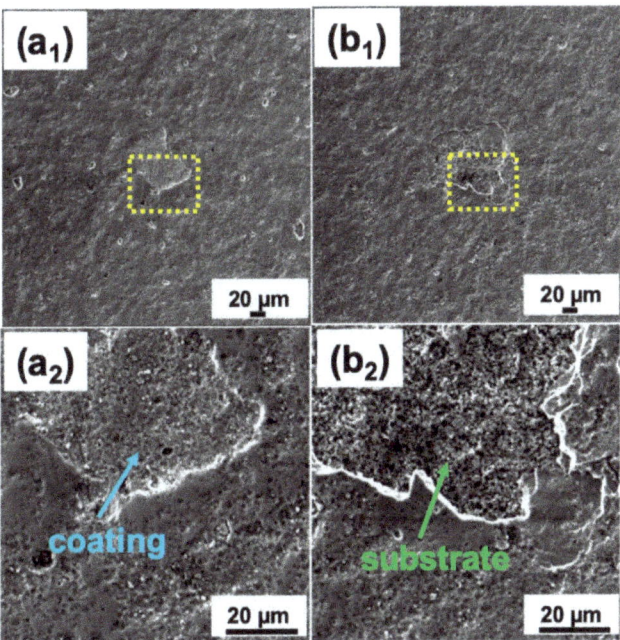

Figure 6. SEM images of the contact damage tests of AlCrSiN coatings deposited on cBN substrates with a Rockwell C tip. The failures in (a_1,a_2) were under 9.8 N and in (b_1,b_2) were at 29.4 N.

3.4. Mechanical Response under Contact Loading

Both materials were indented using a spherical indenter with a monotonic load and cyclic load to investigate the contact mechanical response under different conditions [47–49]. Figure 7 shows the SEM images of the damage failure induced via spherical indentation under monotonic loads. The surface of the cBN substrate sample presented discontinuous cracks under the monotonic loads of 2000 N. When the load increased up to 3000 N, the edge of the indentation formed a complete crack circle. However, for the samples deposited on the WC-Co, the discontinuous cracks and complete ring cracks appeared at larger loads of 5000 N and 6000 N, respectively, which shows the form and shape of the cracks over the load of the first damage. In agreement with previous results, the contact resistance of the coatings deposited on the WC-Co is larger than for the coatings deposited on the cBN substrates.

Based on the results, fatigue tests were conducted at the loads causing first damage, that is 5000 N and 2000 N for the samples on WC-Co and cBN substrates, respectively, as presented in Figure 8. For the cBN substrate, the ring crack was fully developed, and partial delamination could be observed, as seen in Figure 8(b_1). However, for the WC-Co substrate, the ring crack was not fully developed, even after 10^3 cycles under 5000 N, the same load that appeared for first damage, as seen in Figure 8(a_1). The results indicate that the AlCrSiN coatings deposited on the WC-Co substrate performed better under contact fatigue than the same coatings deposited on the cBN substrates.

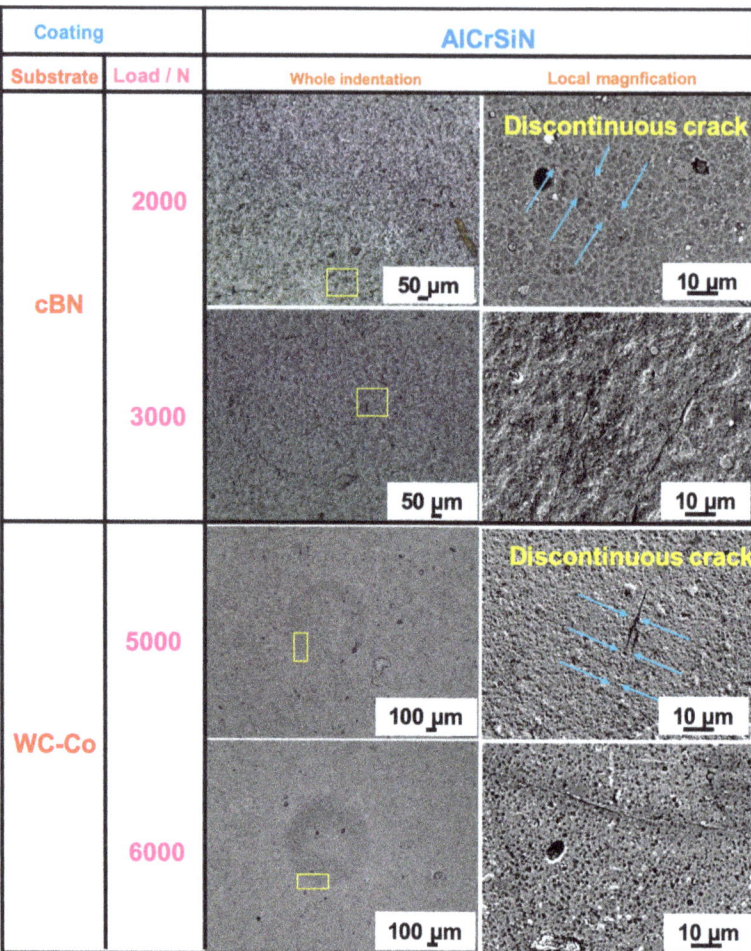

Figure 7. SEM images of damage features of AlCrSiN coatings deposited on WC-Co and cBN substrates induced via spherical indentation under different loads (2000 N, 3000 N, 5000 N, and 6000 N).

Figure 8. SEM images of spherical indentation cyclic loads with 10^3 cycles. (a_1) AlCrSiN coating deposited on WC-Co substrate under the load of 5000 N; (b_1) AlCrSiN coating deposited on cBN substrate under the load of 2000 N.

In order to further explore the direction of crack development caused by fatigue tests, the cross-section at the complete circle crack were performed using FIB, as shown in Figure 9. The ring cracks of the surface penetrated both the AlCrSiN coatings and both substrates. For the cBN substrate, a part of each coating was removed from the surface, as seen in Figure 9(b_1), with a delaminated area of coating forming a valley shape. However, as seen in Figure 8(b_1), the coatings were completely spalled from the cBN substrate, indicating that cohesive and adhesive failure of the coatings coexisted under contact fatigue loading. As seen in Figure 9(b_2), the ceramic binder of the cBN substrate suffered microcracking due to the cyclic loads. For the samples deposited on the WC-Co substrate, only some transverse cracks appeared in the coating and substrate, which kept their original morphologies, as seen in Figure 9(a_1,a_2); the overall shape of coating–substrate system remained intact, and transverse cracks only appeared inside the coatings but not at the interfaces. Another interesting phenomenon can be observed in Figure 9(b_1,b_2), where the interface between the AlCrSiN coatings and cBN substrate was not flat but rather kept the tortuous morphology of the surface of the cBN substrate. Multiple cracks were observed in the ceramic binder of the cBN substrate, and the substrate appeared to be crushed along with the periphery of the boron nitride particles.

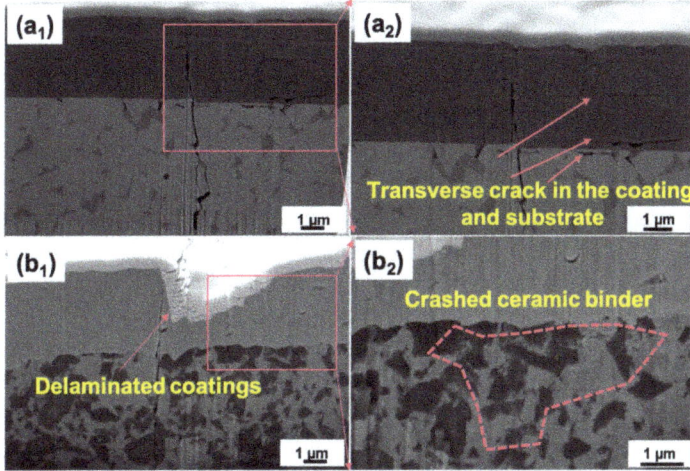

Figure 9. Cross-section SEM images of indentations and morphologies of fatigue cracks. (a_1,a_2) present the sample of AlCrSiN coating deposited on WC-Co substrate under the cyclic load of 6000 N with 10^3 cycles, (b_1,b_2) present the sample of AlCrSiN coating deposited on cBN substrate under the cyclic load of 3000 N with 10^3 cycles. (a_2,b_2) are the enlarged views of the corresponding areas in (a_1,b_1), respectively.

4. Discussion

From the results presented above, it is clear that coatings deposited on cBN have a lower mechanical performance as compared with the same coatings deposited on WC-Co substrates. Adhesion strength, as measured by scratch testing, is lower, implying that the coating delamination is easier in this substrate. In addition, different fracture features are evident for both substrates.

This lower adhesion can be partially attributed to the difference in chemical nature of each substrate. However, it is also seen that the structural integrity of the coating is also more affected when deposited on the cBN substrate under contact loadings, albeit having the same composition and intrinsic mechanical properties such as hardness and elastic modulus. In this sense, the differences in mechanical properties of the two substrates induce different contact damage responses and damage evolutions with increasing loads (as seen in Figures 7 and 8). As seen in Table 2, cBN is harder than WC-Co. If the contact loading is performed at relatively high loads (such as the contact damage and scratch tests) deformation will be a combined response of both coating and substrate. Because of the cBN substrate's higher hardness, its deformation will be concentrated close to the surface, which will lead to a larger amount of microcracking in order to accommodate the deformation [50]. These results are in agreement with Sveen et al. [27], who also observed much lower scratch resistance and different damage mechanisms for the cBN substrate. In the case of contact loads, the results are also in agreement with the work of S. Gordon et al. [21]. who studied TiN/TiAlN-coated cBN under contact loads and found that failure appears near the interface between the AlCrSiN coating and the cBN substrate. In Figure 9, in the FIB cross-section, one can see the existence of microcracking of the substrate, which can explain the failure of the substrate observed by these authors and in this work as well.

From the above results, it is clear that the contact resistance of coated hard materials is a multifaceted issue, in which the microstructures of both coating and substrate, as well as interfacial adhesion, play a key role.

5. Conclusions

The mechanical performances of the AlCrSiN coatings deposited on WC-Co and cBN, two different substrates, have been comparatively characterized. Our main conclusions can be summarized as:

(1) Nano-indentation results show the same value of mechanical properties and thicknesses of the coatings when deposited on both substrates as well as chemical compositions. The crystal structure of AlCrSiN coatings deposited on different substrates was mainly (Cr,Al)N solid solution. Si was not detected using XRD due to its relatively small amounts or the formation of an amorphous phase. Therefore, the coating structure is not affected by the substrate.

(2) Through microscratch tests, AlCrSiN coatings deposited on the WC-Co substrate presented better adhesion than those deposited on the cBN substrate. This is further evidenced by the contact damage tests, during which a part of an AlCrSiN coating totally detached from the cBN substrate at 29.4 N, but the delamination of the AlCrSiN coating on the cBN substrate appeared at 294 N.

(3) The results of mechanical response under contact Hertzian loads presented that AlCrSiN coatings deposited on the WC-Co substrate perform better resistance to monotonic loads and cyclic loads than those coatings deposited on the cBN substrate. Using SEM and FIB, cohesive and adhesive failures of AlCrSiN coatings were observed under cyclic loads when deposited on the WC-Co and cBN substrates, respectively.

The differences in mechanical performances of the same coating deposited on different substrates depend on both adhesion strengths and distinct mechanical properties of the substrates.

Author Contributions: Conceptualization, J.L., E.J.-P. and J.F.d.A.; experiments, J.L., M.S., S.G., E.A. and J.F.d.A.; analysis of the results, all authors; writing—original draft preparation, J.L.; writing—review and editing, E.J.-P., L.L., J.F.d.A. and J.L.; funding acquisition, E.J.-P. All authors have read and agreed to the published version of the manuscript.

Funding: This research was funded by The Spanish Ministry of Science, Innovation and Universities through grants PGC2018-096855-B-C41, PGC2018-096855-A-C44 and PID2021-126614OB-100.

Institutional Review Board Statement: Not applicable.

Informed Consent Statement: Not applicable.

Data Availability Statement: The data that support the findings of this study are available from the corresponding authors upon reasonable request.

Acknowledgments: The authors would like to thank R. M' Saoubi from SECO Tools for kindly providing the substrates. The work was funded by The Spanish Ministry of Science, Innovation and Universities through grants PGC2018-096855-B-C41, PGC2018-096855-A-C44 and PID2021-126614OB-100.

Conflicts of Interest: The authors declare no conflict of interest. The funders had no role in the design of the study; in the collection, analyses, or interpretation of data; in the writing of the manuscript; or in the decision to publish the results.

References

1. Rech, J.; Djouadi, M.; Picot, J. Wear resistance of coatings in high speed gear hobbing. *Wear* **2001**, *250*, 45–53. [CrossRef]
2. Fellah, M.; Aissani, L.; Zairi, A.; Samad, M.A.; Nouveau, C.; Touhami, M.Z.; Djebaili, H.; Montagne, A.; Iost, A. Thermal treatment effect on structural and mechanical properties of Cr–C coatings. *Trans. IMF* **2018**, *96*, 79–85. [CrossRef]
3. Spor, S.; Jäger, N.; Meindlhumer, M.; Hruby, H.; Burghammer, M.; Nahif, F.; Mitterer, C.; Keckes, J.; Daniel, R. Evolution of structure, residual stress, thermal stability and wear resistance of nanocrystalline multilayered Al0.7Cr0.3N-Al0.67Ti0.33N coatings. *Surf. Coat. Technol.* **2021**, *425*, 127712. [CrossRef]
4. Chen, Y.; Zhang, Z.; Yuan, T.; Mei, F.; Lin, X.; Gao, J.; Chen, W.; Xu, Y. The synergy of V and Si on the microstructure, tribological and oxidation properties of AlCrN based coatings. *Surf. Coat. Technol.* **2021**, *412*, 127082. [CrossRef]
5. Li, C.; Bing, Y.; Xu, Y.; Fei, P.; Zhou, L.; Yong, D. Improved thermal stability and oxidation resistance of Al–Ti–N coating by Si addition. *Thin Solid Films* **2014**, *556*, 369–375.
6. Mcintyre, D.; Greene, J.E.; Hakansson, G.; Sundgren, J.E.; Munz, W.D. Oxidation of metastable single-phase polycrystalline Ti0.5Al0.5N films: Kinetics and mechanisms. *J. Appl. Phys.* **1990**, *67*, 1542–1553. [CrossRef]
7. Veprek, S.; Männling, H.-D.; Jilek, M.; Holubar, P. Avoiding the high-temperature decomposition and softening of (Al1−xTix)N coatings by the formation of stable superhard nc-(Al1−xTix)N/a-Si3N4 nanocomposite. *Mater. Sci. Eng. A* **2004**, *366*, 202–205. [CrossRef]
8. Park, J.H.; Chung, W.S.; Cho, Y.-R.; Kim, K.H. Synthesis and mechanical properties of Cr–Si–N coatings deposited by a hybrid system of arc ion plating and sputtering techniques. *Surf. Coat. Technol.* **2004**, *188–189*, 425–430. [CrossRef]
9. Veprek, S.; Veprek-Heijman, M.; Karvankova, P.; Prochazka, J. Different approaches to superhard coatings and nanocom-posites. *Thin Solid Films* **2005**, *476*, 1–29. [CrossRef]
10. Baptista, A.; Silva, F.J.G.; Porteiro, J.; Míguez, J.L.; Pinto, G. Sputtering Physical Vapour Deposition (PVD) Coatings: A Critical Review on Process Improvement and Market Trend Demands. *Coatings* **2018**, *8*, 402. [CrossRef]
11. Leyendecker, T.; Lemmer, O.; Esser, S.; Ebberink, J. The development of the PVD coating TiAlN as a commercial coating for cutting tools. *Surf. Coat. Technol.* **1991**, *48*, 175–178. [CrossRef]
12. Oliver, W.C.; Pharr, G.M. An improved technique for determining hardness and elastic modulus using load and dis-placement sensing indentation experiments. *J. Mater. Res.* **1992**, *7*, 1564–1583. [CrossRef]
13. Vidakis, N.; Antoniadis, A.; Bilalis, N. The VDI 3198 indentation test evaluation of a reliable qualitative control for layered compounds. *J. Mater. Process. Technol.* **2003**, *143–144*, 481–485. [CrossRef]
14. Mishra, S.K.; Ghosh, S.; Aravindan, S. Investigations into friction and wear behavior of AlTiN and AlCrN coatings deposited on laser textured WC/Co using novel open tribometer tests. *Surf. Coat. Technol.* **2020**, *387*, 125513. [CrossRef]
15. Wu, Z.; Huan, J.; Geng, D.; Ye, R.; Wang, Q. Investigation on plastic deformation of arc-evaporated AlCrSiN and AlCrSiON nanocomposite films by indentation. *Surf. Coat. Technol.* **2022**, *441*, 128570. [CrossRef]
16. Berger, L.M. Coatings by Thermal Spray—ScienceDirect. *Compr. Hard Mater.* **2014**, *1*, 471–506.
17. Gant, A.J.; Morrell, R.; Wronski, A.S.; Jones, H.G. Edge toughness of tungsten carbide based hardmetals. *Int. J. Refract. Met. Hard Mater.* **2018**, *75*, 262–278. [CrossRef]
18. Huang, P.; Chou, Y.K.; Liang, S.Y. CBN tool wear in hard turning: A survey on research progresses. *Int. J. Adv. Manuf. Technol.* **2007**, *35*, 443–453. [CrossRef]
19. Cook, M.W.; Bossom, P.K. Trends and recent developments in the material manufacture and cutting tool application of polycrys-talline diamond and polycrystalline cubic boron nitride. *Int. J. Refract. Met. Hard Mater.* **2000**, *18*, 147–152. [CrossRef]
20. Bushlya, V.; Lenrick, F.; Bjerke, A.; Aboulfadl, H.; Thuvander, M.; Ståhl, J.-E.; M'Saoubi, R. Tool wear mechanisms of PcBN in machining Inconel 718: Analysis across multiple length scale. *CIRP Ann.* **2021**, *70*, 73–78. [CrossRef]
21. Gordon, S.; Roa, J.J.; Rodriguez-Suarez, T. Infuence of microstructural assemblage of the substrate on the adhesion strength of coated PcBN grades. *Ceram. Int.* **2022**, *48*, 22313–22322. [CrossRef]
22. Pozuelo, S.G. Microstructural and Mechanical Properties Correlation at the Micro- and Nanometric Length Scale of Different cBN Grade. Master's Thesis, Universitat Politècnica de Catalunya, Barcelona, Spain, July 2018.
23. Gordon, S.; García-Marro, F.; Rodriguez-Suarez, T.; Roa, J.; Jiménez-Piqué, E.; Llanes, L. Spherical indentation of polycrys-talline cubic boron nitride (PcBN): Contact damage evolution with increasing load and microstructural effects. *Int. J. Refract. Met. Hard Mater.* **2023**, *111*, 106115. [CrossRef]
24. Gordon, S.; Besharatloo, H.; Wheeler, J.; Rodriguez-Suarez, T.; Roa, J.; Jiménez-Piqué, E.; Llanes, L. Micromechanical mapping of polycrystalline cubic boron nitride composites by means of high-speed nanoindentation: Assessment of microstructural assemblage effects. *J. Eur. Ceram. Soc.* **2023**, *43*, 2968–2975. [CrossRef]
25. Gao, Y.; Cai, F.; Lu, X.; Xu, W.; Zhang, C.; Zhang, J.; Qu, X. Design of cycle structure on microstructure, mechanical properties and tribology behavior of AlCrN/AlCrSiN coatings. *Ceram. Int.* **2022**, *48*, 12255–12270. [CrossRef]

26. Tuchida, K.; Wathanyu, K.; Surinphong, S. Thermal Oxidation Behavior of TiAlCrSiN and AlCrTiN Films on HastelloyX. In Proceedings of the 2012 International Conference on Nanotechnology Technology and Advanced Materials (ICNTAM 2012), Hong Kong, China, 12–13 April 2012.
27. Sveen, S.; Andersson, J.; M'saoubi, R.; Olsson, M. Scratch adhesion characteristics of PVD TiAlN deposited on high speed steel, cemented carbide and PCBN substrates. *Wear* **2013**, *308*, 133–141. [CrossRef]
28. Mosquera, A.; Mera, L.; Fox-Rabinovich, G.; Martínez, R.; Azkona, I.; Endrino, J. Advantages of nanoimpact fracture testing in studying the mechanical behavior of CrAl (Si) N coatings. *Nanosci. Nanotechnol. Lett.* **2010**, *2*, 352–356. [CrossRef]
29. Oliver, W.C.; Pharr, G.M. Measurement of hardness and elastic modulus by instrumented indentation: Advances in understanding and refinements to methodology. *J. Mater. Res.* **2004**, *19*, 3–20. [CrossRef]
30. Kayali, Y.; Taktak, S. Characterization and Rockwell-C adhesion properties of chromium-based borided steels. *J. Adhes. Sci. Technol.* **2015**, *29*, 2065–2075. [CrossRef]
31. Zhu, L.-H.; Song, C.; NI, W.-Y.; Liu, Y.-X. Effect of 10% Si addition on cathodic arc evaporated TiAlSiN coatings. *Trans. Nonferrous Met. Soc. China* **2016**, *26*, 1638–1646. [CrossRef]
32. Ho, W.Y.; Hsu, C.H.; Chen, C.W.; Wang, D.Y. Characteristics of PVD-CrAlSiN films after post-coat heat treatments in nitrogen atmosphere. *Appl. Surf. Sci.* **2011**, *257*, 3770–3775. [CrossRef]
33. Gui, Y.; Zhao, J.; Chen, J.; Jiang, Y. Preparation and Characterization of Ni Spines Grown on the Surface of Cubic Boron Nitride Grains by Electroplating Method. *Materials* **2016**, *9*, 153. [CrossRef] [PubMed]
34. Liang, J.; Almandoz, E.; Ortiz-Membrado, L.; Rodríguez, R.; de Ara, J.F.; Fuentes, G.G.; Llanes, L.; Jiménez-Piqué, E. Mechanical Performance of AlCrSiN and AlTiSiN Coatings on Inconel and Steel Substrates after Thermal Treatments. *Materials* **2022**, *15*, 8605. [CrossRef]
35. *ASTM C1624-22; Standard Test Method for Adhesion Strength and Mechanical Failure Modes of Ceramic Coatings by Quantitative Single Point Scratch Testing.* ASTM International: West Conshohocken, PA, USA, 2010.
36. Zhang, X.; Tian, X.-B.; Zhao, Z.-W.; Gao, J.-B.; Zhou, Y.-W.; Gao, P.; Guo, Y.-Y.; Lv, Z. Evaluation of the adhesion and failure mechanism of the hard CrN coatings on different substrates. *Surf. Coat. Technol.* **2019**, *364*, 135–143. [CrossRef]
37. Zhang, S.; Sun, D.; Fu, Y.; Du, H. Effect of sputtering target power on microstructure and mechanical properties of nano-composite nc-TiN/a-SiNx thin films. *Thin Solid Films* **2004**, *447*, 462–467. [CrossRef]
38. Kabir, M.S.; Munroe, P.; Zhou, Z.; Xie, Z. Scratch adhesion and tribological behaviour of graded Cr/CrN/CrTiN coatings synthesized by closed-field unbalanced magnetron sputtering. *Wear* **2017**, *380–381*, 163–175. [CrossRef]
39. Wang, Q.; Zhou, F.; Zhou, Z.; Li, L.K.-Y.; Yan, J. An investigation on the crack resistance of CrN, CrBN and CrTiBN coatings via nanoindentation. *Vacuum* **2017**, *145*, 186–193. [CrossRef]
40. Bull, S.; Rickerby, D. New developments in the modelling of the hardness and scratch adhesion of thin films. *Surf. Coat. Technol.* **1990**, *42*, 149–164. [CrossRef]
41. Huang, Y.-C.; Chang, S.-Y.; Chang, C.-H. Effect of residual stresses on mechanical properties and interface adhesion strength of SiN thin films. *Thin Solid Films* **2009**, *517*, 4857–4861. [CrossRef]
42. Chang, S.-Y.; Tsai, H.-C.; Chang, J.-Y.; Lin, S.-J.; Chang, Y.-S. Analyses of interface adhesion between porous SiOCH low-k film and SiCN layers by nanoindentation and nanoscratch tests. *Thin Solid Films* **2008**, *84*, 5334–5338. [CrossRef]
43. Wang, Q.; Zhou, F.; Yan, J. Evaluating mechanical properties and crack resistance of CrN, CrTiN, CrAlN and CrTiAlN coatings by nanoindentation and scratch tests. *Surf. Coat. Technol.* **2016**, *285*, 203–213. [CrossRef]
44. Patnaik, L.; Maity, S.R.; Kumar, S. Evaluation of Crack resistance and Adhesive Energy of AlCrN and Ag doped a-C Films deposited on Chrome Nitrided 316 LVM Stainless Steel. *Adv. Mater. Process. Technol.* **2022**, *8*, 1048–1069. [CrossRef]
45. Casas, B.; Anglada, M.; Sarin, V.K.; Llanes, L. TiN coating on an electrical discharge machined WC-Co hardmetal: Surface integrity effects on indentation adhesion response. *J. Mater. Sci.* **2006**, *41*, 5213–5219. [CrossRef]
46. Yang, J.; Odén, M.; Johansson-Jõesaar, M.P.; Llanes, L. Influence of substrate microstructure and surface finish on cracking and delamination response of TiN-coated cemented carbides. *Wear* **2016**, *352*, 102–111. [CrossRef]
47. Tarrés, E.; Ramírez, G.; Gaillard, Y.; Jiménez-Piqué, E.; Llanes, L. Contact fatigue behavior of PVD-coated hardmetals. *Int. J. Refract. Met. Hard Mater.* **2009**, *27*, 323–331. [CrossRef]
48. Llanes, L.; Tarrés, E.; Ramírez, G.; Botero, C.A.; Jiménez-Piqué, E. Fatigue susceptibility under contact loading of hardmetals coated with ceramic films. *Procedia Eng.* **2010**, *2*, 299–308. [CrossRef]
49. Zheng, Y.; Fargas, G.; Lavigne, O.; Serra, M.; Coureaux, D.; Zhang, P.; Yao, Z.; Zhang, Q.; Yao, J.; Llanes, L. Contact fatigue of WC-6%wtCo cemented carbides: Influence of corrosion-induced changes on emergence and evolution of damage. *Ceram. Int.* **2022**, *48*, 5766–5774. [CrossRef]
50. Kong, Y.; Tian, X.; Gong, C.; Chu, P.K. Enhancement of toughness and wear resistance by CrN/CrCN multilayered coatings for wood processing. *Surf. Coat. Technol.* **2018**, *344*, 204–213. [CrossRef]

Disclaimer/Publisher's Note: The statements, opinions and data contained in all publications are solely those of the individual author(s) and contributor(s) and not of MDPI and/or the editor(s). MDPI and/or the editor(s) disclaim responsibility for any injury to people or property resulting from any ideas, methods, instructions or products referred to in the content.

Article

Thermal Fracture of Functionally Graded Coatings with Systems of Cracks: Application of a Model Based on the Rule of Mixtures

Vera Petrova *, Siegfried Schmauder and Alexandros Georgiadis

Institute for Materials Testing, Materials Science and Strength of Materials (IMWF), University of Stuttgart, Pfaffenwaldring 32, 70569 Stuttgart, Germany
* Correspondence: vera.petrova@imwf.uni-stuttgart.de; Tel.: +49-711-685-63339

Abstract: This paper is devoted to the problem of the thermal fracture of a functionally graded coating (FGC) on a homogeneous substrate (H), i.e., FGC/H structures. The FGC/H structure was subjected to thermo-mechanical loadings. Systems of interacting cracks were located in the FGC. Typical cracks in such structures include edge cracks, internal cracks, and edge/internal cracks. The material properties and fracture toughness of the FGC were modeled by formulas based on the rule of mixtures. The FGC comprised two constituents, a ceramic on the top and a metal as a homogeneous substrate, with their volume fractions determined by a power law function with the power coefficient λ as the gradation parameter for the FGC. For this study, the method of singular integral equations was used, and the integral equations were solved numerically by the mechanical quadrature method based on the Chebyshev polynomials. Attention was mainly paid to the determination of critical loads and energy release rates for the systems of interacting cracks in the FGCs in order to find ways to increase the fracture resistance of FGC/H structures. As an illustrative example, a system of three edge cracks in the FGC was considered. The crack shielding effect was demonstrated for this system of cracks. Additionally, it was shown that the gradation parameter λ had a great effect on the fracture characteristics. Thus, the proposed model provided a sound basis for the optimization of FGCs in order to improve the fracture resistance of FGC/H structures.

Keywords: thermal fracture; system of cracks; functionally graded coatings; fracture toughness; rule of mixtures

1. Introduction

Functionally graded coatings (FGCs) are widely used in different engineering structures, e.g., for thermal barrier coatings (TBCs) in gas turbine engine blades and vanes to protect metal components from overheating and melting [1]. Functionally graded materials (FGMs) are a special type of composite, consisting of a graded pattern of material composition and/or microstructures, so that the properties of FGMs vary continuously across spatial coordinates through the depth of the layer. FGCs with a gradual compositional variation from heat-resistant ceramics to fracture-resistant metals have been proposed in order to reduce thermal residual stresses causing delamination and the debonding of interfaces, enhance coating toughness, and improve the long-term performance of TBCs. However, cracks can occur because of initial defects or microcracks that appear during manufacturing or operation. Therefore, the study of the fracture of FGCs is important for a better understanding of the fracture resistance of graded coatings.

At present, great progress has been achieved in the study of the fracture behavior of FGMs and FGM structures. An overview of important trends in the study of the fracture behavior of FGMs is presented in [2], where one can find useful references on a wide range of problems, including the thermal and thermo-mechanical fracture of FGMs and related structures. A recent review [3] includes work on FGM fracture under mechanical loads, especially dynamic and fatigue loads, and discusses crack propagation and crack growth trajectory in FGMs, as well as the effects of material gradation on crack tip fields and crack growth.

Finite element methods and extended finite element methods are extensively used for the computational modeling of FGMs and cracks in FGMs (see references in [2]). Recently, the methods of peridynamics [4] and the phase field approach [5] have been effectively applied in modeling the complex crack propagation paths in FGMs. In [4], systems of interacting cracks in FGMs, namely a macrocrack and various microcrack configurations, were studied with respect to the shielding effect of microcracks (and improved fracture toughness) or the opposite effect of crack propagation acceleration due to microcracks. In [5], 2D and 3D cracks in FGMs were investigated by the phase field formulation. Finite–discrete element methods are also applied in fracture problems for FGMs. For example, in [6], such a method was used to analyze a coupled thermo-mechanical fracture problem for a system of two interacting edge cracks in an FGM.

Analytical and semi-analytical solutions are important tools in fracture studies for FGMs. They can be used as a separate approach for a study, or as a part of a numerical modeling process. An overview of some of the analytical methods used to solve crack problems in FGMs can be found in [2]. Analytical solutions for steady-state heat transfer in a finite cylinder and a hollow sphere made of an FGM (both undamaged) are presented in [7,8], respectively. For the thermal conductivity coefficient, a power function of two spatial coordinates was used. The Robin boundary condition (also called the third type boundary condition) [9] was adopted for this problem. The choice of this FGM model and these boundary conditions made it possible to derive the exact general analytical solution. The influence of material constants and the conductivity ratio was investigated to shed light on the material selection with respect to the temperature distribution. It should be noted that in [10], the use of the Robin boundary condition in the stationary problem of heat conduction helped to derive explicit representations of the singular terms of the asymptotic expansion of the heat flow in the vicinity of the crack tips in an FGM plate with a crack.

Functional gradation opens up new directions for optimizing both material and component structures to achieve high performance and material efficiency. At the same time, it posits many challenging mechanics problems, including the prediction and measurement of the effective properties, thermal stress distributions, and fracture of FGMs. The modeling and evaluation of the effective properties of FGMs has been considered in many works [11–13]. In [14], a comprehensive experimental and numerical study of the deformation and fracture of aluminum matrix–carbide particle composites was carried out. The applied numerical procedure could be useful for studying the fracture of ceramic/metal FGM structures. The evaluation of the fracture toughness of FGMs is also a very important problem for studying FGC fractures [15–18]. In [19], the analysis of the fracture parameters with respect to critical loads for a system of edge and internal cracks in FGCs showed the importance of taking into account the variation in fracture toughness through the thickness of the FGCs.

In our previous works [19–23], the theoretical study of the thermal and thermomechanical fracture of FGC/H structures was carried out under thermo-mechanical loading. These simulations were performed for different crack systems and geometries (edge crack systems [20], edge and internal cracks [21], edge and internal cracks imitating a curved interface [22]). Models for FGMs and some special models for cracks, such as partially thermally permeable cracks and cracks with contact, which could be used in the problem considered herein, were discussed in [23]. In [19–23], an exponential model of material properties was used. In the present work, the problem of the thermal fracture of FGM/H structures was investigated with the application of functions based on the rule of mixtures. To the best of the authors' knowledge, the problem of the interaction of arbitrarily located cracks in FGC/H structures with material properties described by functions based on the rule of mixtures has not be previously addressed.

Considering the possibility of optimizing an FGC in terms of improving its fracture toughness, material models based on the rule of mixtures are preferable. These models contain a gradation parameter, which can be used to evaluate the profile of changes in the properties over the coating thickness. The rule of mixtures, originally applied to conventional composites, has been successfully used in FGMs [13,15]. The present work was devoted to the problem of the thermal fracture of FGC/H structures using functions based on the rule of mixtures. Emphasis was placed on the determination of critical stresses and the energy release rate near crack tips. One illustrative example is discussed for a system of interacting edge cracks in an FGC/H (PSZ/steel) structure.

2. Formulation of the Problem

In this theoretical work, the problem was considered for structures consisting of a functionally graded coating (FGC) of thickness h and an underlying homogeneous substrate (H), that is, for FGC/H structures, as shown in Figure 1. For thermal barrier coatings (TBCs), the top of the FG layer is made of a ceramic, a material with low thermal conductivity, and the homogeneous substrate is made of a metal. The FGC/H structure is cooled by ΔT, $\Delta T > 0$ (e.g., cooling from operating temperatures), and a tensile load p is applied parallel to the surface (see Figure 1).

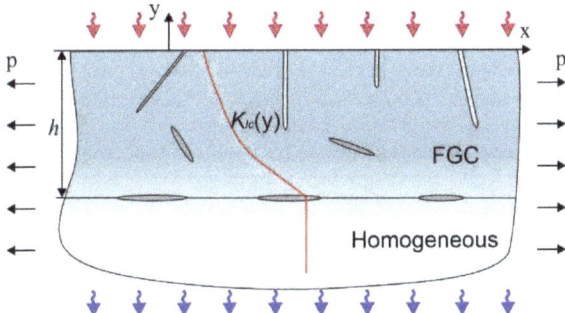

Figure 1. FGC/H structure with a system of cracks under the influence of thermal and mechanical loads. The fracture toughness $K_{Ic}(y)$ changes with the y-coordinate through the FGC thickness.

Pre-existing systems of cracks (length $2a_k$) in the FGCs were considered. A global coordinate system (x, y) with the x-axis lying on the FGC's surface was introduced, and the local coordinates (x_k, y_k) refer to each crack, with the x_k-axis on the crack lines, as shown in Figure 2a. The position of cracks was determined explicitly by their midpoint coordinates (x_k^0, y_k^0) (or in the complex form $z_k^0 = x_k^0 + i y_k^0$, where i is the imaginary unit) and the inclination angles α_k to the x-axis, or β_k for edge cracks ($\beta_k = -\alpha_k$), see Figure 2a.

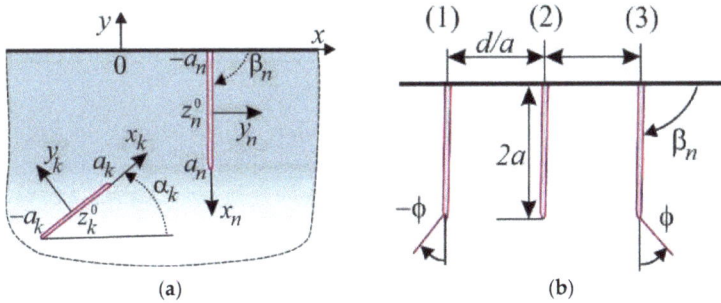

Figure 2. (a) Global (x, y) and local (x_k, y_k) coordinate systems; (b) three edge cracks.

The following assumptions were used for this problem. (1) The uncoupled, quasi-static thermo-elasticity theory was applied, i.e., that the temperature distribution is independent of the mechanical field. (2) The thermal and mechanical properties of an FGC are continuous functions of the thickness coordinate y. (3) The non-homogeneity of the functionally graded material is revealed in the form of the corresponding inhomogeneous stress distributions on the surfaces of cracks [17,24,25]. These assumptions were also used in our previous works (e.g., see [19–23]).

3. Material Properties for FGCs and Residual Stresses

Due to the applications of TBCs and their requirements, in this study, only (ceramic/metal)/metal coatings were considered, that is, coatings in which the material composition varies with the y-coordinate from ceramic at the top of the FGC to metal in the substrate. Consequently, the thermal and mechanical properties of the FGC also vary continuously with the thickness coordinate y and can be mathematically described by a continuous function. It should be noted that this method is applicable to different material combinations.

In our previous works [19,21], an exponential form of Young's modulus and the thermal expansion coefficient was used, while in [23], a linear model was applied. The Poisson's ratio was assumed to be constant and equal to the value of the homogeneous substrate. Along with the change in these mechanical properties, the change in fracture toughness was also taken into account [19,21]. The importance of considering fracture toughness variation when determining critical stresses and assessing fractures was demonstrated in [19].

Another possibility for modeling FGC properties is the rule of mixtures, which with its various modifications has long been used for conventional composites. In contrast to conventional composites, in FGMs, the volume fraction of one material in relation to the other varies; thus, the effective properties for FGMs depend on the volume fraction of one material in relation to the other. In the present study, the thermal and mechanical properties as well as the fracture toughness (K_{Ic}) of the FGC were modeled by the rule of mixtures. The FGC consisted of two constituents, a ceramic on the top and a metal as a homogeneous substrate, with their volume fractions V_c and V_m, respectively, determined by a power law function:

$$V_m(y) = \left[\frac{y}{h}\right]^\lambda = \left[\frac{1}{h}\left(h + \text{Im}(z_n^0) - x_n \sin(\beta_n)\right)\right]^\lambda, \tag{1}$$

$$V_c(y) = 1 - V_m(y) \tag{2}$$

with the power coefficient λ as the gradation parameter for the FGC. This parameter could be set to different values to realize different volume fractions as desired. The indices c and m refer to the volume of ceramic and metal, respectively. The case $\lambda = 0$ corresponds to a homogeneous material.

The material properties (the coefficient of thermal expansion and Young's modulus) of the functionally graded coating were assumed to take the following forms:

$$\alpha_t(y) = \frac{\alpha_{tm} V_m(y) E_m/(1-\nu) + \alpha_{tc} V_c(y) E_c/(1-\nu)}{V_m(y) E_m/(1-\nu) + V_c(y) E_c/(1-\nu)}, \tag{3}$$

$$E(y) = E_c \left[\frac{E_c + (E_m - E_c) V_m(y)^{(2/3)}}{E_c + (E_m - E_c)\left(V_m(y)^{(2/3)} - V_m(y)\right)}\right] \tag{4}$$

These expressions for the effective properties were used by Noda et al. [13]. In Equation (3), ν is Poisson's coefficient.

The fracture toughness (K_{Ic}) for FGCs also varies with the coordinate y and can be determined using one of the models described in [13,15]. This change in $K_{Ic}(y)$ is schematically shown in Figure 1. In [19,21,22], the fracture toughness for an FGM was determined theoretically and used for calculations. In the present work, to determine the fracture toughness, function (4) was used, where E should be replaced by K_{Ic}.

The method of superposition was used to solve this problem, so that loads at infinity are reduced to the corresponding loads on the crack faces. Thus, the tensile load is reduced to the load p_n on the crack surfaces (see, e.g., [26]):

$$p_n = \sigma_n - i\tau_n = p(1 - \exp(2i\beta_n))/2 = pf(\beta_n) \quad (n = 1, 2, ..., N) \tag{5}$$

Here, N is the number of cracks. Additionally, as the temperature changes, e.g., when the structure is cooled by ΔT, residual stresses arise due to the mismatch in the thermal expansion coefficients (σ_n^T). Furthermore, the change in E leads to residual stresses σ_n^e. Thus, in standard FGCs, the full load on the n-th crack consists of p_n, σ_n^T, and σ_n^e, where the index "n" denotes that the functions are written in the local coordinate system (x_n, y_n) connected with the n-th crack (see, [24]):

$$p_n^* = p_n + \sigma_n^e + \sigma_n^T = pf(\beta_n) + pf(\beta_n)[E(y) - 1] + Q[\alpha_t(y) - 1]E(y), \quad (n = 1, 2, ..., N). \tag{6}$$

$$Q = \alpha_{t1}\Delta T E_1 \tag{7}$$

It is assumed that $p = Q$; otherwise, the additional loading parameter p/Q should be considered. E_1 and α_{t1} are the material parameters of the substrate, and $\alpha_t(y)$ and $E(y)$ are obtained from Equations (3) and (4), respectively.

4. Solution and Determination of Fracture Characteristics

The method of singular integral equations was used. The integral equations were formulated using the method of complex potentials (see, [26]). On the right side of the equations were the known functions for the loads (see Equation (6)), and the unknowns were the derivative of the displacement jumps on the crack lines. The equations were solved numerically using the method of mechanical quadrature, based on the Chebyshev polynomials [26]. As a result of this method, the system of singular integral equations was reduced to a system of algebraic equations, from which the unknown derivatives of the displacements jumps on the crack lines were determined. Then, the stress intensity factors near the crack tips were calculated.

A complete description of the singular integral equations for this problem, as well as their numerical solutions, is provided in [21,22] and is not repeated here. The present equations differ from those previously described only in the right-hand sides containing the load functions. In the present problem, these load functions were defined by Equations (6) and (7).

In the considered problem, the cracks are mainly under mixed-mode loading conditions. The mixed-mode loading is due not only to the applied thermal and mechanical loads, but also to the interaction of cracks and the gradation of the material. That is, both stress intensity factors, Mode I and Mode II, are generally not equal to zero. In this case, the cracks will deviate from their initial paths. For predicting crack growth and determining its direction, the criterion of maximum circumferential stresses [27] was applied in this study. According to this criterion, the crack deviation angle ϕ (or the so-called fracture angle, see Figure 2b) and the critical stress intensity factors were calculated using the following relations:

$$\phi_n = 2\arctan\left[\left(K_{In} - \sqrt{K_{In}^2 + 8K_{IIn}^2}\right)/4K_{IIn}\right], \tag{8}$$

$$K_n^{eq} \equiv \cos^3(\phi_n/2)(K_{In} - 3K_{IIn}\tan(\phi_n/2)) = K_{Ic,\,tip} \text{ or } K_n^{eq} = K_{Ic,\,tip} \tag{9}$$

Equation (9) shows that crack growth occurs as soon as the equivalent stress intensity factor K_n^{eq} reaches the fracture toughness K_{Ic}. From Equation (9), the critical loads were obtained as follows:

$$\frac{P_{crn}}{p_0} = \frac{K_{Ic,n\,tip}}{\cos^3(\phi_n/2)(k_{In} - 3k_{IIn}\tan(\phi_n/2))} \frac{\sqrt{a}}{\sqrt{a_n}}, \quad (10)$$

where $p_0 = K_{Ic1}/\sqrt{\pi a}$ is the critical load for a single reference crack subjected to a load p normal to the crack line with the stress intensity factor $K^0 = p\sqrt{\pi a}$ and $a = \max_{n=1,\ldots,N} a_n$ (n = 1,2, ..., N, where N is the number of cracks). For an FGM, K_{Ic} is defined by an expression similar to Equation (4).

In Equations (8)–(10), the dimensional and non-dimensional stress intensity factors (SIFs) are related as follows:

$$K_{In} - iK_{IIn} = p\sqrt{\pi a_n}(k_{In} - ik_{IIn}) \quad (11)$$

The fracture angle ϕ_n is shown in Figure 2b. First, the angle of the crack propagation (fracture angle, Equation (8)) was obtained. Then, the critical loads were calculated near the crack tips (Equation (10)). Finally, the weakest crack was defined by the following conditions:

$$P_{cr} = \min_n p_{cr\,n}/p_0 \quad (n = 1, 2, ..., N).$$

In the present work, attention was mainly paid to the determination of critical loads for systems of interacting cracks in an FGC. Crack path predictions (based on fracture angles) for other material models were investigated, e.g., in [22].

Another fracture characteristic that is particularly useful for studying fracture in FGCs is the energy release rate:

$$\frac{G_n}{G_0} = \left[k_{In}^2 + k_{IIn}^2\right]\frac{a_n}{a} \cdot \frac{E_1'}{E'(y)} \quad (12)$$

where E' represents E for the plane stress and $E/(1-v^2)$ for the plane strain state; E_1 is the Young's modulus of the substrate; $E(y)$ is given in Equation (4); the energy release rate G_0 refers to a single reference crack; and the non-dimensional stress intensity factors k_I and k_{II} are defined in Equation (11).

5. Results

As an illustrative numerical example, consider three edge cracks as shown in Figure 2b. The results were obtained for the following parameters: d/a = 4.0, h/a = 4.0, 2a—crack size; ΔT = 300 °C. The thermal load Q is defined by Equation (7). The material parameters are listed in Table 1 and were taken from [28], where further useful references can be found.

Table 1. Material parameters.

Material Property	Top Coat (Ceramic)—PSZ	Bottom Coat (Metal)—Steel
Young's modulus (GPa)	48.0	207.0
Fracture toughness (MPa m$^{1/2}$)	7.0	50.0
Thermal exp. coeff. (10^{-6} K^{-1})	9.0	15.0

The normalized critical loads p_{cr}/p_0 as functions of the inclination angle β ($\beta_n = \beta$, n = 1, 2, 3) are shown in Figure 3. Figure 3a,b present the results for three FGM models: exponential, linear, and the rule of mixtures. Figure 3c,d present the results for the rule of mixtures for different λ values (0.1, 0.9, 1.9, and 2.9). Figure 3a,c refer to crack 1, and Figure 3b,d refer to the middle crack 2 (see geometry in Figure 2b). The fracture characteristics for crack 3 were similar to those of crack 1 and are not shown in the figures. The largest value for p_{cr}/p_0 was for crack 2, and the smallest was for crack 1; thus, a shielding effect was observed for crack 2. The lowest p_{cr}/p_0 value was for the rule of mixtures model (for $\lambda = 1$), as seen in Figure 3a,b. The p_{cr}/p_0 increased with increasing λ, and for $\lambda = 2.9$, the values were close to p_{cr}/p_0 for the exponential model (compare Figure 3a,c for crack 1, and Figure 3b,d for crack 2).

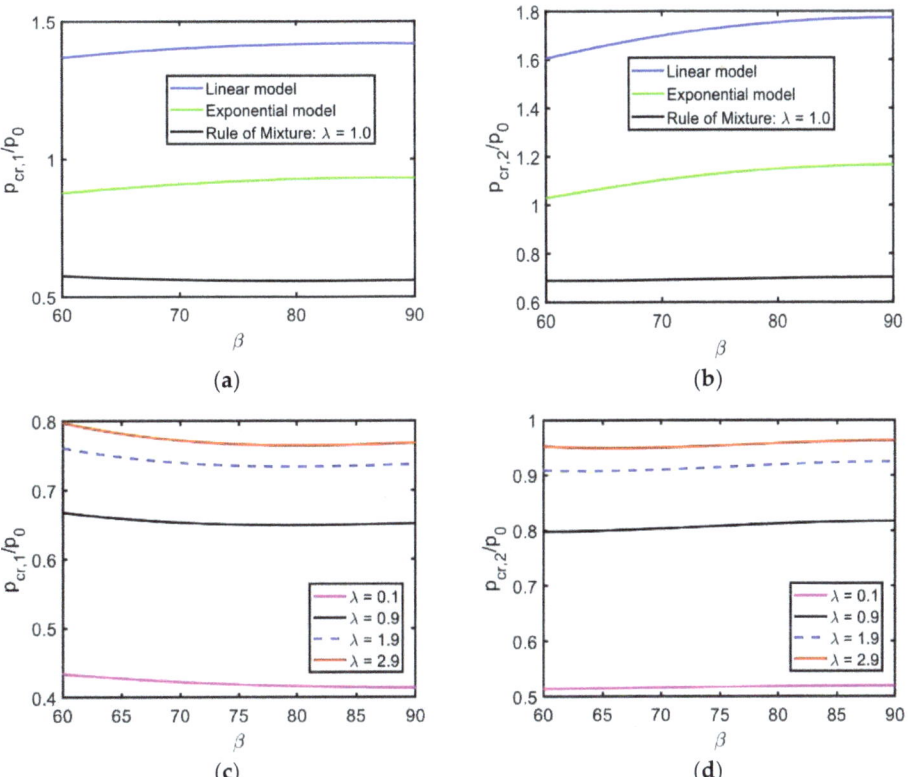

Figure 3. Critical loads as a function of β: (**a**,**b**) for three FG models; (**c**,**d**) for the rule of mixtures for different values of λ (0.1, 0.9, 1.9, and 2.9); (**a**,**c**) for crack 1 and (**b**,**d**) for middle crack 2.

Different values of the critical load for different material models (as seen in Figure 3) but the same crack showed a difference in the fracture toughness values determined by these models, which in turn showed different concentrations of ceramic (metal) near the crack tip. Different models (or different values of the gradation parameter λ in the rule of mixtures model) provided different profiles of material change, which meant different K_{Ic} values near the crack tip. As can be seen from Equation (10), p_{cr}/p_0 depended on K_{Ic}.

Ceramics in coatings create a thermal barrier and protect metal parts from overheating. The metal content in an FGC increases the strength of the FGC but reduces the thermal insulation. Thus, it is necessary to balance the content of ceramic and metal in coatings to improve the thermal insulation properties of FGCs.

Figure 4 depicts the effect of the thermal load Q on the energy release rate (ERR) (Figure 4a,b) and critical loads p_{cr}/p_0 (Figure 4c,d) for a wide range of grading parameter λ values (from 0.1 to 3.0) in the rule of mixtures model. The results are presented for crack 1 (Figure 4a,c) and middle crack 2 (Figure 4a,c); the cracks were inclined on 90° to the surface. The parameter ranges of thermal loading Q corresponded to a temperature change ΔT (°C) from 300° to 400°.

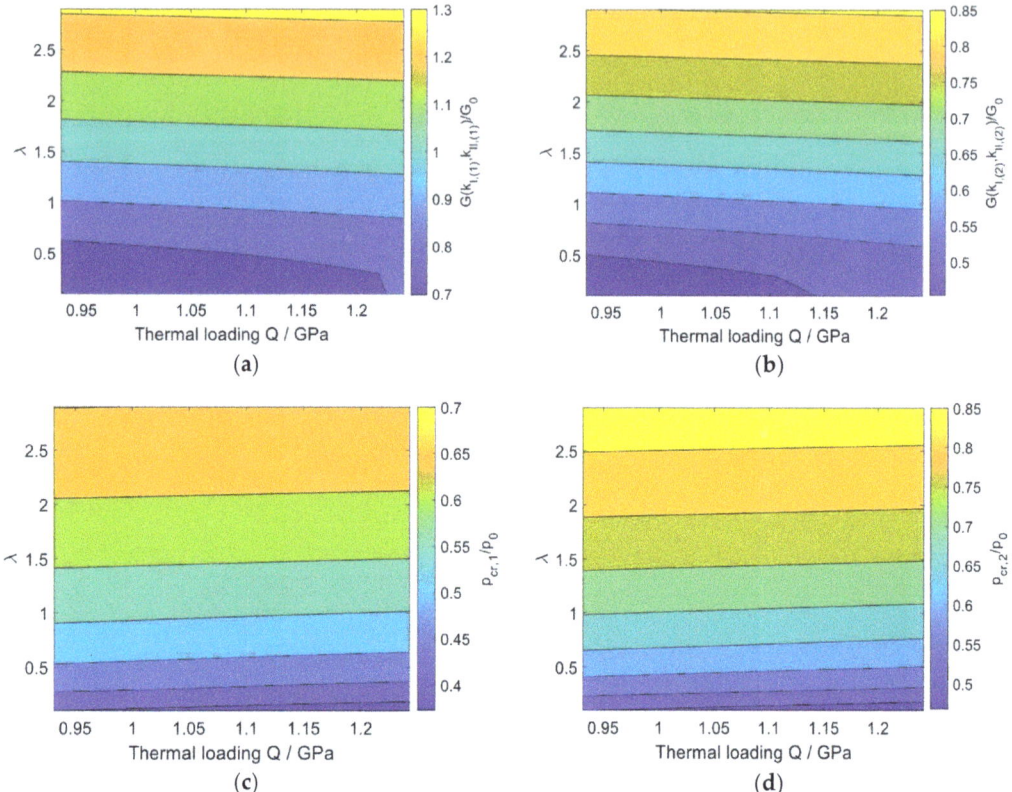

Figure 4. Influence of thermal loading Q and gradation parameter λ on ERR (**a**,**b**) and critical load (**c**,**d**); (**a**,**c**) for crack 1 and (**b**,**d**) for middle crack 2.

An increase in the thermal loading Q caused an increase in the energy release rate but decreased the critical load; that is, a reduction in the resistance to crack propagation was observed. This was due to the higher loading (additional residual stress) acting on the cracks.

The dependence of the fracture characteristics on λ was as follows: with an increase in the gradation parameter λ, both ERR and p_{cr}/p_0 increased. A higher gradation parameter corresponded to a higher metal content and a lower ceramic content in the FGC.

6. Conclusions

A theoretical study based on the method of singular integral equations was performed for FGC/H structures under thermo-mechanical loads. For the modeling of the material gradation of FGCs, functions based on the rule of mixtures were used. The structural variation of fracture toughness in FGCs was also taken into account and was shown to play an important role in evaluating fracture characteristics, especially critical loads. An illustrative example of three edge cracks in an FGC was studied in detail to investigate the influence of the material and geometrical parameters of the problem on the fracture characteristics of interacting cracks. The weakest cracks in the considered system of three edge cracks were the outer edge cracks in the FGC/H (ceramic/metal)/metal structure. A shielding effect was observed for the middle crack. For the rule of mixture model, it was observed that the variation in gradation parameter λ led to a significant change in the fracture parameters, in particular, the critical loads and energy release rates.

The above results for critical loads and energy release rates demonstrate their strong dependence on the gradation parameter λ for different combinations of materials in coatings. Thus, this model is a good candidate for a theoretical evaluation of the desired material properties of FGCs for their further development in order to improve the thermal fracture resistance of coatings.

Author Contributions: Conceptualization, V.P. and S.S.; methodology, V.P.; software, V.P. and A.G.; validation, V.P. and A.G.; formal analysis, V.P. and A.G.; investigation, V.P. and A.G.; writing—original draft preparation, V.P.; writing—review and editing, S.S. and V.P.; visualization, A.G. and V.P.; supervision, V.P. and S.S. All authors have read and agreed to the published version of the manuscript.

Funding: This research was funded by the German Research Foundation, grant number SCHM 746/209-1.

Institutional Review Board Statement: Not applicable.

Informed Consent Statement: Not applicable.

Data Availability Statement: Not applicable.

Conflicts of Interest: The authors declare no conflict of interest.

References

1. Clarke, D.; Oechsner, M.; Padture, N. Thermal-barrier coatings for more efficient gas-turbine engines. *MRS Bull.* **2012**, *37*, 891–941. [CrossRef]
2. Shanmugavel, P.; Bhaskar, G.B.; Chandrasekaran, M.; Mani, P.S.; Srinivasan, S.P. An overview of fracture analysis in functionally graded materials. *Eur. J. Sci. Res.* **2012**, *68*, 412–439.
3. Bhandari, M.; Purohit, K. Dynamic fracture analysis of functionally graded material structures—A critical review. *Compos. Part C Open Access* **2022**, *7*, 100227. [CrossRef]
4. Ozdemir, M.; Imachi, M.; Tanaka, S.; Oterkus, S.; Oterkus, E. A comprehensive investigation on macro–micro crack interactions in functionally graded materials using ordinary-state based peridynamics. *Compos. Struct.* **2022**, *287*, 115299. [CrossRef]
5. Hirshikesh; Natarajan, S.; Annabattula, R.K.; Martínez-Pañeda, E. Phase field modelling of crack propagation in functionally graded materials. *Compos. Part B Eng.* **2019**, *259*, 239–248. [CrossRef]
6. Han, D.; Fan, H.; Yan, C.; Wang, T.; Yang, Y.; Ali, S.; Wang, G. Heat Conduction and cracking of functionally graded materials using an FDEM-based thermo-mechanical coupling model. *Appl. Sci.* **2022**, *12*, 12279. [CrossRef]
7. Amiri Delouei, A.; Emamian, A.; Karimnejad, S.; Sajjadi, H. A closed-form solution for axisymmetric conduction in a finite functionally graded cylinder. *Int. Commun. Heat Mass Transf.* **2019**, *108*, 104280. [CrossRef]
8. Delouei, A.A.; Emamian, A.; Karimnejad, S.; Sajjadi, H.; Jing, D. Two-dimensional analytical solution for temperature distribution in FG hollow spheres: General thermal boundary conditions. *Int. Commun. Heat Mass Transf.* **2020**, *113*, 104531. [CrossRef]
9. Tikhonov, N.; Samarskii, A.A. *Equations of Mathematical Physics*; Dover Publications: Garden City, NY, USA, 1990; p. 765.
10. Glushko, A.V.; Ryabenko, A.S.; Petrova, V.E.; Loginova, E.A. Heat distribution in a plane with a crack with a variable coefficient of thermal conductivity. *Asymptot. Anal.* **2016**, *98*, 285–307. [CrossRef]
11. Bao, G.; Wang, L. Multiple cracking in functionally graded ceramic/metal coatings. *Int. J. Solids Struct.* **1995**, *32*, 2853–2871. [CrossRef]
12. Zuiker, J.R. Functionally graded materials: Choice of micromechanics model and limitations in property variation. *Compos. Eng.* **1995**, *5*, 807–819. [CrossRef]
13. Noda, N.; Ishihara, M.; Yamamoto, N. Two-crack propagation paths in a functionally graded material plate subjected to thermal loadings. *J. Therm. Stress.* **2004**, *27*, 457–469. [CrossRef]
14. Balokhonov, R.; Romanova, V.; Kulkov, A. Microstructure-based analysis of deformation and fracture in metal-matrix composite materials. *Eng. Fail. Anal.* **2020**, *110*, 104412. [CrossRef]
15. Jin, Z.-H.; Batra, R. Some basic fracture mechanics concepts in functionally graded materials. *J. Mech. Phys. Solids* **1996**, *44*, 1221–1235. [CrossRef]
16. Tohgo, K.; Suzuki, T.; Araki, H. Evaluation of R-curve behavior of ceramic–metal functionally graded materials by stable crack growth. *Eng. Fract. Mech.* **2005**, *72*, 2359–2372. [CrossRef]
17. Tohgo, K.; Iizuka, M.; Araki, H.; Shimamura, Y. Influence of microstructure on fracture toughness distribution in ceramic–metal functionally graded materials. *Eng. Fract. Mech.* **2008**, *75*, 4529–4541. [CrossRef]
18. Zhang, Y.; Guo, L.; Wang, X.; Shen, R.; Huang, K. Thermal shock resistance of functionally graded materials with mixed-mode cracks. *Int. J. Solids Struct.* **2019**, *254*, 202–211. [CrossRef]
19. Petrova, V.; Schmauder, S. Analysis of interacting cracks in functionally graded thermal barrier coatings. *Procedia Struct. Integr.* **2020**, *28*, 608–618. [CrossRef]
20. Petrova, V.; Schmauder, S. Modeling of thermo-mechanical fracture of FGMs with respect to multiple cracks interaction. *Phys. Mesomech.* **2017**, *20*, 241–249. [CrossRef]

21. Petrova, V.; Schmauder, S. A theoretical model for the study of thermal fracture of functionally graded thermal barrier coatings with a system of edge and internal cracks. *Theor. Appl. Fract. Mech.* **2020**, *108*, 102605. [CrossRef]
22. Petrova, V.; Schmauder, S. Thermal fracture of functionally graded thermal barrier coatings with pre-existing edge cracks and multiple internal cracks imitating a curved interface. *Contin. Mech. Thermodyn.* **2021**, *33*, 1487–1503. [CrossRef]
23. Petrova, V.; Schmauder, S. Fracture of functionally graded thermal barrier coating on a homogeneous substrate: Models, methods, analysis. *J. Phys. Conf. Ser.* **2018**, *973*, 012017. [CrossRef]
24. Afsar, A.M.; Sekine, H. Crack Spacing Effect on the Brittle Fracture Characteristics of Semi-infinite Functionally Graded Materials with Periodic Edge Cracks. *Int. J. Fract.* **2000**, *102*, L61–L66. [CrossRef]
25. Afsar, A.M.; Song, J.I. Effect of FGM coating thickness on apparent fracture toughness of a thick-walled cylinder. *Eng. Fract. Mech.* **2010**, *77*, 2919–2926. [CrossRef]
26. Panasyuk, V.; Savruk, M.; Datsyshin, A. *Stress Distribution Near Cracks in Plates and Shells*; Naukova Dumka: Kiev, Ukraine, 1976; p. 270. (In Russian)
27. Erdogan, F.; Sih, G.C. On the crack extension in plates under plane loading and transverse shear. *J. Basic Eng.* **1963**, *85*, 519–527. [CrossRef]
28. Zhou, Y.C.; Hashida, T. Coupled effects of temperature gradient and oxidation on thermal stress in thermal barrier coating system. *Int. J. Solids Struct.* **2001**, *38*, 4235–4264. [CrossRef]

Disclaimer/Publisher's Note: The statements, opinions and data contained in all publications are solely those of the individual author(s) and contributor(s) and not of MDPI and/or the editor(s). MDPI and/or the editor(s) disclaim responsibility for any injury to people or property resulting from any ideas, methods, instructions or products referred to in the content.

Article

Structural-Phase Change of Multilayer Ceramics Zr-Y-O/Si-Al-N under High Temperature

Marina Fedorischeva [1,*], Mark Kalashnikov [1,2], Irina Bozhko [2], Tamara Dorofeeva [1] and Victor Sergeev [1,2]

1 Institute of Strength Physics and Materials Science SB RAS, 634055 Tomsk, Russia; kmp1980@mail.ru (M.K.); dorofeeva@ispms.ru (T.D.); vs@ispms.tsc.ru (V.S.)
2 Department of Materials Science, National Research Tomsk Polytechnic University, 634050 Tomsk, Russia; bozhko_irina@mail.ru
* Correspondence: fed_mv@mail.ru; Tel.: +79-627-763-406

Abstract: To increase the thermocyclic resistance of material, multilayer coatings with alternating layers of Zr-Y-O and Si-Al-N were obtained via magnetron sputtering. It was established that a coating layer based on Zr-Y-O has a columnar structure; the height of the columns is determined by the thickness of the layer. The Si-Al-N-based layer is amorphous. There were monoclinic and tetragonal phases with a large lattice parameter in the composition of the Zr-Y-O-based coating layer. After high-temperature annealing, a tetragonal phase with a small lattice parameter appeared in the microscope column. In the "in situ" mode, a change in the structural state of the Zr-Y-O coating layer was detected in the temperature range of 450–500 °C; namely, a change in the grain size and coherent scattering regions, and an increase in internal elastic stresses. It was found that the thermocyclic resistance increased by more than two times for multilayer samples compared to the single-layer ones we studied earlier.

Keywords: structure-phase state; multilayer coatings; grain size; curvature torsion; phase transition; thermal cyclic resistance

1. Introduction

Heat-protective coatings based on ZrO_2-Y_2O_3 are used in the production of gas turbine engines (GTE) to protect from the effects of the high temperature of the main components: combustion chambers, nozzle and turbine blades, etc. Such coatings also effectively protect the metal base of the GTE blades from high-temperature overloads, oxidizing, erosive and corrosive effects of the aggressive environment (gases) formed during the combustion of the fuel, and also allow one to effectively deal with the problems that determine the reliability, operability and the working life of structures, equipment, etc. [1–6].

Ceramic materials based on partially stabilized zirconium dioxide have unique properties: high bending strength and crack resistance in combination with chemical inertia and hardness, which makes them widely in demand as structural and functional ceramics [7,8].

It is known that zirconium dioxide ZrO_2 has three stable crystal structures that depend on temperature: a monoclinic structure below 1200 °C, a tetragonal structure between 1200 °C and 2370 °C and a cubic structure above 2370 °C. The mechanical properties of ceramics based on zirconium dioxide are known to be the function of the phase composition and the structure. Tetragonal zirconium dioxide has high strength and toughness. Various methods and technologies have been improved for the production and stabilization of the tetragonal phase in zirconium dioxide materials. The most common method has become the formation of tetragonal zirconium dioxide by adding stabilizing impurities to it. Analysis of the literature data has shown that yttrium oxide Y_2O_3 is most often used as a stabilizing impurity, which inhibits the transformation of the tetragonal phase into the monoclinic one during cooling.

The ZrO$_2$-Y$_2$O$_3$ ceramics system based on the tetragonal modification of zirconium dioxide (t-ZrO$_2$) has the highest mechanical characteristics, which, in accordance with the state diagram at normal pressure, exists in the temperature range of 1200–2370 °C [8,9]. Alloying of zirconium dioxide with 3–5 mol.% Y$_2$O$_3$ makes it possible to preserve the high-temperature tetragonal phase in the metastable state up to room temperature [10–14]. The high-temperature phases of zirconium dioxide have high physical and mechanical properties, which contribute to the great interest in this material. The presence of the monoclinic–tetragonal transition is accompanied by a sharp increase in volume, which leads to the destruction of ZrO$_2$ products. To ensure the possibility of using ZrO$_2$, it is necessary to stabilize high-temperature modifications in the entire temperature range up to room temperature. It is also known that diffusion and diffusion-free phase transformations occur in zirconium dioxide ceramics depending on the stabilizing impurities, the pressure, the temperature, the ambient gas environment, the humidity and the time of their exposure [15–20], which lead to significant changes in the phase composition, the structure and the mechanical properties of the above material.

For example, stabilization of zirconium dioxide with up to 6 mole% erbium (Er$_2$O$_3$), samarium (Sm$_2$O$_3$) or neodymium (Nd$_2$O$_3$) leads to stable t'-ZrO$_2$. The compounds Cr$_2$O$_3$-ZrO$_2$ and Al$_2$O$_3$-ZrO$_2$ decompose into m- and t-phases after heat treatment, and Cr$_2$O$_3$-ZrO$_2$ retain a stable t'-ZrO$_2$ phase [21]. In the work of Shen [22], the rare earth elements erbium and yttrium were used for alloying. The use of these elements in various ratios has led to an increase in the number of the cycles during thermocyclic tests. It is interesting to note that the number of cycles during thermal cycling for the Y-ZrO$_2$ coating remains the maximum. This value is comparable with the YEr-ZrO$_2$ coating. The authors explain the improvement of the thermal conductivity and the thermal cyclical resistance through the ability of the alloying elements to lead to a non-transformable tetragonal t' ZrO$_2$ phase. In other words, the joint alloying of Y and Er leads to a columnar microstructure and a relatively good thermal insulation ability, which may be responsible for the long service life of the Y-Er-ZrO$_2$ TBS under thermal shock. In [23], a thermal barrier coating doped with ZrO$_2$-YO$_{1.5}$-NbO$_2$ is considered. The addition of niobium contributed to a significant decrease in the thermal conductivity by about 52% compared to zirconium dioxide with a molar content of yttrium of 7.6%.

A study of the above phenomena allows one to better understand the nature of phase transformations, as well as to learn how to control its properties and even predict the ceramics' behavior.

The aim of this study was to study the structural and phase states of multilayer nanocomposite coatings based on alternating heterogeneous Zr-Y-O and Si-Al-N layers and to trace the processes occurring in the coating structure under heating in the "in situ" mode via TEM and X-ray, and also to investigate the thermal cycling of the multilayer coating at high temperatures.

2. Materials and Methods

The deposition of the coating was produced using the «KVANT-03MI» equipment [24] with two magnetrons using mosaic zirconium–yttrium and aluminum–silicon targets. The surface of the substrate was polished up to the roughness Ra = 0.16 μm before ionic treatment by Ti ions. The magnetron was powered from a pulse source with a frequency of 50 kHz. The sample was placed in the chamber on a rotating table, which could be moved in different directions—in front of the ion source for ion bombardment, and then in front of the magnetron for deposition of a coating. The substrate bias potential was −900 V under the bombardment of the surface layer of the substrate and −100 V during the deposition of a coating.

The fine structure of the multilayer coatings was investigated via transmission electron microscopy (TEM) using the JEM-2100 device (Japan). A foil was prepared via the «cross-section» method using the ION SLISER-EM-09100IS installation. The grain structure of the coatings was investigated in the temperature range of 25–900 °C in the microscope column and the vacuum camera of the XRD-7000S device in the «in situ» mode. The secant method was used to determine the grain size [25]. Another method for determining the grain size which was used in this study was the method of estimating the average grain size of ultrafine materials by the number of reflections on the Debye rings of the microdiffraction patterns obtained by TEM. This technique is described in detail in [26,27]. The authors propose using the correlation curve to relate the continuity of the diffraction rings to the average size of the grains that form the diffraction reflections.

The curvature torsion (indicating the presence of internal elastic stresses) was determined using the parameters of the bending extinction contours observed in the TEM images of the coating material. This technique is described in detail in [27–30].

The structural-phase state was studied via X-ray using the DRON-7 device (Burevestnik, Russia, "NANOTECH" ISPMS SB RAS). X-ray investigation of the coatings was carried out under continuous 2θ scanning with Bragg–Brentano focusing in Co Kα radiation. The high-temperature X-ray analysis was carried out using the XRD-7000S X-ray diffractometer (Shimadzu, Japan) with a high-temperature prefix. The sample was heated in a high-temperature chamber for 25 min to the desired temperature. Then, the sample was cooled spontaneously together with the chamber to RT. A tungsten–rhenium thermocouple was used to control the temperature. The database JCPDS PDF-4 was used to interpret the diffractograms. Structural characteristics, such as the crystalline lattice parameter, the size of the coherent scattering regions, and the phase composition were determined by the methods described in [31]. The resistance of the coatings to cracking and peeling when changing the temperature was determined from the results of thermal cycling of the samples according to the following regime: heating the sample to 1000 °C for 1 min, then forced cooling for 1 min to a temperature of 20 °C, photographing the surface of the sample on the side of the coating using a special Microscope DCM500 camera on an optical microscope BMG-160 (Carl Zeiss, Oberkochen, Germany), the data being transferred directly to the computer, then heating again. The total duration of each cycle, including all the stages of the process—heating, cooling, photographing—was 5 min. Photographing of the coating was also carried out before testing for thermal cycling resistance. For the criterion of the thermal stability of the coatings, the number of the cycles until the detachment of 50% of the coating area from the sample surface was selected. After that, the tests were stopped. The thermal cycling resistance is described in detail in [32].

3. Results

Figure 1 shows the TEM image of the cross-section of a multilayer coating based on the Zr-Y-O/Si-Al-N. The layers of the multilayer coating were found to consist of the elements Zr, Y, O, Si, Al, and N. Figure 1b shows the results of the EDS in the selected area of the transverse cross-sections of the coating and the distribution of the elements over the multilayer coating (the area of study is highlighted with a red rectangle). We can see that the thickness of the single layer is about 1000 ± 100 nm.

In Figure 2, one can see the TEM image of the structure of a multilayer coating consisting of the alternating Si-Al-N and Zr-Y-O layers of equal thickness. The darker contrast has a layer on the basis of Zr-Y-O. The lighter contrast has a layer based on the Si-Al-N.

The results of TEM studies of the structural-phase state of the Zr-Y-O coatings in cross-section are presented in Figure 2. The light-field images (Figure 2d) clearly show that the coating has a crystalline columnar structure. The crystals in the form of columns are oriented perpendicular to the surface of the substrate and extend over the entire thickness of the formed coatings. This columnar structure is one of the characteristic features of coatings obtained via pulsed magnetron deposition.

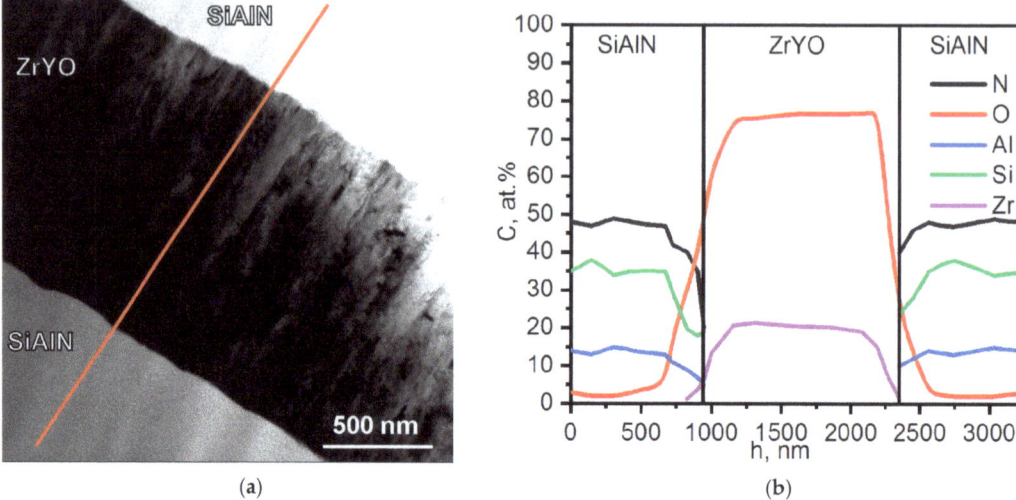

Figure 1. Cross-sections of the TEM images of the multilayer coatings on the basis of Zr-Y-O/Si-Al-N with a thickness of individual layers about 1000 ± 100 nm: bright-field image (**a**), depth distribution of the elements (**b**).

Microdiffraction patterns (Figure 2c) of the Zr-Y-O coatings are a set of ring reflexes. The identification of the microdiffraction patterns presented in Figure 2j–l shows that under pulsed magnetron deposition, zirconium dioxide ZrO_2 is formed in the coatings of the Zr-Y-O system in the monoclinic and tetragonal modifications. It should be noted that the ring reflexes in the indicated microdiffraction pattern consist of a large number of closely spaced point reflexes. This type of microdiffraction pattern allows one to assert that the Zr-Y-O coating has a crystal structure with a small size of structural elements. A detailed study of the bright-field and dark-field electron microscopic images reveals that as the Zr-Y-O coating grows, the transverse dimensions of its crystallites increase. Thus, the results of the measurements have allowed us to establish that the average transverse size of the crystallites of this coating near the substrate is about 25–30 nm, and near the next layer, the average size of the crystallites increases and is about 80 nm.

As for the layer based on Si-Al-N, it is amorphous. This is evidenced by the microdiffraction patterns in Figure 2a which are typical for an amorphous material.

The same picture shows the TEM images of the multilayer coatings based on Zr-Y-O in the initial state (d), at a temperature of 900 °C (e) and after cooling in a high-temperature chamber (f), the microdiffraction patterns in the nanodiffraction mode corresponding to the selected areas (g–i) and the ring microdiffraction from the visible area (j–l).

It was found using TEM that the structure of the layer based on Zr-Y-O in the Zr-Y-O/Si-Al-N multilayer coating contains the tetragonal and monoclinic phases in the initial state and at a temperature of 900 °C. In the Zr-Y-O coating layer after cooling, the tetragonal phase t' with a small crystal lattice parameter (a = 3.64 Å, c = 5.27 Å) was found, which was not observed in the initial state and at 900 °C. It is known that the above phase, in contrast to the tetragonal t-phase, does not undergo reversible phase transformations [33] of the tetragonal and monoclinic phases.

It is also known that the stability of the tetragonal phase, which is the most preferable for operations under extreme conditions, depends on the grain size. The larger the grain size the faster the phase transition from the high-temperature tetragonal phase to the monoclinic one [33]. Therefore, the grain size is an important characteristic of a material operating at high temperatures. The effect of temperature on the grain size of the ZrO_2 layer in the Si-Al-N–Zr-Y-O multilayer coating is discussed further in detail.

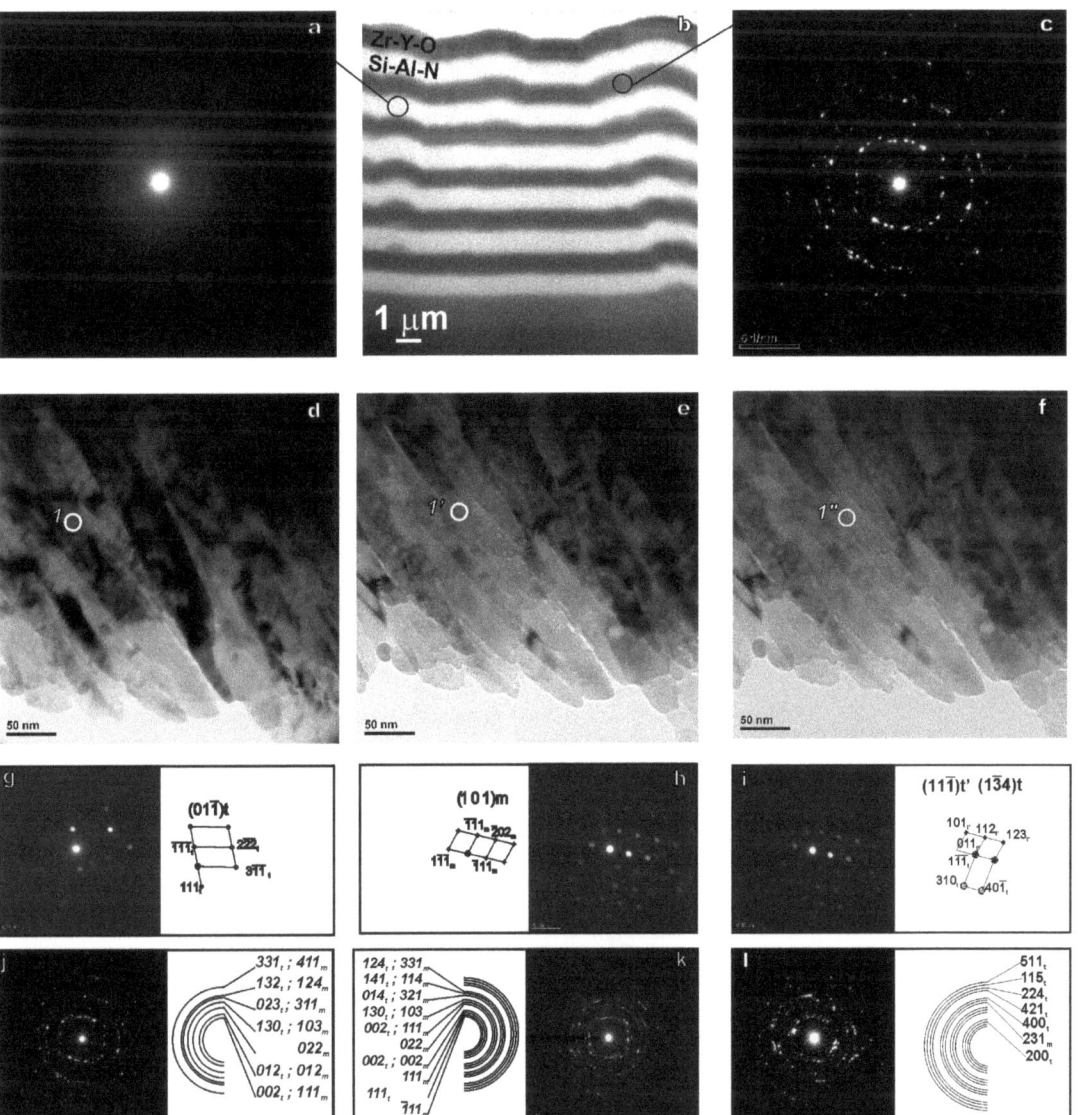

Figure 2. TEM image of the cross-section of the protective coatings of the Zr-Y-O-Si-Al-N system obtained by pulsed magnetron sputtering: alternating layers (**b**); the microdiffraction patterns of the amorphous layer on the basis of Si-Al-N (**a**); the ring microdiffraction of the layer Zr-Y-O (**c**); the bright field of the layer Zr-Y-O in the initial state (**d**) and at a temperature of 900 °C (**e**) and after cooling (**f**); the microdiffraction patterns in the nanodiffraction mode from the circled area (**g,h,i**), from the visible area (**j,k,l**) and the indexing scheme, respectively.

Further, the changes occurring at temperatures from room temperature to 900 °C were investigated. First, the sample was heated from room temperature to 400 °C for 30 min. Similar heatings were carried out at temperatures of 450 °C, 475 °C, 600 °C and 900 °C.

Figure 3 presents the microdiffraction patterns and the grain size distributions of the coating layer based on Zr-Y-O for the above temperatures. As can be seen, all the grain size distributions for all temperatures are unimodal and close to normal. The average grain size in the initial sample is about 15 nm. When it was heated up to 400 °C, the distribution became noticeably blurred and the most probable values shifted to the region of the large grain size, with the average grain size increased up to 20 nm.

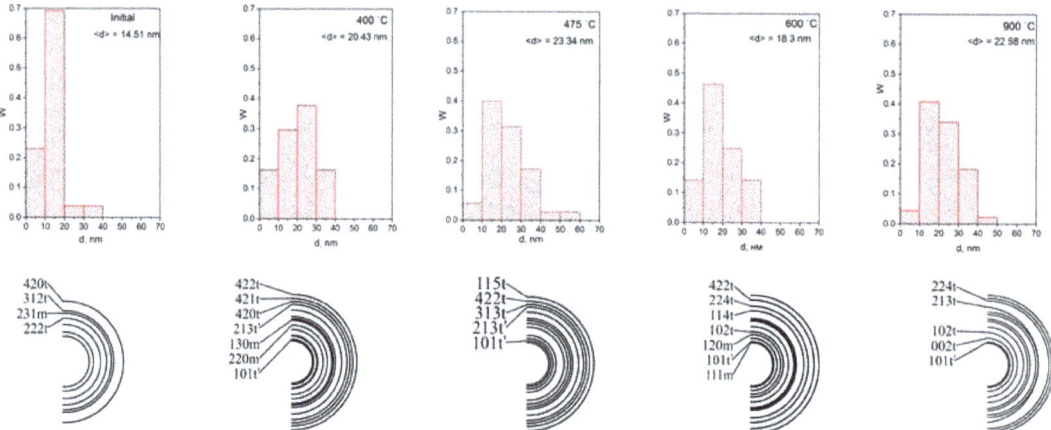

Figure 3. The grain size distribution of the coating based on Zr-Y-O for different temperatures (shown in the figures), the corresponding microdiffraction patterns and the indexing scheme.

The number of small grains significantly reduced. With a further increase in temperature, the fraction of the small grain size (less than 10 nm) decreased and the regions with a large (up to 60 nm) grain size appeared. The distribution had an asymmetric shape: there was a small "tail" in the region of the large grain size. In this case, the most probable grain size of about 20 nm remained at the level of the initial state almost up to 900 °C. The proportion of the grains with a minimum grain size of less than 10 nm was the smallest at temperatures of 475 °C and 900 °C.

Let us try to interpret the results in Figure 3. Judging by the blurring of the image of the distribution at 400 °C, recrystallization first occurs with an increase in the average grain size to about 20 nm. A further increase in temperature to 475 °C leads to a maximum increase in the average grain size to 23 nm, and in this case, the maximum blurring of the grain size distribution is observed, which suggests that at the above temperature, the martensitic phase transition from the tetragonal phase to the monoclinic one occurs most actively [1]. It is known that a method that stabilizes the high-temperature phase at room temperature is the introduction of impurity ions (in our case, yttrium) into the crystal lattice, which produces local compression regions. In this case, solid solutions are formed with stabilizer oxides, whose cations are larger than the zirconium ions and cause local compressive stresses [34–37].

A decrease in the concentration of the stabilizing ion in the bulk of the material can be responsible for the phase transformation of the tetragonal phase into the monoclinic one. It is known that with an increase in temperature, the rate of atoms diffusing, including those of the alloying elements, increases, with the atom migrating to the grain boundaries [37]. This leads to the fact that the number of local compression regions in the solid solution decreases and, consequently, the stabilizing effect of yttrium on the crystal lattice of the ZrO_2 tetragonal phase decreases.

An increase in the amount of the monoclinic phase, which has a larger volume (according to various sources, from 4% to 9%) compared with the tetragonal one, leads to an increase in the internal elastic stresses [33,37,38]. As is known, the internal elastic stresses lead to a phase transition (from a tetragonal to a monoclinic one) in this system. The relaxation of these stresses takes place, according to the dislocation model, due to the formation of dislocations followed by the formation of a grain–subgrain structure. This is evidenced by a decrease in the internal elastic stresses and a decrease in the grain size of the structure at 600 °C and high temperature [39].

Table 1 shows the values of the curvature torsion of the crystal lattice, which characterizes the internal elastic stresses of the coating layer in the Zr-Y-O/Si-Al-N multilayer coating. The components of the internal stress tensor were reconstructed from the parameters of the bending extinction contours observed in the TEM images of the coating layer. It can be seen that the maximum value of the curvature torsion of the material occurs at a temperature range of 450 °C–475 °C.

Table 1. Grain size determined by secant and diffraction patterns, coherent scattering blocks and values of the curvature torsion of the crystal lattice in the coating layer based on Zr-Y-O at different annealing temperatures.

Temperature, (°C)	Average Value of the Transverse Grain Size, nm (TEM)	Average Grain Size by Microdiffraction, nm (TEM)	Size of Coherent Scattering Units (X-ray)	Average Value of Curvature of Torsion χ, (cm^{-1}) (TEM)
Initial	15 ± 2	20 ± 2	26 ± 2	2.51
400	20 ± 2	24 ± 2		3.28
450	22 ± 2	26 ±2	37 ± 2	4.27
475	23 ± 2	22 ± 2		3.82
600	19 ± 2	19 ± 2		2.02
900	23 ± 2	22 ± 2		1.82
1000	-	-	25 ± 2	
1100			26 ± 2	
1300			35 ± 2	
1400			37 ± 2	
25			36 ± 2	

It is known that in a substrate coating system (in our case, coating–coating layers), internal elastic stresses always arise. During heating, the internal energy of the system changes. This process is caused by changes in any system and is accompanied by the change in the phase composition, grain growth, etc. With increasing temperature, two competing processes take place: the appearance of thermal stresses and their relaxation [33].

Presumably, in our case, up to temperatures of 400–500 °C, the internal energy is accumulated, which leads to an increase in the transverse size of the grains, and, consequently, to the curvature torsion of the crystal lattice (emergence of stresses). Next, the martensitic transition of the tetragonal phase to the monoclinic one (the relaxation stage) starts. In [33,37], the authors observed a martensitic phase transition at a temperature of 500 °C for YSZ coatings with an yttrium content of up to 3 mol%.

Further, the grain size decreases until the temperature of 600 °C is reached, and then it gradually increases up to 1400 °C (X-ray), remaining unchanged after cooling. Evidently, a large quantity of t′-ZrO$_2$ is contained in the Zr-Y-O coating, which stabilizes coating structure.

The data of the X-ray studies confirm the data of the electron microscopy. Figure 4a shows the fragments of the X-ray patterns of the coating layer based on Zr-Y-O. It can be seen that with an increase in the heating temperature, the reflexes of the X-ray diffractograms shift to the left which indicates an increase in the internal elastic stresses (Figure 4a). At room temperature after cooling together with the camera, the X-ray peaks return to their initial position. At the same time, the width at half-height also changes, which indicates a change in the coherent scattering regions. Indeed, their size is maximum at a temperature of 450 °C, i.e., the main trend remains: the grain size measured by two methods (TEM) and the size of the coherent scattering regions (X-ray) increases in the temperature range of 400–500 °C. In this regard, the curvature torsion of the crystal lattice (χ) increases, indicating an increase in the internal elastic stresses (TEM) at the same temperatures. Figure 4b shows the dependences of the grain size determined by two methods: the size of the coherent scattering regions and the curvature of the torsion of the crystal lattice. It can be seen that all the features (a sharp increase in the grain size, an increase in the internal elastic stresses) take place within the temperature range of 400–500 °C.

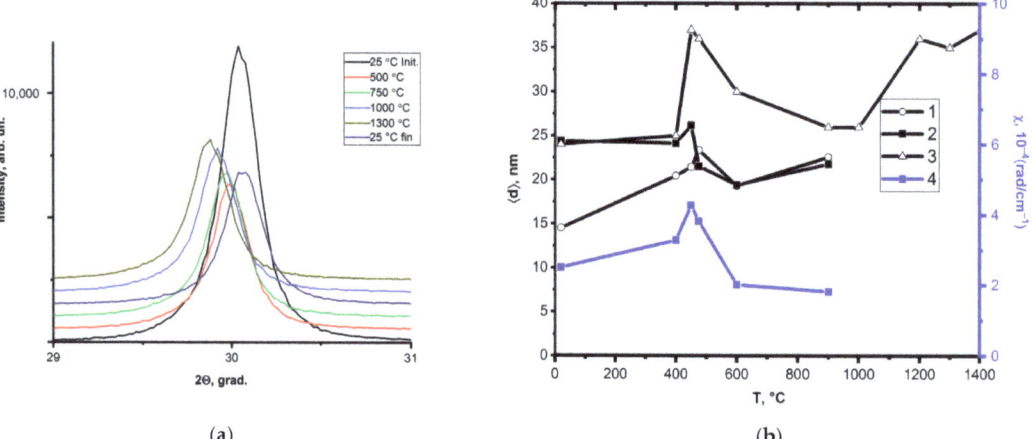

Figure 4. Fragments of the X-ray patterns obtained at temperatures from room temperature to 1300 °C (**a**), the dependence of the coherent scattering blocks (curve 1), the grain size (curves 2, 3) and the curvature torsion of the crystal lattice on temperature (curve 4) (**b**).

Thermocyclic resistance of the multilayer Zr-Y-O/Si-Al-N coatings was researched.

For the criterion of the thermal stability of the coatings, the number of the cycles before the destruction of 50% of the coating area of the sample surface was selected [32]. After that, the tests were stopped. Figure 5 shows that half of the coating will peel off only after 95 cycles in the multilayer coatings.

This value is significantly higher than that for the single-layer coating studied in [32].

The higher values of the thermocyclic resistance of a multilayer coating based on Zr-Y-O/Si-Al-N compared to a single-layer Zr-Y-O are caused by the presence of the Si-Al-N intermediate layer.

The intermediate layer is the main carrier of mechanical loads and relaxes the amplitude stresses arising during thermal cycling in a multi-level coating. It has high strength, high crack resistance and resistance to thermal shock, high relaxation ability and a low coefficient of thermal expansion. Si-Al-N-based ceramics meet all these requirements. The introduction of the intermediate layer under thermal cycling allows for a "multilayer coating-substrate" system, reducing the amplitude values of thermal compression–tension stresses in the upper functional layer. As a result, the decrease in the crack braking effect observed at high temperatures due to weakening of the mechanism of transformational hardening of yttrium-stabilized zirconium dioxide can be partially or completely compensated [40].

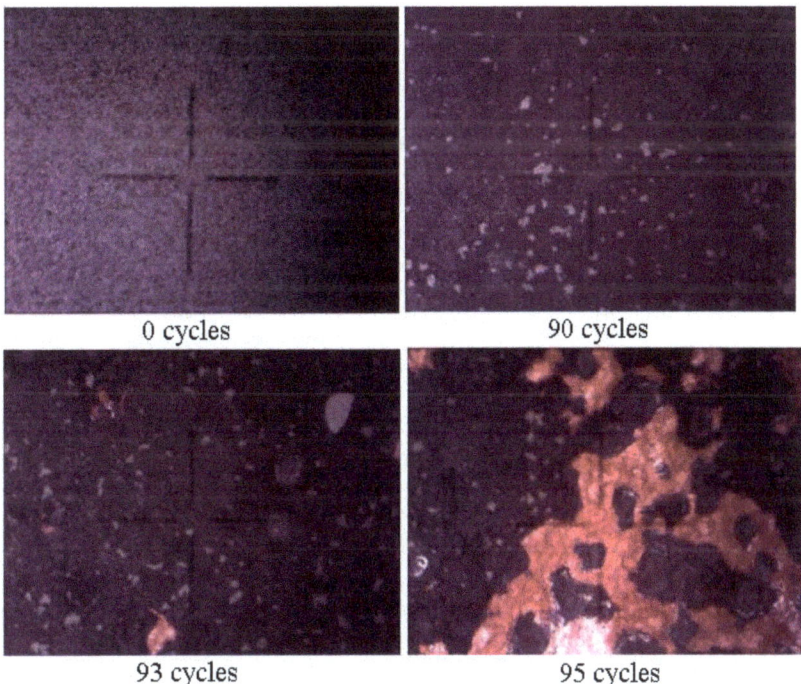

Figure 5. Optical images of the heat-shielding surface coatings on the basis of the Zr-Y-O/Si-Al-N coating obtained under thermocyclic resistance test (the dimension of the observed field is 2.09 × 1.56 mm).

4. Conclusions

Thus, pulsed magnetron sputtering was used to form multilayer nanostructured coatings based on the Zr-Y-O/Si-Al-N systems with alternating layers with a total coating thickness of ~10 μm. It was established by TEM and X-ray methods that the Zr-Y-O layers contain the ZrO_2 phases in the tetragonal and monoclinic modifications. The layers based on Zr-Y-O in the multilayer coating of Zr-Y-O/Si-Al-N have a columnar structure, with the size of the columns in the cross-section reaching 80 nm and the height of the columns being about 1000 nm, which, in this case, corresponds to the thickness of the applied layer. The structure of the Si-Al-N-based layers is amorphous.

During high-temperature TEM studies in the "in situ" mode in the Zr-Y-O layer of a multilayer coating based on Zr-Y-O/Si-Al-N in the temperature range of 400–500 °C, the grain size and the coherent scattering regions change. As a result, internal elastic stresses appear, leading to the martensitic phase transformation.

After cooling, the tetragonal phase t′-ZrO_2 with a small crystal lattice parameter (a = 3.64 Å c = 5.27 Å) was detected in the Zr-Y-O coating layer, which was not observed in the initial state and at 900 °C.

Thermocyclic resistance of the multilayer Zr-Y-O/Si-Al-N coatings is about two times higher than that of a single-layer coating. This is due to the presence of the Si-Al-N intermediate layer, which has a high crack resistance and a resistance to thermal shock, a high relaxation ability and low coefficient of thermal expansion.

Author Contributions: Conceptualization, V.S. and M.F.; methodology, M.F.; validation, V.S. and M.F.; investigation, T.D., M.K. and M.F.; resources, V.S.; writing—original draft preparation, I.B.; writing—review and editing, M.F. and I.B.; funding acquisition, V.S. All authors have read and agreed to the published version of the manuscript.

Funding: This research was performed under the government statement of work for ISPMS Project No FWRW-2021-0003.

Institutional Review Board Statement: Not applicable.

Informed Consent Statement: Not applicable.

Data Availability Statement: Not applicable.

Conflicts of Interest: The authors declare no conflict of interest.

References

1. Yasuda, K.; Goto, Y.; Takeda, H. Influence of Tetragonality on Tetragonal-to-Monoclinic Phase Trans-formation during Hydrothermal Aging in Plasma-Sprayed Yttria-Stabilized Zirconia Coatings. *J. Am. Ceram. Soc.* **2020**, *84*, 1037–1042.
2. Langjahr, P.A.; Oberacker, R.; Hoffmann, M.J. Long-Term Behavior and Application Limits of Plasma-Sprayed Zirconia Thermal Barrier Coatings. *J. Am. Ceram. Soc.* **2001**, *84*, 1301–1308. [CrossRef]
3. Gruninger, M.F.; Boris, M.V. Thermal Barrier Ceramics for Gas Tur-bine and Reciprocating Heat Engine Applications. In *Thermal Spray: International Advances in Coatings Technology*; Berndt, C.C., Ed.; ASM International: Almere, The Netherlands, 1992; pp. 487–492.
4. Soechting, F.O. A Design Perspective on Thermal Barrier Coatings. In Proceedings of the Thermal Barrier Coating Workshop, NASA CP-33, Westlake, OH, USA, 27–29 March 1995; Volume 12, pp. 1–15.
5. Bose, S.; Demasi-Marcin, J. Thennal Barrier Coating Experience inGas Turbine Engine at Pratt & Whitney. In Proceedings of the Thermal Barrier Coating Workshop, NASA CP-3312, Westlake, OH, USA, 27–29 March 1995; pp. 63–73.
6. Tanaka, M. Ion- and electron-beam-induced structural changes in cubic yttria stabilized zirconia. *Appl. Phys. A* **2018**, *124*, 647. [CrossRef]
7. Sinitsyn, D.Y.; Anikin, V.N.; Eremin, S.A.; Yudin, A.G. Protective coatings based on ZrO_2-Y_2O_3 and Al_2O_3–TiO_2 systems with modifying additives on CCCM. *Refract. Ind. Ceram.* **2017**, *58*, 194–201. [CrossRef]
8. Zhigachev, A.O.; Golovin, Y.I.; Umrikhin, A.V.; Korenkov, V.V.; Tyurin, A.I.; Rodaev, V.V.; Dyachek, T.A. *Ceramic Materials Based on Zirconium Dioxide*; Golovin, Y.I., Ed.; Technosfera: Moscow, Russia, 2018; 358p.
9. Stubican, V.S.; Hink, R.C.; Ray, S.P. Phase Equilibria and Ordering in the System ZrO_2-Y_2O_3. *J. Am. Ceram. Soc.* **1978**, *6*, 17–21. [CrossRef]
10. Lyakishev, N.P. (Ed.) *Diagrams of Binary Metallic Systems*; Mashinostroenie: Moscow, Russia, 1997.
11. Kulkov, S.N.; Buyakova, S.P. Phase composition and features of structure formation based on stabilized zirconium dioxide. *Russ. Nanotechnol.* **2007**, *2*, 119–132.
12. Lughi, V.; Sergo, V. Low Temperature Degradation -Aging- of Zirconia: A Critical Review of the Relevant Aspects in Dentistry. *Dent. Mater.* **2010**, *8*, 807–820. [CrossRef]
13. Chevalier, J.; Gremillard, L.; Virkar, A.V.; Clarke, D.R. The Ttetragonal-Monoclinic Transformation in Zirconia: Lessons Learned and Future Trends. *J. Am. Ceram. Soc.* **2009**, *92*, 1901–1920. [CrossRef]
14. Eichler, J.; Rodel, J.; Ulrich, E.; Mark, H. Effect of Grain Size on Mechanical Properties of Submicrometer 3Y-TZP: Fracture Strength and Hydrothermal Degradation. *J. Am. Ceram. Soc.* **2007**, *90*, 2830–2836. [CrossRef]
15. Hannink, R.H.; Kelly, P.M.; Muddle, B.C. Transformation Toughening in Zirconia-Containing Ceramics. *J. Am. Ceram. Soc.* **2000**, *83*, 461–487. [CrossRef]
16. Zhou, K.; Xie, F.; Wu, X.; Wang, S. Fretting wear behavior of nano ZrO_2 doped plasma electrolytic oxidation. *Surf. Coat. Technol.* **2021**, *421*, 127429. [CrossRef]
17. Borik, M.A.; Kulebyakin, A.V.; Myzina, V.A.; Lomonova, E.E.; Milovich, F.O.; Ryabochkina, P.A.; Sidorova, N.V.; Shulga, N.Y.; Tabachkova, N.Y. Mechanical characteristics, structure, and phase stability of tetragonal crystals of ZrO_2-Y_2O_3 solid solutions doped with cerium and neodymium oxides composite coatings on TC21 titanium alloy. *J. Phys. Chem. Solids* **2021**, *150*, 109908. [CrossRef]
18. Pang, E.L.; Olson, G.B.; Schuh, C.A. Schuh, The mechanism of thermal transformation hysteresis in ZrO_2-CeO_2 shape-memory ceramics. *Acta Mater.* **2021**, *213*, 116972. [CrossRef]
19. Kablov, E.N.; Muboyadzhyan, S.A. Heat-resistant and heat-shielding coatings for turbine blades high-pressure promising gas turbine engines. *Aviat. Mater. Technol.* **2012**, *1*, 60–70.
20. Tamarin, Y. *Protective Coatings for Turbine Blades USA*; ASM International: Almere, The Netherlands, 2002; pp. 3–300.
21. Boissonnet, G.; Chalk, C.; Nicholls, J.R.; Bonnet, G.; Pedraza, F. Phase stability and thermal insulation of YSZ and erbia-yttria co-doped zirconia EB-PVD thermal barrier coating systems. *Surf. Coat. Technol.* **2020**, *389*, 125566. [CrossRef]

22. Shen, Z.; Liu, Z.; Mu, R.; He, L.; Liu, G. Y–Er–ZrO$_2$ thermal barrier coatings by EB-PVD: Thermal conductivity. *Appl. Surf. Sci. Adv.* **2021**, *3*, 100043. [CrossRef]
23. Takahashi, J.M.K.; Assis, F.; Piorino Neto, D.A.P. Reis Thermal conductivity study of ZrO$_2$-YO$_{1.5}$-NbO$_{2.5}$ TBC. *J. Mater. Res. Technol.* **2022**, *19*, 4932–4938. [CrossRef]
24. Sergeev, V.P.; Yanovsky, V.P.; Paraev, Y.N.; Kozlov, S.A.; Zhuravlyov, S.A. Installation of ion magnetron sputtering of nanocrystalline coatings "KVANT". *Phys. Mesomech.* **2004**, *7*, 333–336.
25. Saltykov, S.A. *Stereometric Metallography*; Metallurgy: Moscow, Russia, 1970.
26. Morris, D.G.; Morris, M.A. Microstructure and Strength of Nanocrystalline Copper Alloy prepared by mechanical Alloying. *Acta Met.* **1991**, *39*, 1763–1770. [CrossRef]
27. Ivanov, Y.F.; Kozlov, E.V. Electron microscopic analysis of nanocrystalline materials. *Phys. Met. Met. Sci.* **1991**, *7*, 206–208.
28. Korotaev, A.D.; Tyumentsev, A.N. Physical Design Principles of Thermally Stable Multicomponent Nanocomposite Coatings. *Phys. Mesomech.* **2023**, *26*, 137–151. [CrossRef]
29. Kozlov, E.V. Structure and Resistance to Deformation of UFG Metals and Alloys. In *Severe Plastic Deformation*; Altan, B.S., Ed.; Nova Science Publishers, Inc.: New York, NY, USA, 2005; pp. 295–332.
30. Panin, V.E.; Panin, A.V.; Elsukova, T.F.; Popkova, Y.F. Fundamental role of crystal structure curvature in plasticity and strength of solids. *Phys. Mesomech.* **2015**, *18*, 89–93. [CrossRef]
31. Gorelik, S.S.; Rastorguev, L.N.; Skakov, Y.A. *X-ray and Electron-Optical Analysis*; Metallurgy: Moscow, Russia, 1994; pp. 124–127.
32. Fedorischeva, M.; Kalashnikov, M.; Bozhko, I.; Sergeev, V. Influence of the structural-phase state of a copper substrate upon modification with titanium ions on the thermal cyclic resistance of a coating based on Zr-Y-O. *Metals* **2022**, *12*, 65. [CrossRef]
33. Akimov, G.Y.; Timchenko, V.M.; Gorelik, I.V. Specific features of phase transformations in finely dispersed zirconium dioxide deformed by high hydrostatic pressure. *FTT* **1994**, *36*, 3582–3585.
34. Trunec, M. Effect of Grain Size on Mechanical Properties of 3Y-TZP Ceramics. *Ceram. Silik.* **2008**, *52*, 165–171.
35. Scott, H.G. Phase relationships in the zirconia-yttria system. *J. Mater. Sci.* **1975**, *10*, 1527–1535. [CrossRef]
36. Zhu, W.; Nakashima, S.; Marin, E.; Gu, H.; Pezzotti, G. Microscopic mapping of dopant content and its link to the structural and thermal stability of yttria-stabilized zirconia polycrystals. *J. Mater. Sci.* **2020**, *55*, 524–534. [CrossRef]
37. Akimov, G.Y.; Marinin, G.A.; Kameneva, V.Y. Evolution of the phase composition and physical and mechanical properties of ceramics ZrO$_2$+4mol.% Y$_2$O$_3$. *Phys. Solid State* **2005**, *47*, 2060–2062. [CrossRef]
38. Stark, D. *Diffusion in Solid*; Trusov, Energy: Moscow, Russia, 1980; p. 239.
39. Perevalova, O.B.; Konovalova, E.V.; Koneva, N.A.; Kozlov, E.V. *Effect of Atomic Ordering on Grain Boundary Ensembles of FCC Solid Solutions*; Portnova, T.C., Ed.; NTL Publisher: Tomsk, Russia, 2014; p. 250.
40. Sergeev, V.P. *Kinetics and Mechanism of the Formation of Nonequilibrium States of Surface Layers under Conditions of Magnetron Sputtering and Ion Bombardment*; Lyachko, N.Z., Psahie, S.G., Eds.; Nanoengineering Surface; Publishing House of the SB RAS: Novosibirsk, Russia, 2008; pp. 227–276.

Disclaimer/Publisher's Note: The statements, opinions and data contained in all publications are solely those of the individual author(s) and contributor(s) and not of MDPI and/or the editor(s). MDPI and/or the editor(s) disclaim responsibility for any injury to people or property resulting from any ideas, methods, instructions or products referred to in the content.

Article

Characterization of Thermal Barrier Coatings Using an Active Thermography Approach

Francesca Curà [1,*], Raffaella Sesana [1], Luca Corsaro [1] and Riccardo Mantoan [2]

1. DIMEAS Department of Mechanical and Aerospace Engineering, Politecnico di Torino, Corso Duca degli Abruzzi 24, 10129 Torino, TO, Italy
2. ATLA, Via Secondo Caselle 10, 10023 Chieri, TO, Italy
* Correspondence: francesca.cura@polito.it

Abstract: The aim of this paper is to define and set up an experimental procedure, based on active thermography, for the characterization of coatings for industrial applications. This procedure is intended to be a fast and reliable method, alternative to the consolidated one described in International Standards. In more detail, a classical active thermographic set up, and not a dedicated apparatus, was used for that aim, and data processing techniques referred to the analytical approach described in Standards. The active thermography procedure provided the measurement of the surface temperature of specimens undergoing a thermal excitation, applied by means of a laser pulse (Pulsed Technique). Temperature data processing, according to and adapting the Standard procedures, allowed to obtain thermal conductivity and diffusivity information. In particular, two coating processes (Atmospheric and Suspension Plasma Spray) applied to the same base material, Inconel 601, and the same coating material were investigated. These results were compared in terms of thermal properties variation with respect to base and coated materials, and in terms of different coating processes (APS and SPS). Obtained results were also compared to those available in literature.

Keywords: thermal barrier coatings; atmospheric plasma spray; suspension plasma spray; thermal properties; active thermography

Citation: Curà, F.; Sesana, R.; Corsaro, L.; Mantoan, R. Characterization of Thermal Barrier Coatings Using an Active Thermography Approach. *Ceramics* 2022, 5, 848–861. https://doi.org/10.3390/ceramics5040062

Academic Editors: Amirhossein Pakseresht and Kamalan Kirubaharan Amirtharaj Mosas

Received: 19 September 2022
Accepted: 21 October 2022
Published: 25 October 2022

Publisher's Note: MDPI stays neutral with regard to jurisdictional claims in published maps and institutional affiliations.

Copyright: © 2022 by the authors. Licensee MDPI, Basel, Switzerland. This article is an open access article distributed under the terms and conditions of the Creative Commons Attribution (CC BY) license (https://creativecommons.org/licenses/by/4.0/).

1. Introduction

Suspension Plasma Spray (SPS) and Atmospheric Plasma Spray (APS) coatings represent a challenging field of research, both for academic interest and commercial applications. In the aerospace environment, coatings are widely applied, and a typical application is Thermal Barrier Coating (TBC) for underneath metallic layers. Improvement to TBCs allows gas turbines to operate at higher temperatures, so the thermal conductivity of the coating dictates the temperature drop across the thermal barrier [1].

The increased coating requirements, such as low thermal conductivity and high level of thermal cycling resistance, can be reached using the new SPS technology; SPS can generate columnar microstructures typical of Electron-Beam Physical Vapor Deposition (EB-PVD) TBC, but at a much lower investment cost [2,3]. In more detail, SPS coating combines the advantages of APS and EB-PVD systems: its structure is columnar like EB-PVD, with columns, and its characteristics in terms of uniformity and density provide a higher thermal resistance compared to EB-PVDs.

From the technological point of view, SPS coatings may be produced with micro and nano scale features by using ultrafine particle feedstock (less than 5 microns). To facilitate the use of these ultrafine powders, the particles are suspended in solution, providing them with enough momentum to carry them into the plasma gun and make a coating on the target surface. Suspension feedstocks can cover a range of material compositions, including ceramics and metal alloy blends. Since the powder feedstocks used are ultrafine, the resulting coatings have a wider range of columnar structures.

SPS is a process that enables the use of thermal spray feedstocks too small for conventional plasma spray processes. These feedstocks come in the form of a slurry, with micron and submicron sized particles suspended in water, alcohol, or other solvents. Generally, the concentration of the fine particles in the slurry can be controlled and ranges between 5–80% by weight. This allows for the best combination of coating microstructure and deposition rate to be achieved.

During the SPS process, the thermal spray slurry is pumped to the outlet of the thermal spray torch and injected into the thermal spray jet. Once entrained in the plasma, the droplets fragment and the liquid phase evaporates, leaving ultrafine particles accelerating towards the substrate. By using these ultrafine particles sizes, it is possible to generate uniform coatings as thin as 25 microns thick. Practically, SPS combines the versatility and rapid deposition rate of APS with the ability to produce advanced microstructures from submicron and nanoscale powders. At the same time, it maintains excellent resistance to thermal shock, typical of coatings with a columnar structure.

Therefore, SPS succeeds in satisfying both fundamental requirements of the main aero-engine manufacturers: increasing turbine inlet temperature (TIT), with consequent increase in performance and efficiency, and ensuring resistance to thermal variations due to in-flight maneuvers.

In support of this 4th generation coating, material selection plays a key role, both for the bond coat and the TBC.

In general, preparation methods for coatings and corresponding key process parameters substantially influence their performance. In this context, the characterization of thermal diffusivity and conductivity properties is critical for process and material selection.

Over the years, many experimental techniques, such as optical microscopy, scanning electron microscopy (SEM), field emission scanning electron microscopy (FE-SEM), and X-ray diffraction, were used to measure physical characteristics in TBCs and their microstructure. An example of the screening of experimental variables leading to formation of a columnar microstructure in suspension plasma sprayed zirconia coatings is widely described in [4]. In that case, the thermal diffusivity of coatings was characterized with the use of NDT methods (IR thermography, flash method) and thermal conductivities of coatings were then determined. Recently, a novel approach was presented to characterize microstructural features as porosity of atmospheric-plasma-sprayed (APS) thermal barrier coatings using terahertz spectroscopy [5].

Anyhow, characterization of thermal diffusivity and conductivity in TBCs, already in the design phase, plays an important role in the performance definition.

Available and commercialized methods to measure the thermal conductivity can be classified [6] into steady-state conditions methods (guarded hot plate, heat flowmeter) and transient conditions methods (transient plane source, transient hot wire, laser flash apparatus, modulated DSC, 3ω method, thermocouple method).

Steady state methods may directly measure the thermal conductivity, based on Fourier's law, while in transient methods, the thermal diffusivity is firstly needed to calculate the thermal conductivity if both specific heat and density are known.

An example of the transient method is the so-called flash method (or flash diffusivity technique [7]), widely used to verify the design parameters of TBCs and to assess both in-plane and in cross-plane the thermal diffusivities of the top coats in the blades of gas turbines, pointing out the anisotropy of thermal conduction [8]. Flash-pulse thermography was recently introduced for quantitative inspection of the thickness of thermally sprayed coatings [9] and for porosity verification [10].

Pulsed active thermography in reflection configuration was applied in [11] to monitor the thermal diffusivity variation due to ageing of ceramic thermal barrier coatings and to detect the presence of cracks at the interface of Atmospheric Plasma Spray (APS) TBCs [12]. Thermography, in passive configuration, was also used to analyze the delamination crack growth in a thermal barrier coating [13].

During the last decade, thermal characterization of coatings was classically performed by the laser-flash method and following the procedure described in detail in a dedicated ASTM E1461-13 Standard [14] and in Standards [15–17]. Tests for thermal characterization were usually carried on by dedicated equipment, well described in [14–17]. Standards [14–17] provide both technical requirements for experimental set up and measurements and multi-layer analytical models for the calculation of thermal diffusivities.

In a very recent paper [18], the thermal diffusivity of coatings was measured by the dedicated laser-flash apparatus described in ASTM E1461-13 Standard [14], and calculated by an analytical two layers model. That model was applied for the determination of the thermal diffusivity of individual layers by separating its contribution from the one of the overall sample [18].

Active thermographic techniques were developed to characterize mechanical and thermal properties of components with coatings, regardless of the approaches involved in Standards [14–17]. In more detail, a thermal excitation was applied to the sample by laser sources or flash lamps and then the corresponding thermal response was detected by an infrared thermal camera. As an example, pulsed and lock-in infrared thermography, both in reflection mode and by using flash lamps, were used in [19] to characterize the variation in thickness of a topcoat.

The aim of this paper is to define and set up an experimental procedure, based on active thermography, for the characterization of coatings for industrial applications.

This procedure is intended to be a fast and reliable method, alternative to the consolidated one described in International Standards [14–17]. In more detail, a classical active thermographic set up, and not a dedicated apparatus, was used for that aim, and data processing techniques referred to the analytical approach described in Standards.

The active thermography procedure provided the measurement of the surface temperature of specimens undergoing a thermal excitation, applied by means of a laser pulse (Pulsed Technique).

Temperature data processing, according to and adapting the Standard procedures [15–17], allowed to obtain thermal conductivity and diffusivity information.

In particular, two coating processes (Atmospheric and Suspension Plasma Spray) applied to the same base material, Inconel 601, and the same coating material were investigated.

These results were compared in terms of thermal properties (both for base and coated materials) with those available in the literature.

2. Active Thermography Approach for Thermal Characterization of Coatings and Experimental Setup

Standards [15–17] are intended to provide methodologies for the thermal characterization of materials. In particular, these Standards describe the fundamental procedures based on the laser flash method (testing apparatus and formulas) for determining the thermal diffusivity of thermal barrier coatings [15], of monolithic ceramics [17] and plastics [16]. Thermal conductivity may be obtained from thermal diffusivity values by its physical equation.

The typical apparatus for measuring thermal diffusivity, according to the flash method, is well described in [15,17] and mainly consists of a specimen holder, a flash source (laser) to generate a temperature rise on one of the two major surfaces of the specimen, and a transient IR detector with its associated electronics. The detector acquires the thermal heating profile of the surface opposite to that heated by the laser flash. This setup is the so called "transmission configuration" and it allows to record the thermal heating behavior due to the energy transmitted through the sample, which will be affected by the proper thermal properties. In the case of multilayer materials, as thermal barrier coatings, the heated surface is the base material while the detected surface is the coated one. Standards [15–17] describe shape, dimensions, and thicknesses of samples to be tested.

More in detail, Standard [17] describes the procedure to compute the thermal property of fine and advanced ceramic materials. The so-called half rise time method is proposed to compute the thermal diffusivity of the material. In [15], a multi-layer analytical model is provided for determining thermal diffusivities. In particular, the areal heat diffusion time method allows to compute the apparent thermal diffusivity of the sample and the real thermal diffusivity and thermal conductivity of each layer. For both Standards [15,17], appropriate corrections are proposed to obtain the real thermal properties of the analyzed material.

The Active Thermography technique (AT) presented in this paper aims to substitute the consolidated equipment described in International Standards [15–17], but based on the same theoretical approach.

This AT technique provides the measurement of the samples' surface temperature undergoing an external thermal excitation, applied by means of a laser pulse (Pulsed Technique). In more detail, the external thermal excitation is a laser beam, able to give a higher quantity of energy in a small application surface (maximum power 50 W) and the response thermal sensor is an IR thermal camera FLIR A6751sc (Wilsonville, OR, USA) (sensitivity lower than 20 mK and 3–5 µm spectral range) (see Figure 1). Temperature data processing, according to and adapting the Standard procedures and analytical models [15,17], allows thermal diffusivity and conductivity information.

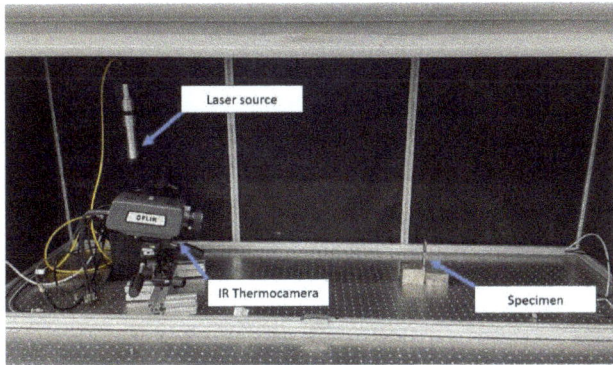

Figure 1. Experimental equipment.

Shape and dimensions of specimens, thickness values of base material and coating are chosen following Standard indication [15]. As a matter of fact, coating thermal diffusivity and thermal conductivity can be obtained by referring to the substrate material. This means that, for experimental characterization, the base material sample is always necessary, as well as the coated sample.

The calculation of the coating thermal diffusivity is based on the temperature-rise of the base material (see Figure 2) by using the half-rise time method described in [17] for monolayer materials and the areal heat diffusion time method, specific for metallic and other inorganic coatings [15].

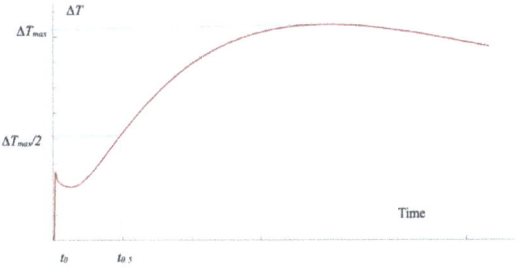

Figure 2. Transient temperature curve of a specimen after light pulse heating [17].

In more detail, for monolayer materials, the thermal diffusivity α [m²/s] may be obtained by the following equation, based on the half-rise time method:

$$\alpha = (0.1388\, d^2)/t_{0.5} \qquad (1)$$

where $t_{0.5}$ is the time delay when the temperature of the rear face reaches one-half of the maximum temperature rise, ΔT_{max}, after the front face was heated by the laser pulse (see Figure 2) and d is the sample thickness.

In order for the rise-time to be validly applied, both duration of the laser pulse and sampling time have to be less than 1% of the time delay $t_{0.5}$ (first condition [17]).

Furthermore, the data acquisition unit requires a sampling time faster than 1% of the half rise-time (second condition [17]).

The calculation of thermal diffusivity using Equation (1) may be modified if the duration time of the heating laser pulse does not respect initial and boundary conditions, as indicated in [17] by several correction methods (Centroid method, determination of chronological centroid of laser pulse and Triangular pulse approximation). In the present work, the Centroid method was applied to satisfy the laser square pulse condition.

In the same Standard [17], some correction factors may be introduced if radiative heat losses cannot be neglected. In particular, Standard [17] proposes Cape and Lehman's equation [20], Clark and Taylor's method [21], the Takahashi method [22], and Cowan method [23]. The Cowan method was chosen here, due to its effect on the stabilization zone of the thermal profile.

A specific approach for metallic and other inorganic coatings is described in Standard [15]. In particular, the areal heat diffusion time method provides both thermal diffusivity of each layer α_i (and therefore of the substrate too) and apparent thermal diffusivity of the multilayer sample αapp (substrate and coatings). The method refers to the areal heat diffusion time represented in Figure 3 and to the analytical multi-layer model shown in Figure 4, both reported in [15].

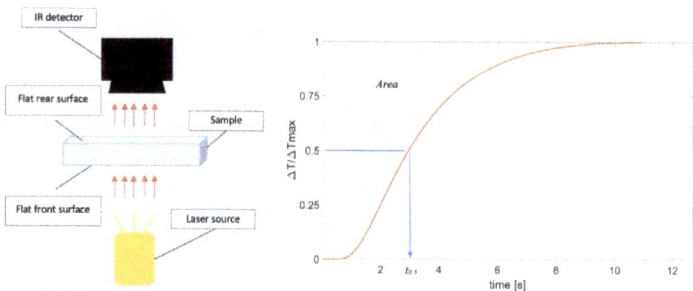

Figure 3. Temperature-rise curve under ideal conditions according to Standard [15].

The thermal diffusivity of a single layer belonging to a multi-layered system is computed as follows:

$$\alpha_i = (d_i)^2/\tau_i \qquad (2)$$

where d_i is the thickness of the layer and the heat diffusion time τ_i is evaluated with a specific formulation for each layer of the specimen.

Physical parameters required to compute the heat diffusion time of each layer (τ_1, τ_2, ..., τ_n) are illustrated in Figure 4 [15].

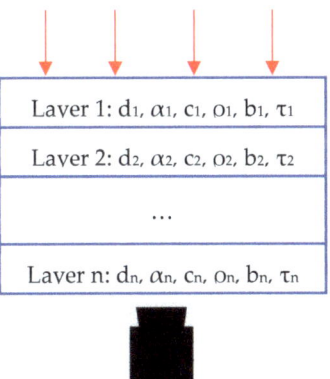

Figure 4. Multi-layer model.

In Standard [15], the analytical formulations for calculating the heat diffusion time of both layer 1 (τ_1) (the substrate) and layer 2 (τ_2) (the coating) are described.

In particular, τ_1 is defined as:

$$\tau_1 = 6 \, A_{1-S} \qquad (3)$$

where A_{1-S} is the area (see Figure 3) obtained from the temperature profile generated with the base material specimen, τ_2 is evaluated as:

$$\tau_2 = [6 \, A_{2-S} \, (c_1 \, \rho_1 \, d_1 + c_2 \, \rho_2 \, d_2) - (c_1 \, \rho_1 \, d_1 + 3 \, c_2 \, \rho_2 \, d_2) \, \tau_1)]/(3 \, c_1 \, \rho_1 \, d_1 + c_2 \, \rho_2 \, d_2) \qquad (4)$$

and A_{2-S} (see Figure 3) is the area obtained from the temperature profile generated with the two-layer specimen (base material and coating).

Standard [15] also suggests a specific formulation for the evaluation of the apparent thermal diffusivity of a multilayer specimen:

$$\alpha_{app} = (d_s + d_c)^2 / (6 \, A_{2-S}) \qquad (5)$$

where d_s and d_c are, respectively, substrate and coating thicknesses.

Equation (5) represents an alternative formulation with respect to Equation (1) to evaluate the apparent thermal diffusivity.

Finally, the thermal conductivity can be obtained from the thermal diffusivity α_{app} values by the following relationship:

$$k = \alpha \, \rho \, c \qquad (6)$$

where ρ and c are, respectively, density and heat capacity of the material useful for the purpose.

3. Materials, Samples, and Testing Procedure

Samples adopted for this study were made of two materials. Inconel 601 was the substrate material, Yttria-Stabilized Zirconia (YSZ) based powder was selected for the ceramic coating. Two types of deposition process were utilized, respectively, APS and SPS.

Figure 5 shows the samples adopted for measurements. Samples were prepared with the aim to respect the geometrical characteristics requested in Standards and to create uniform heating on the surface sample. Substrate geometry was a flat sample with 2 mm thickness and 5×5 mm square surface. Three specimens per samples were tested and at least three replications for each measurement were performed.

Figure 5. Samples.

A SEM analysis was performed to evaluate both structure and thickness of coatings, generated by APS and SPS deposition processes. Thickness values were necessary to calibrate experimental set up parameters. Figure 6 shows SEM images of APS and SPS specimens, emphasizing both stiffnesses of Inconel 601 and coatings and the structure of coatings. SEM images refer to different zones of the same specimen (1 and 2 for APS, 3 and 4 for SPS).

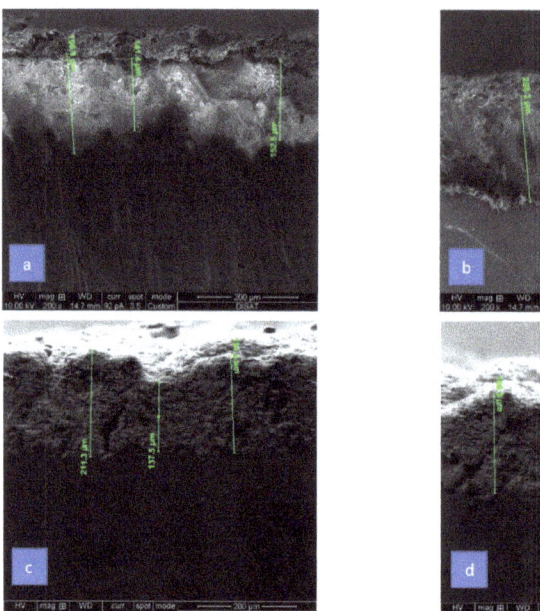

Figure 6. SEM images: (**a**,**b**) APS samples; (**c**,**d**) SPS samples.

Physical characteristics values (density ρ and specific heat c) adopted for the estimation of thermal conductivity may be found in [24] for substrate (Inconel 601) and in [25] for coatings (see Table 1). Inconel 601 data are useful to check this procedure.

Table 1. Physical characteristics of materials.

Material	ρ [kg/m^3]	c [J/kg °C]
Inconel 601 (substrate)	8110	448
Yttria Stabilized Zirconia (coating)	5200	467

A dedicated experimental configuration and apparatus were setup to carry out the experimental tests following the procedure described in [15,17], but by means of different equipment, as was stated in the previous sections. The experimental equipment was composed by a thermal camera, a laser excitation source, and a PC control unit.

A transmission-mode configuration was utilized, where samples were positioned between the laser source and thermal camera. Distances between the thermal camera and sample and between the laser source and sample were 0.8 m and about 0.1 m, respectively. Samples were clamped in a device with two polymeric supports to reduce the possible heat transmission. Figure 7 shows the schematic experimental configuration.

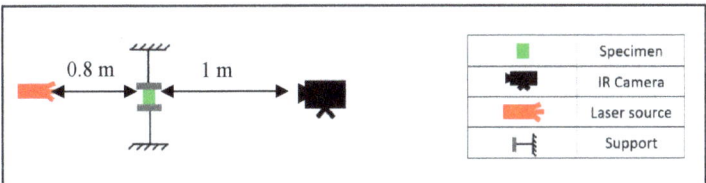

Figure 7. Schematic experimental configuration.

Laser excitation and acquisition parameters (frame rate) were tuned both for the substrate and coatings. The substrate material (Inconel 601) was heated with a pulse period of 50 ms at the maximum power, while coated samples were heated with a pulse period of 5 ms at the maximum power. As a matter of fact, the substrate material requires more energy to increase its surface temperature, due to the high conductivity characteristic. Acquisition rates were chosen at 785.67 Hz for the substrate and 600 Hz for the specimen with the coating, respectively.

To extract the temperature profile from thermograms, the evaluation of the emissivity of each sample was done, according to Standard ISO 18434 [26].

The extraction of temperature profiles was done by using ResearchIR Software. A Region Of Interest (ROI) was chosen to extract the temperature profiles from the corresponding thermogram. As an example, Figure 8 illustrates the thermal image and the corresponding ROI for the Inconel 601 sample. In general, the ROI is located at the specimen center in order to reduce possible external contour influences.

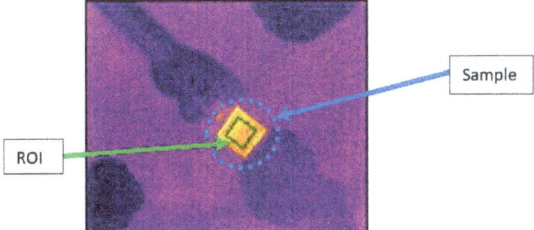

Figure 8. Inconel 601: thermal image and ROI.

As already observed, tests performed on Inconel 601 allowed to characterize the base material from the thermal point of view (thermal conductivity) and to verify the methodology reliability (experimental set up and data processing).

4. Results

The experimental activity carried on by the above-described dedicated methodology provided an estimation of thermal diffusivity and conductivity values of both substrate and coating materials, based on Standards approaches [15,17]. Temperature profiles obtained for all specimens, substrate and coating, allowed to obtain the thermal characteristics of interest.

A filtering process (Gaussian filter) was applied to raw signals in order to reduce the effects of high acquisition frame rates. This way, the temperature increment after the laser excitation was correctly identified, without altering the original profile. An example of comparison between unfiltered and filtered temperature profiles (row and denoised signals) for Inconel 601 is illustrated in Figure 9.

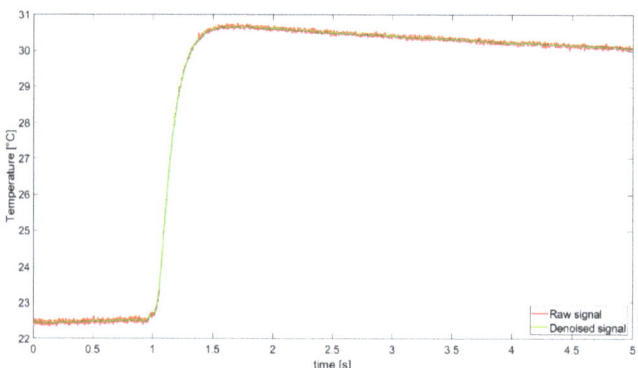

Figure 9. Inconel 601: raw and denoised signals.

Figure 10 shows denoised signals corresponding to Inconel 601 relative temperatures ∆*Temperature* (difference between measured temperature T referred to the ambient temperature T_{amb}, ∆*Temperature* = $T - T_{amb}$) [°C].

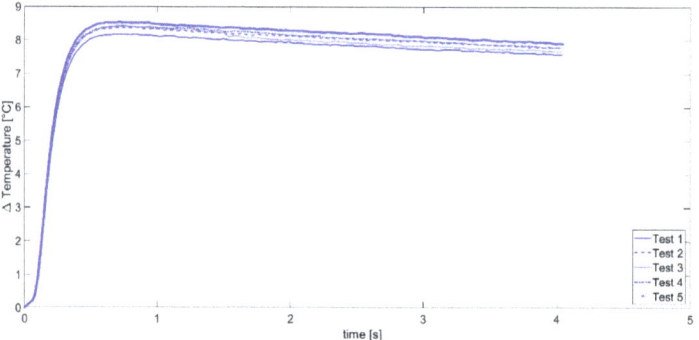

Figure 10. Inconel 601: relative temperature ∆Temperature.

Five test replications were performed for the base material (Inconel 601).

Relative temperature profiles ∆*Temperature*, according to Standards [15–17], were processed in a time range up to its maximum relative temperature rise.

Figure 11 shows the normalized temperature rise [15] for Inconel 601. In particular, this plot represents the same relative temperature profiles ∆*Temperature* shown in Figure 10, here normalized with respect to maximum relative value for each curve ($T_{max} - T_{amb}$). In the same Figure 11, the parameters of interest (area and half rise time $t_{0.5}$) useful for thermal characterization are illustrated.

Thermal diffusivity and conductivity were computed by using Equations (1)–(3) and (6), reported in [15,17], and by applying the corrections suggested in Standard [17], as the Centroid method (to correct the finite pulse period of the excitation source) and the Cowen method (to correct the heat losses) [15,17].

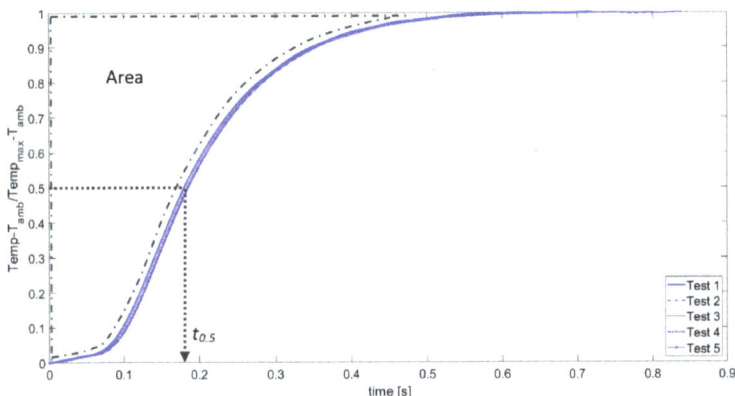

Figure 11. Inconel 601: normalized temperature rises and Standard parameters [15,17].

Table 2 resumes all results related to the thermal characterization of the substrate (Inconel 601) (averaged values of all replications) by applying Standards [15,17] formula.

Table 2. Inconel 601: thermal diffusivity and thermal conductivity.

Thickness: 2 mm	ISO18755 [17]								ISO18555 [15]		
	α_{ideal} [m²/s]	k_{ideal} [W/m °C]	α_{tc} [m²/s]	k_{tc} [W/m °C]	α_{Cowen5} [m²/s]	k_{Cowen5} [W/m °C]	$\alpha_{Cowen10}$ [m²/s]	$k_{Cowen10}$ [W/m °C]	$\alpha_{1\text{-}ideal}$ [m²/s]	$k_{1\text{-}ideal}$ [W/m °C]	τ_1 [s]
Inconel 601	3.31×10^{-6}	12.02	3.82×10^{-6}	13.90	3.84×10^{-6}	13.96	3.83×10^{-6}	13.90	3.53×10^{-6}	12.82	1.25
Standar deviation	4.4×10^{-8}	0.175	6.4×10^{-8}	0.232	1.0×10^{-7}	0.372	6.5×10^{-8}	0.236	4.4×10^{-8}	0.161	1.5×10^{-2}

In particular, Table 2 can be thought as divided into two parts: the left side concerns results obtained by applying to the substrate a monolayer ceramic material model (ISO18755 [17]), while the right one concerns the results that refer to a multilayer material model (ISO18555 [15]).

In more detail, on the left side (ISO18755 [17]), the calculated values of ideal thermal diffusivity and conductivity (α_{ideal} e k_{ideal}) as they are defined in [17], which is the diffusivity and conductivity calculated by means of the half rise method, and the corresponding values corrected by the Cetroid Method (α_{tc} e k_{tc}) and by the Cowen Method are presented (α_{Cowen5}, k_{Cowen5}, $\alpha_{Cowen10}$, and $k_{Cowen10}$) (respectively, 5 and 10 times the half rise time "$t_{0.5}$").

On the right side (ISO18555 [15]), the calculated values of ideal thermal diffusivity and conductivity obtained by considering the Inconel 601 as a monolayer material ($\alpha_{1\text{-}ideal}$ and $k_{1\text{-}ideal}$) are shown.

Finally, Table 2 reports the calculated heat diffusion time of the substrate (τ_1), useful for thermal characterization of APS and SPS coatings.

From the analysis of Table 2, it can be observed that for Inconel 601, both Standard analytical formula ([15,17]) provide similar diffusivity and conductivity values. These values are in a good agreement with the corresponding provided by the material producer (k =11.2 to 12.7 W/m °C).

Table 3 shows a comparison between thermal conductivity values of Inconel 601 shown in Table 2 and those available in the literature [24].

Table 3. Inconel 601: thermal conductivity values.

	k_{ideal} [W/m °C]	k_{tc} [W/m °C]	k_{Cowen5} [W/m °C]	$k_{Cowen10}$ [W/m °C]	$k_{1-ideal}$ [W/m °C]	k [22] [W/m °C]
Inconel-601	12.02	13.90	13.96	13.90	12.82	11.2–12.7

Figures 12–15 resume all thermal profiles related to APS and SPS coatings.

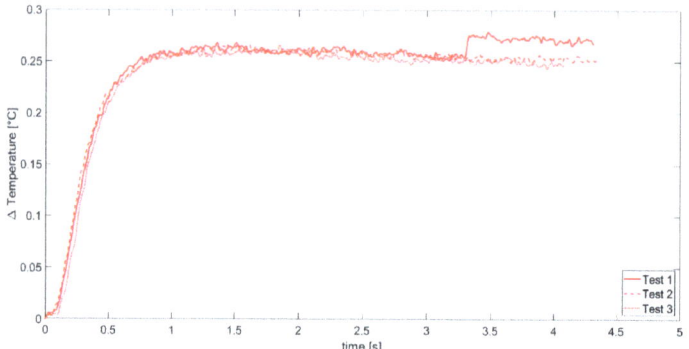

Figure 12. APS: relative temperature $\Delta Temperature$.

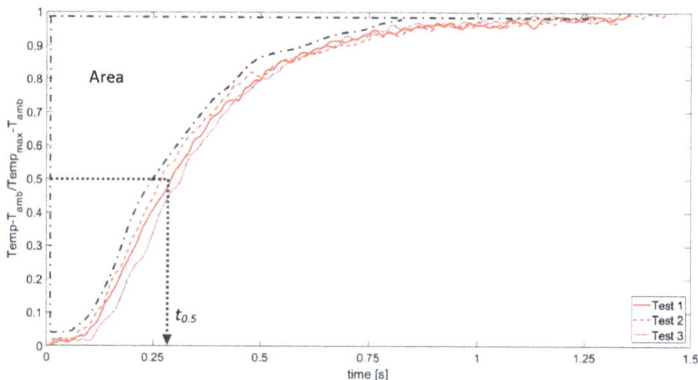

Figure 13. APS: normalized temperature rises and standard parameters [13,17].

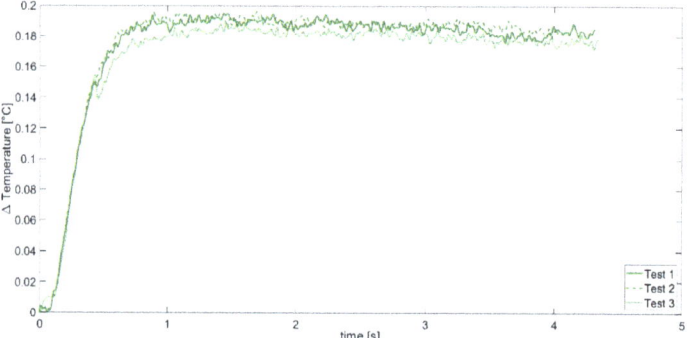

Figure 14. SPS: relative temperature $\Delta Temperature$.

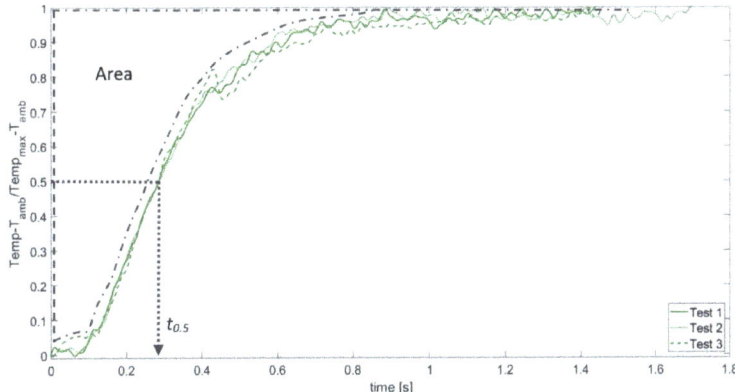

Figure 15. SPS: normalized temperature rises and standard parameters [15,17].

In particular, Figures 12 and 14 show the denoised relative temperature profiles, respectively, for APS and SPS coatings.

Figures 13 and 15 report the normalized temperature rises with respect to maximum relative value for each curve ($T_{max} - T_{amb}$) and the parameters of interest (area and half rise time $t_{0.5}$) for APS and SPS coatings.

Finally, Table 4 resumes all thermal diffusivity and conductivity values for APS and SPS coatings, organized as Table 2.

Table 4. APS and SPS: thermal diffusivity and thermal conductivity.

	Mean Thickness [mm]	ISO18755 [17]				ISO18555 [15]		
		$\alpha_{app\text{-}ideal}$ [m^2/s]	$\alpha_{app\text{-}tc}$ [m^2/s]	$\alpha_{app\text{-}Cowen5}$ [m^2/s]	$\alpha_{app\text{-}Cowen10}$ [m^2/s]	$\alpha_{app\text{-}ideal}$ [m^2/s]	$\alpha_{2\text{-}ideal}$ [m^2/s]	$k_{2\text{-}ideal}$ [W/m °C]
APS	0.1868	3.18 × 10^{-6}	3.20 × 10^{-6}	3.09 × 10^{-6}	3.17 × 10^{-6}	3.11 × 10^{-6}	1.31 × 10^{-7}	0.32
Standard deviation		2.5 × 10^{-7}	2.6 × 10^{-7}	2.0 × 10^{-7}	2.2 × 10^{-7}	1.0 × 10^{-7}	1.2 × 10^{-8}	1.3 × 10^{-1}
SPS	0.1759	2.64 × 10^{-6}	2.67 × 10^{-6}	2.63 × 10^{-6}	2.66 × 10^{-6}	2.58 × 10^{-6}	1.24 × 10^{-7}	0.30
Standard deviation		9.3 × 10^{-9}	9.5 × 10^{-9}	7.5 × 10^{-8}	1.9 × 10^{-8}	4.6 × 10^{-8}	6.3 × 10^{-9}	1.5 × 10^{-2}

Real and apparent (substrate and coating respectively, two layers material) thermal diffusivities and thermal conductivities of APS and SPS coatings were obtained from Equations (1), (2), and (4)–(6) ([15,17]), taking into account the mean value of calculated variables as for Inconel 601. Computation of coatings thermal characteristics required the thickness information, obtained by the SEM analysis as averaged thickness values. The thermal conductivity evaluation is based on an estimation of the physical properties of the coating (density and specific heat), as already described.

Tests performed on these specimens allowed us to characterize both SPS and APS coatings, once the heat diffusion time (τ_1) of the substrate was calculated (see Table 2).

All results related to coated specimens (substrate and TBCs) and averaged on all replications are resumed in Table 4, organized with the same layout of Table 2 (left side and right side refer, respectively, to ISO18755 Standard [17] and ISO18555 Standard [15] approaches).

In particular, respectively, for APS and SPS coated specimens, the following results are shown on the left side of Table 4: ideal apparent thermal diffusivity values ($\alpha_{app\text{-}ideal}$), apparent thermal diffusivity values corrected by both Centroid method ($\alpha_{app\text{-}tc}$) and Cowen Method ($\alpha_{app\text{-}Cowen5}$ and $\alpha_{app\text{-}Cowen10}$, respectively, 5 and 10 times the half rise time "$t_{0.5}$").

The right side of the same table shows: ideal apparent thermal diffusivity values ($\alpha_{app\text{-}ideal}$) of substrate + coatings, ideal thermal diffusivity and conductivity values of coatings ($\alpha_{2\text{-}ideal}$ and $k_{2\text{-}ideal}$), obtained based on the heat diffusion time of Inconel 601 (τ_1) (see Table 2).

As already commented for Table 2, a similar consideration can be stated referring to TBCs results, based on the analysis of Table 4. In more detail, it can be concluded that obtained thermal parameters values for SPS and APS coatings are comparable to those available in the literature [3] (k = 0.6 to 1 W/m °K) as order of magnitude. A great tuning of data is difficult to be reached due to the variability of TBC thickness that strongly influences the obtained values.

5. Conclusions

The results obtained in the present work allow us to draw the following conclusions.

The experimental procedure defined and set up the thermal characterization of industrial TBCs, based on active thermography, provided very good results, comparable to those available in the literature. As wished and hypothesized, this methodology can be thought as a fast and reliable method, alternative to the consolidated one described in International Standards ISO18755 and ISO18555, for which is required dedicated equipment.

The classical active thermography apparatus utilized in this research activity, consisting of a thermal camera and a fast thermal excitation source, was able to measure thermal profiles suitable to be processed and analyzed following the International Standards requirements.

Moreover, the developed technique, together with its processing algorithms, provided a robust method to be easily used in the industrial environment.

From the coatings point of view, it can be conclude that, despite the difficulty in defining samples with quite constant thickness, obtained thermal diffusivity and conductivity values are representative of the actual behavior of TBCs obtained by different production processes. Similar considerations may be drawn for base materials characterization.

Finally, particularly for concerns of the present industrial application, it can be concluded that SPS and APS coatings show similar thermal behavior and the choice to employ one of these can be left to production and business choices.

Author Contributions: Conceptualization, F.C. and R.S.; methodology, F.C.; software, L.C.; validation, L.C., F.C. and R.S.; formal analysis, F.C.; investigation, L.C.; data curation, R.S.; writing—original draft preparation, F.C.; writing—review and editing, R.M. and R.S.; visualization, L.C.; supervision, R.M. All authors have read and agreed to the published version of the manuscript.

Funding: This research was possible thanks to ATLA Company technical support and material supply.

Institutional Review Board Statement: Not applicable.

Informed Consent Statement: Not applicable.

Data Availability Statement: Data are available on request.

Conflicts of Interest: The authors declare no conflict of interest.

References

1. Clarke, D.R.; Phillpot, S.R. Thermal barrier coating materials. *Materials Today* **2005**, *8*, 22–29. [CrossRef]
2. Tang, Z.; Kim, H.; Yaroslavski, I.; Masindo, G.; Celler, Z.; Ellsworth, D. Novel thermal barrier coatings produced by axial suspension plasma spray. In Proceedings of the International Thermal Spray Conference and Exposition, Hamburg, Germany, 27–29 September 2011.
3. Dautov, S.S.; Shornikov, P.G.; Rezyapova, L.R.; Akhatov, I.S. Increasing thermal and mechanical properties of thermal barrier coatings by suspension plasma spraying technology. *J. Phys. Conf. Ser.* **2019**, *1281*, 012008. [CrossRef]
4. Sokołowski, P.; Kozerski, S.; Pawłowski, L.; Ambroziak, A. The key process parameters influencing formation of columnar microstructure in suspension plasma sprayed zirconia coatings. *Surf. Coat. Tech.* **2014**, *260*, 97–106. [CrossRef]
5. Yea, D.; Wanga, W.; Zhoub, H.; Fanga, H.; Huanga, J.; Lia, Y.; Gongc, H.; Li, Z. Characterization of thermal barrier coatings microstructural features using terahertz spectroscopy. *Surf. Coat. Tech.* **2020**, *394*, 125836. [CrossRef]
6. Palaciosa, A.; Conga, L.; Navarroa, M.E.; Dinga, Y.; Barrenechea, C. Thermal conductivity measurement techniques for characterizing thermal energy storage materials–A review. *Renew Sustain. Energy Rev.* **2019**, *108*, 32–52. [CrossRef]

7. Chi, W.; Sampath, S.; Wang, H. Ambient and High-Temperature Thermal Conductivity of Thermal Sprayed Coatings. *J. Therm. Spray Technol.* **2006**, *15*, 773–778. [CrossRef]
8. Akoshima, M.; Takahashi, S. Anisotropic Thermal Diffusivities of Plasma-Sprayed Thermal Barrier Coatings. *Int. J. Thermophys.* **2017**, *38*, 28–38. [CrossRef]
9. Muzika, L.; Svantner, M.; Houdkov, S.; Sulcov, P. Application of flash-pulse thermography methods for quantitative thickness inspection of coatings made by different thermal spraying technologies. *Surf. Coat. Tech.* **2021**, *406*, 126748. [CrossRef]
10. Cernuschi, F. Can TBC porosity be estimated by non-destructive infrared techniques? A theoretical and experimental analysis. *Surf. Coat. Tech.* **2015**, *272*, 387–394. [CrossRef]
11. Bison, P.; Cernuschi, F.; Grinzato, E. In-depth and In-plane Thermal Diffusivity Measurements of Thermal Barrier Coatings by IR Camera: Evaluation of Ageing. *Int. J. Thermophys.* **2008**, *29*, 2149–2161. [CrossRef]
12. Cernuschi, F.; Bison, P.; Marinetti, S.; Campagnoli, E. Thermal diffusivity measurement by thermographic technique for the non-destructive integrity assessment of TBCs coupons. *Surf. Coat. Tech.* **2010**, *205*, 498–505. [CrossRef]
13. Schweda, M.; Beck, T.; Offermann, M.; Singheiser, L. Thermographic analysis and modelling of the delamination crack growth in a thermal barrier coating on Fecralloy substrate. *Surf. Coat. Tech.* **2013**, *217*, 124–128. [CrossRef]
14. *ASTM E1461-13*; Standard Test Method for Thermal Diffusivity by the Flash Method. ASTM International: West Conshohocken, PA, USA, 2013. [CrossRef]
15. *ISO 18555*; Metallic and Other Inorganic Coatings—Determination of Thermal Conductivity of Thermal Barrier Coatings. International Standard Organisation: Geneva, Switzerland, 2016.
16. *BS EN ISO 22007-4*; Plastics—Determination of Thermal Conductivity and Thermal Diffusivity. Part 4: Laser flash method. International Standard Organisation: Geneva, Switzerland, 2017.
17. *ISO 18755*; Fine Ceramics (Advanced Ceramics, Advanced Technical Ceramics)—Determination of Thermal Diffusivity of Monolithic Ceramics by Laser Flash Method. International Standard Organisation: Geneva, Switzerland, 2005.
18. Abu-warda, N.; Boissonnet, G.; López, A.J.; Utrilla, M.V.; Pedraza, F. Analysis of thermo-physical properties if NiCr HVOF coatings on T24, T92, VM12 and AISI 304 steels. *Surf. Coat. Tech.* **2021**, *416*, 127163. [CrossRef]
19. Shrestha, R.; Kim, W. Infrared Phys. Evaluation of coating thickness by thermal wave imaging: A comparative study of pulsed and lock-in infrared thermography—Part II: Experimental investigation. *Infrared Phys. Technol.* **2018**, *92*, 24–29. [CrossRef]
20. Cape, J.; Lehman, G. Temperature and Finite Pulse-Time Effects in the Flash Method for Measuring Thermal Diffusivity. *J. Appl. Phys.* **1963**, *34*, 1909–1913. [CrossRef]
21. Clark, L.M., III; Taylor, R.E. Radiation loss in the flash method for thermal diffusivity. *J. Appl. Phys.* **1975**, *46*, 714–719. [CrossRef]
22. Takahashi, S.; Akoshima, M.; Tanaka, T.; Endo, S.; Ogawa, M.; Kojima, Y.; Taniguchi, S.; Kobaiashi, Y.; Ono, F. Determination of thermal conductivity of thermal barrier coatings. In Proceedings of the 5th Asean Thermal Spray Conference, Tsukuba, Japan, 26–28 November 2012; pp. 11–12.
23. Cowan, R.D. Pulse Method of Measuring Thermal Diffusivity at High Temperatures. *J. Appl. Phys.* **1963**, *34*, 926. [CrossRef]
24. Special Metals–2005. Available online: www.specialmetals.com (accessed on 20 October 2022).
25. Akoshima, M.; Tanaka, T.; Endo, S.; Baba, T.; Harada, Y.; Kojima, Y.; Kawasaki, A.; Ono, F. Thermal Diffusivity Measurement for Thermal Spray Coating Attached to Substrate Using Laser Flash Method. *Jpn. J. Appl. Phys.* **2011**, *50*, 11RE01. [CrossRef]
26. *ISO 18434*; Condition Monitoring and Diagnostics of Machine Systems—Thermography—Part 2: Image Interpretation and Diagnostics. International Standard Organisation: Geneva, Switzerland, 2019.

Article

Structural, Optical, Magnetic and Electrical Properties of Sputtered ZnO and ZnO:Fe Thin Films: The Role of Deposition Power

Ahmed Faramawy [1,2,*], Hamada Elsayed [3,4], Carlo Scian [1] and Giovanni Mattei [1]

[1] Department of Physics and Astronomy, University of Padova, Via Marzolo, 8, 35131 Padova, Italy
[2] Department of Physics, Faculty of Science, Ain Shams University, Abbassia, Cairo 11566, Egypt
[3] Department of Industrial Engineering, Università Degli Studi di Padova, 35131 Padova, Italy
[4] Refractories, Ceramics and Building Materials Department, National Research Centre, Cairo 12622, Egypt
* Correspondence: ahmed_faramawy@sci.asu.edu.eg

Citation: Faramawy, A.; Elsayed, H.; Scian, C.; Mattei, G. Structural, Optical, Magnetic and Electrical Properties of Sputtered ZnO and ZnO:Fe Thin Films: The Role of Deposition Power. *Ceramics* 2022, 5, 1128–1153. https://doi.org/10.3390/ceramics5040080

Academic Editors: Amirhossein Pakseresht and Kamalan Kirubaharan Amirtharaj Mosas

Received: 14 October 2022
Accepted: 28 November 2022
Published: 1 December 2022

Publisher's Note: MDPI stays neutral with regard to jurisdictional claims in published maps and institutional affiliations.

Copyright: © 2022 by the authors. Licensee MDPI, Basel, Switzerland. This article is an open access article distributed under the terms and conditions of the Creative Commons Attribution (CC BY) license (https://creativecommons.org/licenses/by/4.0/).

Abstract: Structural, optical, magnetic, and electrical properties of zinc oxide (henceforth, ZO) and iron doped zinc oxide (henceforth, ZOFe) films deposited by sputtering technique are described by means of Rutherford backscattering spectrometry, grazing incidence X-ray diffraction, scanning electron microscope (SEM), UV–Vis spectrometer, vibrating sample magnetometer, and room temperature electrical conductivity, respectively. GIXRD analysis revealed that the films were polycrystalline with a hexagonal phase, and all films had a preferred (002) c-axis orientation. The lattice parameters a and c of the wurtzite structure were calculated for all films. The a parameter remains almost the same (around 3 Å), while c parameter varies slightly with increasing Fe content from 5.18 to 5.31 Å throughout the co-deposition process. The optical gap for undoped and doped ZO was obtained from different numerical methods based on the experimental data and it was increased with the increment of the concentration of Fe dopant from 3.26 eV to 3.35 eV. The highest magnetization (4.26×10^{-4} emu/g) and lowest resistivity (4.6×10^7 $\Omega \cdot$cm) values of the ZO films were found to be at an Fe content of 5% at. %. An explanation for the dependence of the optical, magnetic, and electrical properties of the samples on the Fe concentrations is also given. The enhanced magnetic properties such as saturated magnetization and coercivity with optical properties reveal that Fe doped ZO thin films are suitable for magneto-optoelectronic (optoelectronic and spintronics) device applications.

Keywords: diluted magnetic semiconductor; TM doping; sputtering; magnetic properties; optical band gap; resistivity

1. Introduction

Nowadays, studies on ZO, which is one of the most important 3rd generation semiconductors, support a new stage of comprehensive use due to its multi-functional characteristics. It plays an important role in a very wide range of applications in fields such as: biosensors, biomedical and antibacterial active media, in biological fields [1,2]; UV light-emitting devices, optical waveguides, solar cells in electro-optical fields [3]; transparent high power electronics, varistors, thin film transistors in electronics; gas sensors, water purification and solar photocatalytics in the environmental field [1,4,5]. Moreover, since 2001, many studies have been carried out in the field of Diluted Magnetic Semiconducting Oxides (DMSO) which combine two interesting properties—semiconducting and magnetic—whether to look for a new compound which might be ferromagnetic at room temperature or to find materials which might have a large magnetic moment [6]. The interest in ZO is fueled and fanned by its prospects in optoelectronics applications, having the necessary excellent conditions of ultraviolet and blue emissions because of its unique of chemical and physical properties of a large direct energy gap (e.g., ~3.37 eV at 300 K) and a large exciton binding energy (60 meV) at room temperature which makes it being invaluable compared to numerous other nanomaterials [7]. It is also a promising candidate for semiconductor spintronics applications; Dietl et al. [8] predicted a Curie temperature of >300 for Mn-doped ZO. In addition, n-type doping in Fe-, Co-, or Ni-alloyed ZO was also predicted to stabilize high—Curie—temperature ferromagnetism. Moreover, the well-known room temperature ferromagnetism (RTFM) is one of the more studied phenomena in pure ZO nanostructures doped with magnetic ions. This phenomenon has been found at low temperatures as well, and it has been studied both experimentally and theoretically [9].

Furthermore, the remarkable morphological features of ZO make it a striking material that is widely used as a counterpart of a noteworthy scope of applications. As for ZO's morphological perspectives, synthesis and fabrication procedures also play an important role. The various parameters used (namely surfactant, temperature, concentration, and time, etc.) play a crucial role in the growth of various forms of synthetic processes. An enormous number of different methods for the synthesis of surface structured ZO as nanopowder, composites, and films have been published worldwide. There are widely noted techniques for synthesizing different ZO nanomaterials, composites and doped heterostructures, including precipitation, wet-chemical technique, sol-gel procedure, hydrothermal method, solvothermal procedure, sputtering, coating and microwave techniques, etc. [1].

For several of the above-mentioned applications, a stable, high, and reproducible p-doping is mandatory. Though progress has been made, this aspect still represents a major problem. The contribution of the current ZO research is not only on the same topics as earlier, but also includes nanostructures and nanotechnologies, new growth and doping techniques, and focuses more on the application-related aspects of daily life [9,10].

In this study we planned to control and enhance the optical band gap, and the magnetic properties of Zinc Oxide (ZO) thin films doped by a transition element (TM).

Fe is a magnetic transition metal that has several oxidation states, an electronegativity of 1.83 and ionic radii that range from 0.61 to 0.78 Å (depending on the magnetic and oxidation states) [10]. Because the properties of iron are similar to those of aluminum or gallium, it is considered a potentially attractive alternative for the doping of ZO thin films [11]. However, the performance of iron as a dopant is complicated by the existence of 3d electrons in its electronic configuration. Currently, most studies on Fe-doped ZO thin films focus on their wide applications. As we are aware, fewer reports have investigated Fe-doped ZO thin films as transparent conducting oxides [12–14]. Due to the complex electronic structure of iron, it has been difficult to control the different properties of these films. Moreover, iron-doped ZO thin films have been synthesized by several techniques including electrodeposition, spray pyrolysis, spin coating, ion implantation, and the sol–gel method. Sputtering is the most widely used technique for the preparation of Fe-doped ZO thin films [15] because it provides the ability to control the properties of the films through several deposition parameters, such as the sputtering mode, power, sputtering and reactive gas pressures, substrate temperature, and target composition and allows deposition of different materials [16]. Three sputtering techniques have been used to synthesize Fe-doped ZO thin films:

1. simultaneous co-sputtering from two targets (ZO and Fe) [17];
2. direct-current (DC) sputtering using an Fe-doped Zn target [18] or a Zn target with Fe pieces attached to it [19] and;
3. radio-frequency (RF) magnetron sputtering.

Several of the studies using RF-sputtering for Fe-doped ZO thin film have investigated the effect of doping concentration on the properties of the films [17]. Others have studied the influence of the deposition parameters on the properties of films with fixed dopant concentrations [19].

In this present study, a thin film of iron (Fe) doped zinc oxide (ZO), with various Fe-concentrations, is investigated using RF-magnetron sputtering on silica and silicon substrates. We report the influence of the Fe doping on the microstructure, optical, magnetic and electrical resistivity of the ZO films. The results indicated that ZO film with 5.0 at. % doping-concentration can attain a higher energy gap and magnetic properties, and better electrical resistivity.

2. Materials and Methods

Undoped and doped ZO films with Fe were deposited on single-crystal n-type Si (100) wafer and silica (HSQ 100 from Heraeus) glass preliminarily washed in piranha solution for magnetron sputtering. The ZO target (99.95% purity, 100-mm diameter) is placed in a radio-frequency driven (RF) source while setting the power at 100 W and Fe target in a direct current (DC) source providing powers of 3, 5, 10, and 15 W to produce (1, 2.4, 4.8, 5) at. % of Fe, respectively. Moreover, in order to control the sputtering yield of Fe and instead of reducing the DC power (which produced instabilities of the DC source), we performed the DC sputtering through a grid with a different mesh size to block part of the Fe atoms. During the deposition, the substrate was rotated at a low speed to enhance the thickness uniformity of deposited films. The two sources were tilted of an angle 30° with respect to the axis normal to the sample surface allowing a simultaneous deposition of the two targets and the source-target distance was about 15 cm. The thickness and the composition of the films were properly tuned through an optimized combination of the electrical power of the source and the opening time of the shutters placed in front of the targets. The thickness of the films was 150 nm, which was controlled by the deposition time. Other detailed conditions about the deposition process of the films are summarized in Table 1.

Table 1. Typical deposition parameters for sputtering deposition.

Pre-sputtering pressure (mbar)	1.5×10^{-6}
Sputtering pressure (mbar)	50×10^{-4}
Ar gas pressure (mTorr)	0.5
Ar gas flow (sccm)	20
Sputtering time (min)	17
Sputtering voltage (V)	$V_{DC} \approx 350$ V, $V_{RF} \approx 390$ V

Furthermore, the film thickness was measured using a thickness profile-meter and by Rutherford back scattering (RBS), performed with a HVEC 2.5 MeV Van de Graaff accelerator at Legnaro National Laboratory (LNL)—INFN, using 2 MeV α-particles at normal incidence and a scattering angle of 20°, which was also used to determine the contents by atomic percentage of Zn, O and Fe ions. Furthermore, the hexagonal phase wurtzite structure of samples was confirmed by grazing incidence X-ray diffraction (GIXRD) (Philips X'Pert PRO MRD triple axis diffractometer (angular resolution 0.01°, accuracy 0.001°)) and the scanning electron microscope (SEM), Zeiss Field-emission SEM (FE-SEM), was used to determine the surface morphology. The optical properties of the films were measured using UV-Vis optical spectroscopy, Jasco V-670 UV-Vis-NIR Spectrophotometer with working range 190–2500 nm, in order to determine the optical parameters such as band gap, Urbach energy and refractive index for all investigated samples.

Likewise, the magnetic hysteresis loops for the doped and undoped films were investigated by a vibrating sample magnetometer (VSM) (LakeShore model no. 7410 USA) in a maximum applied field of 20 kOe. Finally, the IV characteristics and electric resistivity was also measured for undoped and doped films by a two probe method with a KEITHLEY 6514 system electrometer. For simplicity, throughout the present work, the "balanced stoichiometry" for the Fe doped ZnO will be named ZOFe. The DC power values are indicated as follows: ZOFe/3 W, ZOFe/5 W, ZOFe/10 W and ZOFe/15 W for 3, 5, 10 and 15 Watt for DC deposited power of the Fe source target.

3. Results

3.1. Elemental Analysis

The elemental composition of ZO and ZOFe films was measured by Rutherford backscattering spectrometry (RBS). The RBS data analysis is reported in Figure 1 where the experimental data for all investigated films is superimposed to the simulation. The quality of the fit is the direct proof of the sample homogeneity.

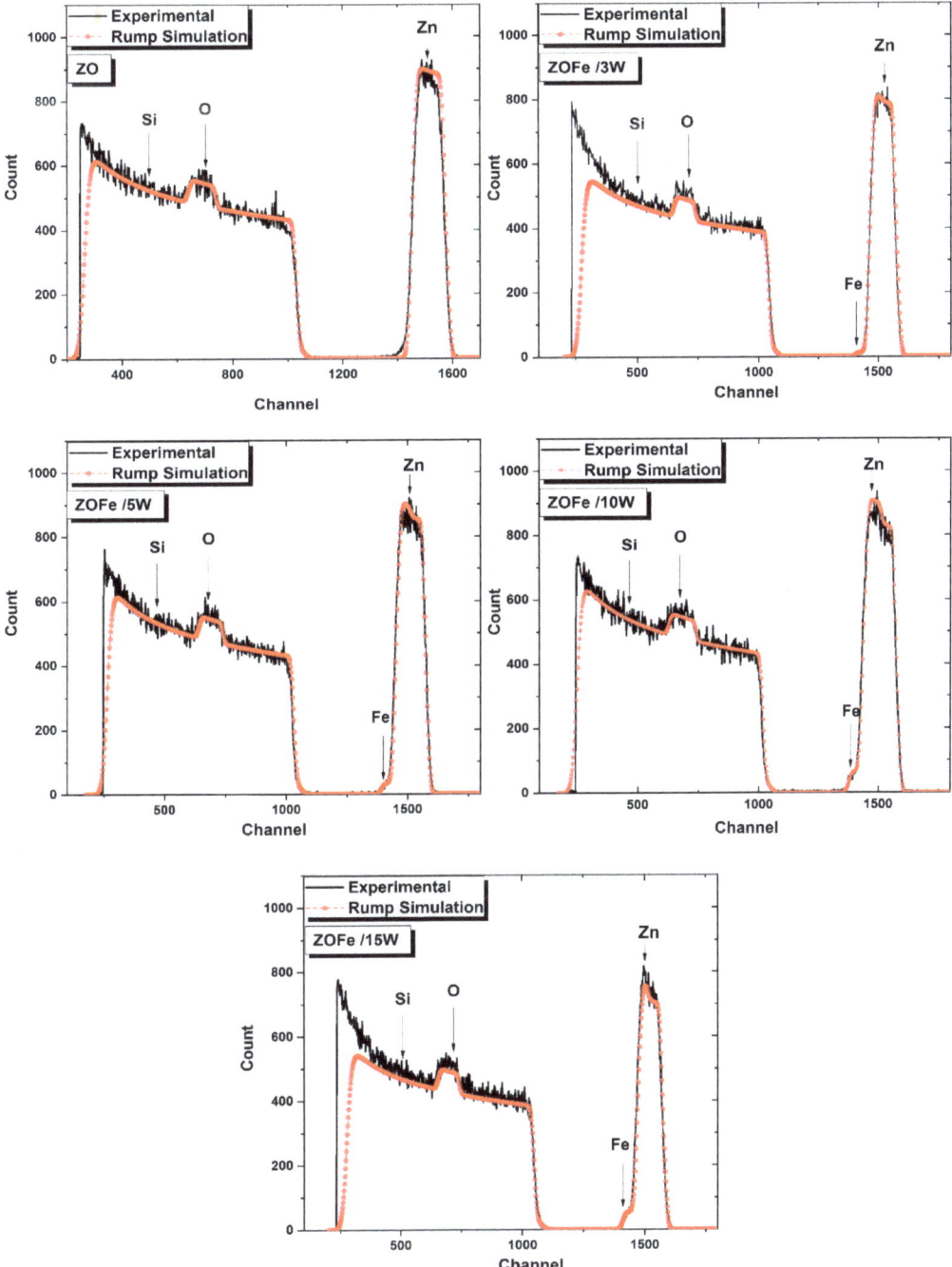

Figure 1. RBS random spectra for the ZO and ZOFe thin films deposited on Si.

One can notice that the surface edge for the different elements (Zn, O, Si and Fe) can be observed clearly from Figure 1. The large peak on the right-hand side located in the high energy range of the spectrum is due to the zinc atoms whereas the low energy peak signal is related to the oxygen, which is superimposed over the continuous spectrum of a Si substrate. Moreover, small peaks were observed on right-hand side of the Zn peak which indicate the Fe. The yield account of these peaks increased with increase in the DC power value in the co-deposited process from 3 W to 15 W.

RBS is not equally sensitive to various elements. The height of the RBS signal depends on the scattering cross section which is proportional to the square of the atomic number of the element. Thus, for light elements, the RBS signal is much lower than for heavy ones [20]. Therefore, it is quite difficult to separate the signal from such elements as nitrogen, carbon or oxygen, from the high background of much heavier substrates as Si. Consequently, it is not easy to evaluate the contribution of light elements. Zinc, oxygen and iron contents in the ZO and ZOFe thin films were evaluated from the RBS random spectra. It has been found that the value O: Zn content is approximately equal 1 and the results are summarized in Table 2.

Table 2. RBS results for the estimation of Zn and Fe content (at. %) in the investigated films.

Film	RBS Fluency (10^{15} at./cm^2)			The Contents Percentage (% at.)	
	Zn	O	Fe	ZnO	Fe
ZO	620	620	0	100	0
ZOFe/3 W	555	555	11	99	1
ZOFe/5 W	630	630	30	97.6	2.4
ZOFe/10 W	650	650	65	95.2	4.8
ZOFe/15 W	473	516	52	95	5

3.2. GIXRD Analysis

Figure 2 depicts the GIXRD pattern of the ZO thin films. One can see that it exhibits a hexagonal crystal phase, which was confirmed with a standard JCPDS card (PDF No. 36–1451). In the GIXRD pattern, the two dominant peaks, (002) and (103) were obtained for all films without any secondary phase.

All films exhibited preferential (002) c-axis orientation, which grew perpendicular to the substrate surface. This can interpreted as follows: the sputtered ZO and ZOFe thin films with a c-axis have a preferential orientation due to the lowest surface free energy of (002) plane in ZO. In the equilibrium state, the films grow with the plane of the lowest surface free energy parallel to the surface if there is no effect from the substrate. Additionally, the sp^3 hybridized orbit in zinc oxide forms a tetrahedral coordination. Because each apex is parallel to the c-axis in the wurtzite structure, ZO films prefer to grow in the (002) direction [21].

Table 3 represents some significant data estimated from the diffracted peaks such as 2θ, d-spacing, (hkl), peak intensity, crystallite size, and the texture coefficient for all deposited films.

The texture coefficient $TC_{(hkl)}$ is another perspective for preferred orientation and/or preferred growth. $TC_{(hkl)}$ can provide quantitative information on the preferred crystal orientation which can be calculated from the expression [22,23]

$$TC_{(hkl)} = \frac{I_{(hkl)}/I_{o(hkl)}}{(1/n)\sum_n I_{(hkl)}/I_{o(hkl)}} \quad (1)$$

where $I_{(hkl)}$ is the measured intensity and $I_{o(hkl)}$ is the standard JCPDS intensity and n is the number of diffraction peaks. If $TC_{(hkl)} \sim 1$ for all the (hkl) planes considered, then the films have a randomly oriented crystallite similar to the JCPDS standard, while values higher than 1 indicate the abundance of grains in a given (hkl) direction.

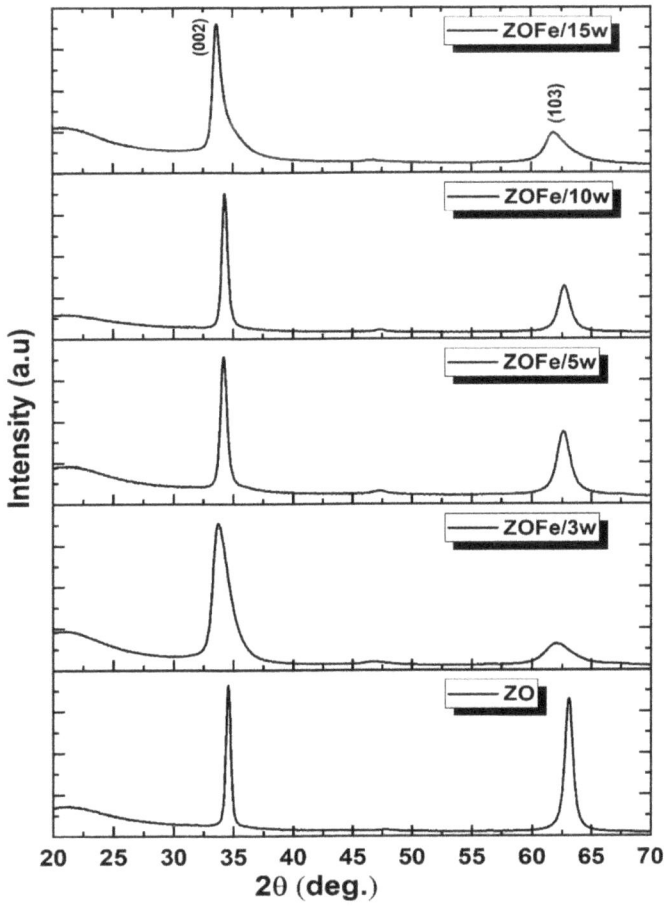

Figure 2. GIXRD patterns of ZO and ZOFe thin films deposited on SiO_2.

Table 3. Planes angle 2θ, d-spacing, intensity I, crystallite size D, and texture coefficient TC for different DC depositing powers of sputtered ZO and ZOFe from GIXRD.

Film	2θ (°)	d-Spacing (Å)	Peak Intensity, I (a.u.)	Crystallite Size, D (nm)	Texture Coefficient
			Plane (002)		$TC_{(002)}$
ZO	34.57	2.59	54,621	40	0.815
ZOFe/3 W	33.28	2.69	88,247	6	1.554
ZOFe/5 W	34.20	2.62	39,271	19	1.120
ZOFe/10 W	34.13	2.61	45,609	25	1.264
ZOFe/15 W	33.67	2.65	72,050	11	1.398
			Plane (103)		$TC_{(103)}$
ZO	63.08	0.52	49,566	-	1.184
ZOFe/3 W	62.08	0.52	15,787	-	0.445
ZOFe/5 W	62.61	0.52	19,252	-	0.879
ZOFe/10 W	62.69	0.52	16,571	-	0.735
ZOFe/15 W	61.87	0.52	19,353	-	0.601

Values 0 < $TC_{(hkl)}$ < 1 indicate the lack of grains oriented in that direction. As $TC_{(hkl)}$ increases, the preferential growth of the crystallites in the direction perpendicular to the *hkl* plane is the greater. Since two diffraction peaks were used ((002) and (103)), the maximum value $TC_{(hkl)}$ possible is two.

The calculated texture coefficient results are displayed in Table 3 and were found to be maximum for the (002) plane and increased by increasing iron content, as shown in Figure 3. Moreover, the (002) polar facet of Fe doped ZO films increased significantly compared to the (103) facet. This also indicates the preferential growth along the c-axis orientation.

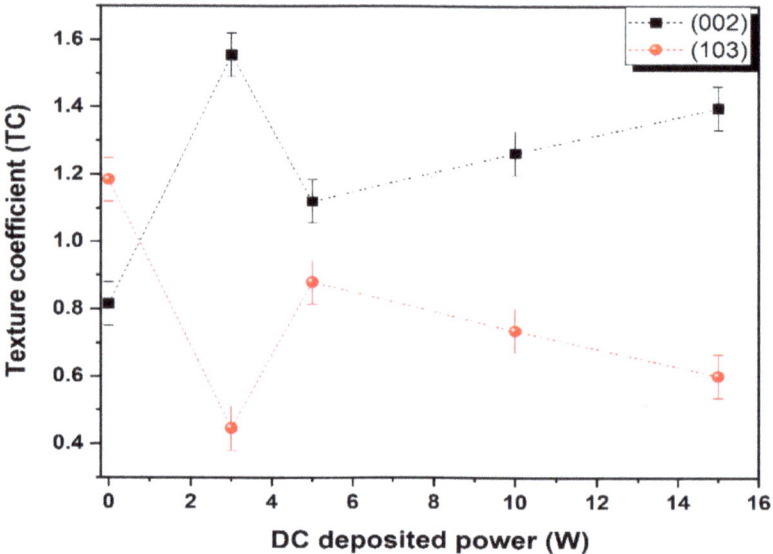

Figure 3. Texture coefficient, TC, for (002) and (103) plans of ZO, Fe-doped ZO deposited at different DC power.

On other side, the crystallite size may be estimated from the full-width at half-maximum (FWHM) of (002) and (103) diffraction peak using the Debye–Scherrer formula [24] which is expressed as:

$$D = \frac{0.9 \lambda}{\beta \cos\theta_{hkl}} \quad (2)$$

where D is the average crystallite size, λ is the wavelength of the X-ray and takes 1.54 Å for Cu Kα, β is the peak width of two peaks at half maximum height (FWHM) in radians, and θ_{hkl} is the diffraction angle of the crystal plane (*hkl*). The values of the average crytallite size are given in Table 3.

In addition, using the interplanar spacing values (d-spacing) and the related hkl parameters for hexagonal systems, the lattice parameters (a, b, and c where $a = b$) of the wurtzite cell of ZO and ZOFe thin films have been calculated by using following formula [22,25]:

$$\frac{1}{d^2} = \frac{4}{3}\left(\frac{h^2 + hk + k^2}{a^2}\right) + \left(\frac{l^2}{c^2}\right) \quad (3)$$

Interplanar d-spacing is calculated according to Bragg's law:

$$n\lambda = 2\, d_{(hkl)} \sin\theta \quad (4)$$

where d is lattice spacing, a and c are lattice constants, h, k, l are miller indices, θ is the angle of corresponding peak and λ is the wavelength of X-ray used (1.5402 Å). Considering Bragg's law, it is possible to rewrite Equation (4) as follows:

$$\frac{4\sin^2\theta}{\lambda^2} = \frac{4}{3}\left(\frac{h^2 + hk + k^2}{a^2}\right) + \left(\frac{l^2}{c^2}\right) \qquad (5)$$

There are two unknowns to be calculated in the formula above. Because of this, while calculating lattice constant a, a peak in the form of (h00) should be used to eliminate c from the equation. As an alternative, the (00l) peak should be used to get an equation with just one unknown for calculating the c constant. Following correct peak selection, the following equations for a and c constants are obtained [25,26].

$$a = \frac{\lambda \cdot \sqrt{h^2 + hk + k^2}}{\sqrt{3}\sin\theta} \qquad (6)$$

$$c = \frac{\lambda \cdot l}{2\sin\theta} \qquad (7)$$

The calculated lattice parameters for all films are listed in Table 4. As can be seen from Table 4, the a parameter remains almost the same, while the c parameter varies slightly due to an increase in Fe content throughout the co-deposition process. As we noticed from XRD, no obvious peaks corresponding to iron metal or iron oxide phase were observed. Furthermore, the wurtzite structure of ZO remained unchanged after iron modification. There is a slight left shift in the (002) and (103) peaks of the ZOFe films which could be interpreted as follows. Fe exists stably in both valence states, i.e., Fe^{2+} and Fe^{3+} state. The peaks would shift towards a lower angle if Fe exists in Fe^{2+} (0.78 Å) state due to its larger ionic radius than Zn^{2+} (0.74 Å), as seen in Figure 4. This confirms that iron has been doped into the ZO lattice. However, if it exists in the Fe^{3+} (0.68 Å) state, the peak shift will occur towards the higher angle side. As a result, we may conclude that the iron in our samples is predominantly in the 2+ state.

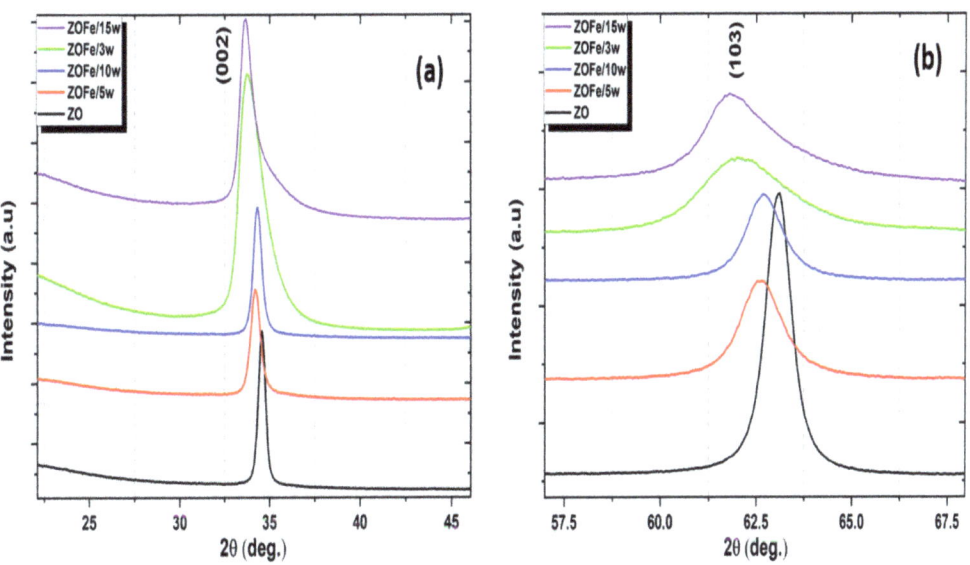

Figure 4. The peak shifts towards lower angle for all deposited films in planes (**a**) (002) (**b**) (103).

Table 4. Values of lattice constants, cell volume, internal parameter, bond length, strain and the compressive stress for the wurtzite hexagonal ZO and ZOFe films deposited at different DC powers.

Film	a (Å)	c (Å)	V (Å³)	U	L (Å)	$e_{zz} \times 10^{-3}$	$\sigma \times 10^9$ (GaP)
ZO	2.993	5.183	40.201	0.361	1.871	−4.152	1.868
ZOFe/3 W	3.015	5.378	41.908	0.361	1.942	3.329	−14.981
ZOFe/5 W	3.024	5.237	41.420	0.361	1.891	6.236	−2.806
ZOFe/10 W	3.015	5.221	41.091	0.361	1.885	3.316	−1.411
ZOFe/15 W	3.071	5.318	43.416	0.361	1.921	2.172	−9.775

As well, the volume V, the internal parameter U, and the bond length L of wurtzite unit cell for hexagonal ZO and ZOFe thin films have been calculated using the following expressions [22,25,27]:

$$V = \frac{\sqrt{3}}{2}a^2 c; \quad U = \frac{1}{3}\left(\frac{a^2}{c^2}\right) + \frac{1}{4}; \quad L = \left[c^2\left(\frac{1}{2}-U\right)^2 + \frac{a^2}{3}\right]^{1/2} \quad (8)$$

The obtained values are tabulated in Table 4 and the variation in a, c and L after Fe incorporation may be attributed to the differences between the atomic radii of the host material (ZO) and the dopant (Fe) [27].

Moreover, the average uniform strain, e_{zz}, in the lattice along the c-axis in the randomly oriented ZO and ZOFe films has been estimated from the lattice parameters as follow:

$$e_{zz} = \frac{c - c_0}{c_0} \quad (9)$$

where c is the lattice parameter of the ZO film calculated from the (002) peak of XRD pattern and the c_0 is the lattice parameter for the ZO bulk. For hexagonal crystals, the stress (σ) in the plane of the film can be calculated using the biaxial strain model [28,29]:

$$\sigma = \left[2C_{13} - \frac{(C_{11} + C_{12})C_{33}}{C_{13}}\right]e_{zz} \quad (10)$$

where C_{ij} are elastic stiffness constants for ZO, (C_{11} = 2.1 × 10¹¹ N/m², C_{33}= 2.1 × 10¹¹ N/m², C_{12} = 2.1 × 10¹¹ N/m², and C_{13} = 2.1 × 10¹¹ N/m²). This yields the following numerical relation for the stress derived from the change in the 'c' lattice parameter:

$$\sigma(\text{Nm}^{-2}) = -4.5 \times 10^{11} e_{zz} \quad (11)$$

The calculated values of stress (σ) in the undoped and doped films are listed in Table 4. The negative sign indicates that the films are in a state of compressive stress which not seen in pure ZO film. The total stress in the film typically consists of two components: intrinsic stress which introduced by impurities, defects and lattice distortions in the crystal, and the extrinsic stress familiarized by the lattice mismatch and thermal expansion coefficient mismatch between the film and substrate. When the thickness of a thin film is large, the extrinsic stress in the thin film normally relaxes. All of the films in our current study have a thickness around 150 nanometers. As a result, extrinsic stress will be absent, and the overall predicted stress values will be dominated by intrinsic stress. Furthermore, in the current scenario of sputtering deposition with rising DC power, this intrinsic stress arises as a result of the bombardment of energetic species. Strong compressive stress is present in films deposited at lower temperatures. In the sputtering process, the highly energetic species could be implanted below the ZO surface, causing high intrinsic stresses by creating the defects [29,30]. The existence of stress in the ZO films was reported by Gupta and Mansingh, who ascribed it to oxygen interstitial defects [31]. It is well known that, because of the higher formation energy, the oxygen interstitials are less expected in ZO films. In the present experiment all the films were deposited in argon atmosphere. So, the possibility of the formation of oxygen interstitials is ruled out and the formation of Zn interstitials and the oxygen vacancies are expected. Thus, in our case, the intrinsic stress in the deposited ZO and ZOFe films seems to arise due to presence of Zn interstitials.

3.3. Microstructural Analysis

Figure 5 shows the morphology of the ZO and Fe doped ZO thin films; the pure zinc oxide film surface seems to be agglomerated, spherical in shape as shown Figure 5a. The surface of the films consists of tightly packed grains forming a smooth surface without any voids and cracks whereas the Fe doped zinc oxide films in Figure 5b,c look like black and white islands of spherical shape. Still, by increasing the DC deposited power for Fe with ZO deposition, the black spots decreased. A SEM image of the sectional view of the samples could help in discovering the shape of the spherical grains and their interface with the substrate as in Figure 5. Moreover, the shape of the grains would be hard to identify because the growth could not be described in a reliable way. Further, the distribution and size in grains is inhomogeneous. Figure 6 shows the EDX spectrum of the ZO and Fe doped ZO thin films. The observed peaks are attributed to zinc, oxygen and iron as principal elemental components with no other impurities being obtained. The Zn:O:Fe atomic percentage ratios of the ZO thin film are listed.

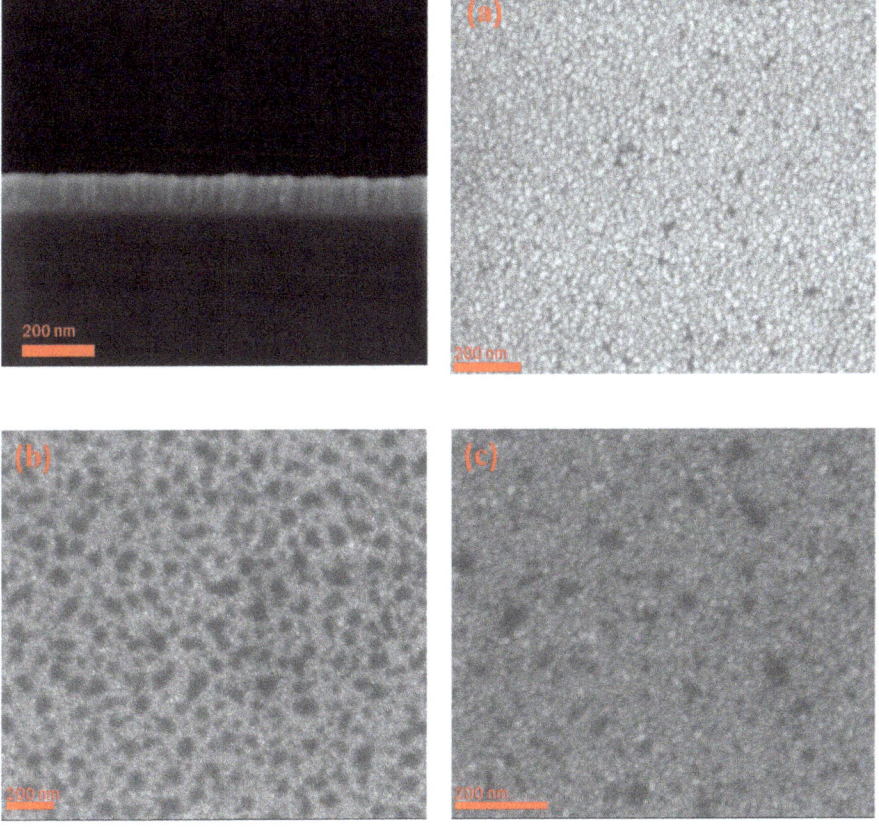

Figure 5. Cross sectional view of ZO thin film deposited in SiO_2 and SEM photos for the (**a**) undoped ZO and Fe doped ZO at DC deposited power (**b**) 3 W and (**c**) 5 W.

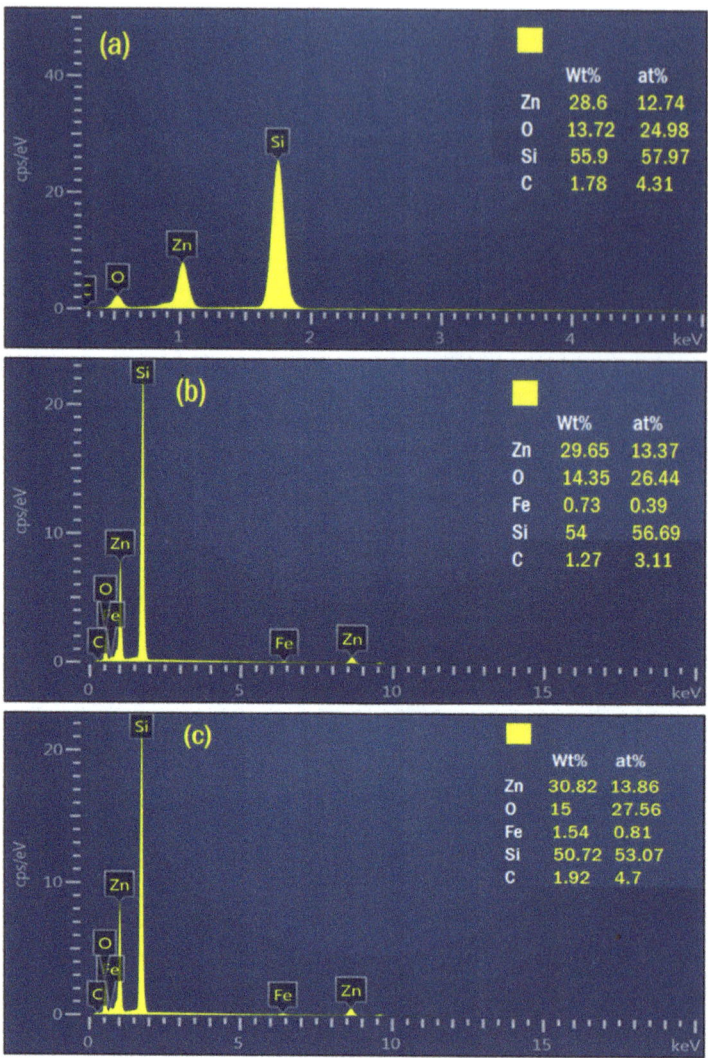

Figure 6. EDX spectrum of (**a**) undoped ZO and Fe doped ZO at DC deposited power (**b**) 3 W and (**c**) 10 W.

3.4. Optical Properties

3.4.1. Transmittance Spectra

Zinc oxide and iron doped zinc oxide samples grown on SiO_2 are transparent to visible and NIR light; thus it was possible to obtain transmittance spectra and make observations on the material properties. The spectra were taken in the range between 300 and 900 nm, as shown in Figure 7. The optical transmittance spectra of the films show a good transmission in the visible region and a sharp fall in the UV region which corresponds to the band gap. The decrease of the transmittance is due to the interaction of the incident long-wavelength radiation with the free electrons in the films [32]. Moreover, the films have good transmission in the visible region of the spectrum terminated at shorter wavelengths while, in the UV region, the obvious absorption could be observed; it increased with increasing Fe content.

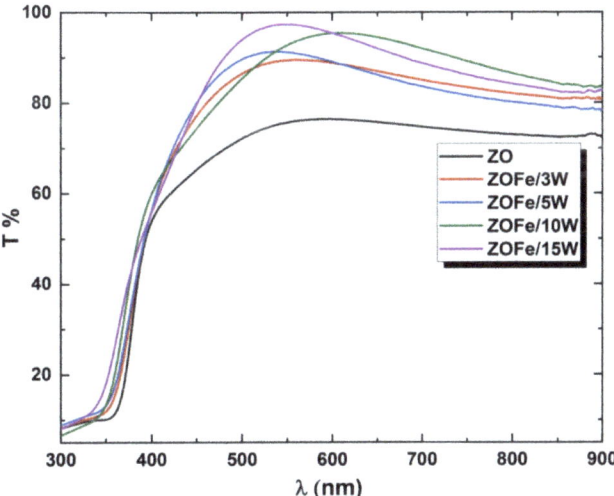

Figure 7. Transmittance of ZO and ZOFe thin films.

3.4.2. Optical Band Gap

The energy gap and refractive index of semiconductors are two essential physical parameters that define their optical and electronic properties. The optical energy gap (E_g) determines the threshold for absorption of photons in a semiconductor. In order to determine the optical energy gap (E_g), there are several numerical methods based on the experimental data. We are used different methods and compared them to obtain (E_g) values which have a good agreement with the literature.

(A) Tauc method:

In most of the literature, the energy gap in the crystalline and amorphous semiconductors is determined by the Tauc method [33]. This method is easy to apply due to fitting of the linear function to the graph for parameter: m = {0.5, 1.5, 2, 3} which represents direct allowed, direct forbidden, indirect allowed and indirect forbidden optical transition, respectively. The energy gap determined using the Tauc method is often subject to high uncertainty. In some cases, it is possible to fit a straight line for all values of parameter m. The energy gap of all the films is determined from the absorption coefficient (α) which can be calculated from the transmittance (T) of the undoped and doped ZO thin films. The absorption coefficient (α) is calculated from Equation:

$$\alpha = \left(\frac{1}{d}\right) ln\left(\frac{1}{T}\right) \qquad (12)$$

where (d) is film thickness and (T) is the transmittance. The absorption edge was analyzed by the following Equation:

$$(\alpha h\nu)^2 = A(h\nu - E_g) \qquad (13)$$

where A is a constant, $h\nu$ is the incident photon energy and E_g is the optical energy band gap. Based on Equation (15), the plots of $(\alpha h\nu)^2$ as a function of incident photon energy ($h\nu$) were obtained for the undoped and doped ZO thin films and are shown in Figure 8. The linear portions of these plots are extrapolated to meet the energy axis and the energy value at $(\alpha h\nu)^2 = 0$ gives the values of the energy gap E_g for the thin films. The E_g values are listed in Table 5.

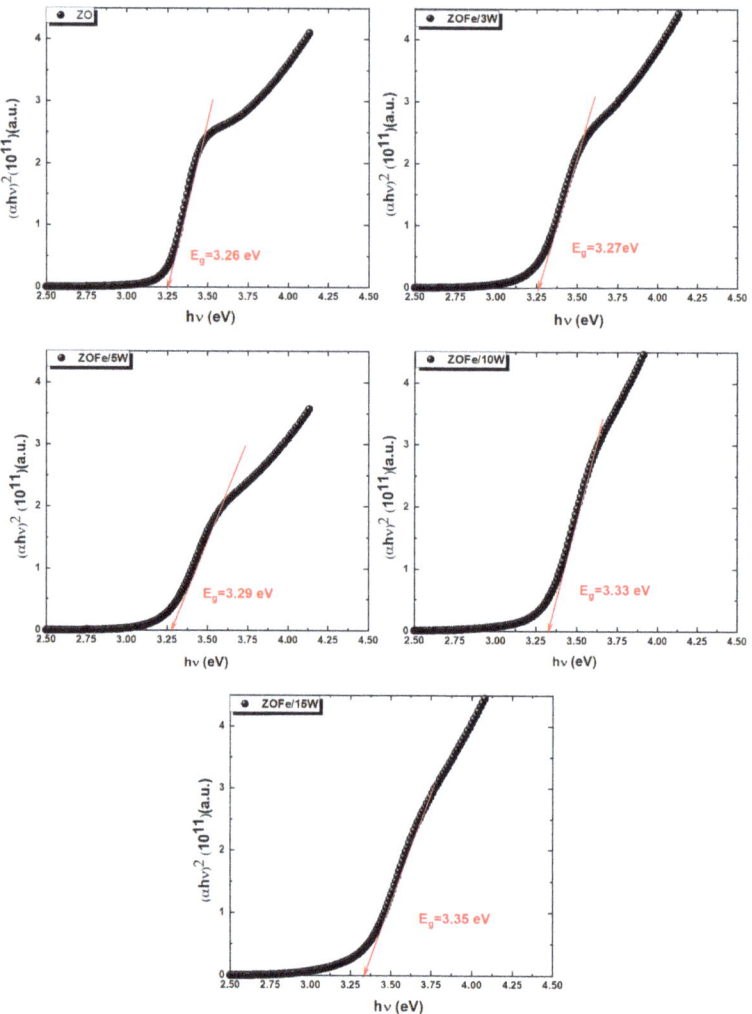

Figure 8. Plot of $(\alpha h\nu)^2$ vs. $(h\nu)$ for the undoped and Fe doped ZO thin films.

Table 5. Energy gap values E_g obtained from different methods for undoped ZO and Fe doped ZO thin films.

Film	Tauc	dT/dE	dT/dλ	LD
ZO	3.26	3.26	3.26	3.26
ZOFe/3 W	3.27	3.27	3.27	3.27
ZOFe/5 W	3.29	3.28	3.29	3.28
ZOFe/10 W	3.33	3.34	3.34	3.33
ZOFe/15 W	3.35	3.39	3.40	3.40

(B) The derivative of transmittance (T) against photon energy $(h\nu)$

This method is applicable for transitions between direct band gaps [34]. The transmission through a film may be approximated as:

$$T(E) \approx [1 - R(E)]^2 e^{\alpha(E)d} \tag{14}$$

where E is the energy of the incident light, R is the reflectance, and d is film thickness.

$$\alpha(E) = \frac{C}{E\eta_r(E)}\sqrt{E - E_g} \quad (15)$$

where C is a constant and $\eta_r(E)$ is the energy-dependent index of refraction. The behavior of the reflectance at the band gap energy must be considered while calculating the transmission derivative. It can be shown that as $E \to E_g$, the quantities $(1 - R(E))$ and $dR(E)/dE$ remain finite at energies in the environs of E_g due to the presence of a band tail that prohibits singularity-type behavior. The reflectance and its derivative are well-behaved values near the band gap energy, according to experimental studies concerning ZO thin film properties [35]. As a result, we can write the transmission through a direct gap semiconductor as

$$T(E) = e^{Cd(1/E\eta_r(E))\sqrt{E-E_g}} \quad (16)$$

After taking the first derivative of $T(E)$ with respect to energy and then at the limit $E \to E_g$, this results in a spike towards negative infinity at $E = E_g$. Due to band tail states, it was assumed that $d\eta_r(E)/dE$ is continuous around E_g. As a result, when plotting dT/dE versus E, a significant singularity will appear at the band gap energy. In realistic instances, absorption tails exist, which soften the divergence and create well-defined peaks around the gap energy, as also can be seen in Figure 9 for ZO and ZOFe/10 W thin films. The values of the band gap of all investigated samples obtained by the derivative method are listed also in Table 5. These values of band gap are very close to the values obtained by Tauc method. Discrepancy between values of band gap obtained by both methods can be explained: the fitting of a line in the linear region of $(\alpha h\nu)^2$ is quite challenging which adds errors in determination of the band gap. Hence, the derivative method is considered more appropriate when compared to Tauc method.

$$\lim_{E \to E_g} \frac{dT}{dE} = -C\left(-0 - 0 + \frac{1}{0}\right) \to -\infty \quad (17)$$

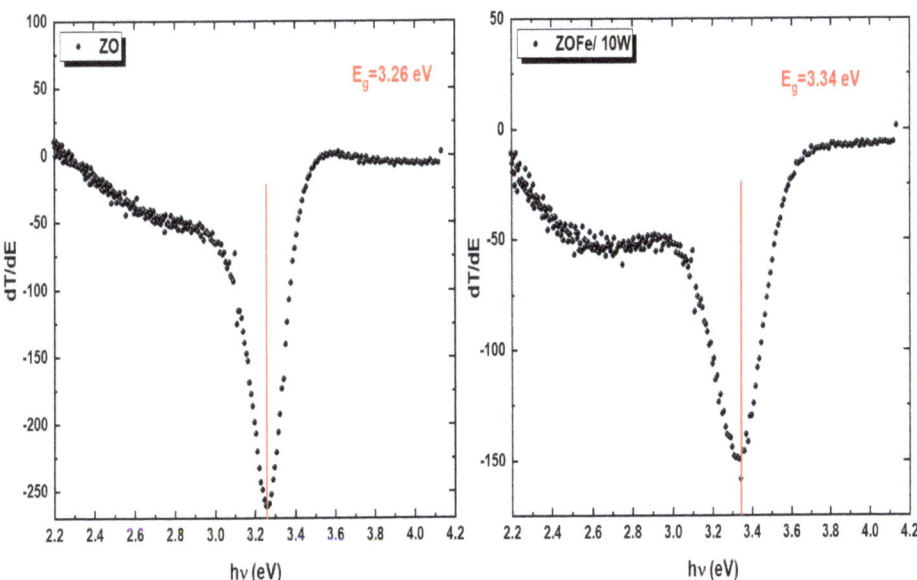

Figure 9. The plots of the derivative of the transmission with respect to energy for the undoped and ZOFe/10 W thin films.

(C) The derivative of transmittance (T) against wavelength (λ)

As we noticed in the transmittance curve of ZO films, there are transmittance tails at the high energy side of the absorption edge. Because a sharp absorption edge is generally observed in the transmittance spectra of direct band gap semiconductor films, we computed the derivative of transmittance (T) against wavelength (λ) in an effort to accurately determine the optical band gap of undoped and doped ZO films. The sharp absorption edge in the transmittance spectra may result in a sharp peak in a plot of $dT/d\lambda$ vs. $h\nu$ from which the optical band gap can be successfully determined. Figure 10 shows plots of $dT/d\lambda$ vs. $h\nu$ for the ZO and ZOFe/10 W thin films. One can see in Figure 10 a broad peak centered around 3.3 eV, corresponding to a band gap of all investigated films, also listed in Table 5.

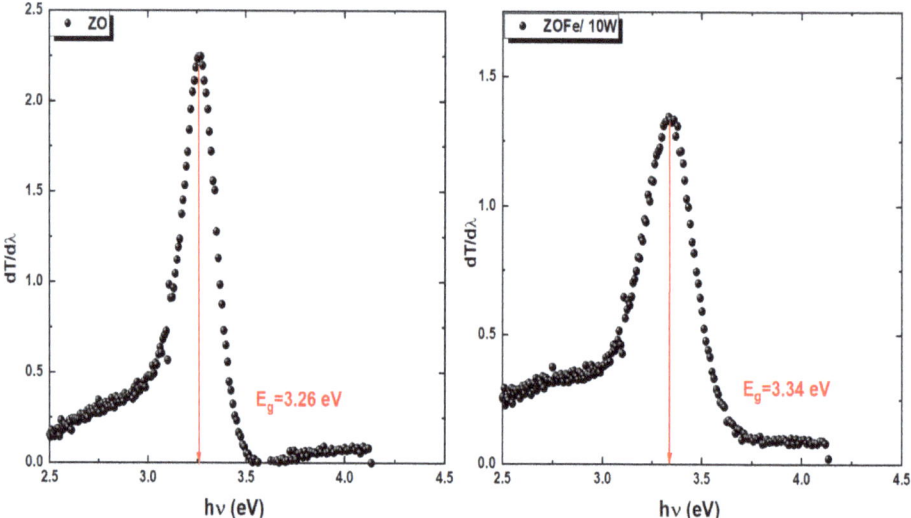

Figure 10. $dT/d\lambda$ vs. photon energy for the undoped and undoped and ZOFe/10 W thin films.

(D) Logarithmic derivative (LD) method:

This approach combines the advantages of both the Tauc [33] and McLean [36] methods: the fit is linear, and the parameter m is calculated as a consequence of the fitting. Significantly, the results of all three approaches are consistent in simple cases.

As we referred above, the Tauc method is based on linear fit for $\alpha h\nu^{1/m}$ data for selected $m = \{0.5, 1.5, 2, 3\}$ values. The zero point of the fitted linear function is then the energy gap E_g. In the McLean method, one can determine m, E_g and parameter A in Equation (13) by fitting a power function to $\alpha h\nu$ data. LD transformation starts from the Tauc equation. For the sake of calculating the natural logarithm, let us suppose that all the quantities in Equation (13) are unitless. Then the natural logarithm will be:

$$ln(\alpha h\nu) = m\, ln(A) + m\, ln(h\nu - E_g) \qquad (18)$$

By differentiation of Equation (18) with respect to $h\nu$

$$\frac{d\, ln(\alpha h\nu)}{d(h\nu)} = m \left(\frac{1}{h\nu - E_g} \right) \qquad (19)$$

The left side of Equation (19) is calculated based on the experimental data. This numerical derivative should be calculated as the difference of transformed measurement data ($h\nu$, $\alpha h\nu$) from subsequent measurement steps.

Now, we can determine the precise value of the optical band gap from the $\frac{d\ln(\alpha h\nu)}{d(h\nu)}$ vs. $h\nu$ curve as in Figure 11. We can see that a peak in the curve appeared at the band gap energy (E_g); i.e., $h\nu = E_g$. The peak at a particular energy value gives the approximate optical band gap, E_g [37].

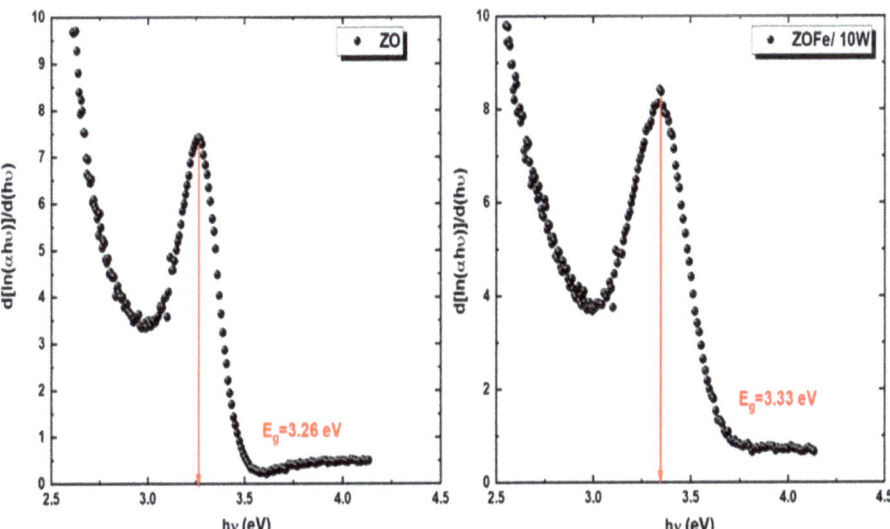

Figure 11. LD of $\alpha h\nu$ as a function of photon energy $h\nu$ with fitted straight lines for the undoped and Fe/10 w doped ZO thin films.

On other hand, the type of the optical transition can be estimated from the slope (m) of the curve of $\ln(\alpha h\nu)$ vs. $\ln(h\nu - E_g)$ using the E_g value of the highest Fe concentration doped ZO (15 at. %), as an example, and to determine the value of (m) which was found about 1/2 from the slope as shown in Figure 12. This indicates that the optical transition through the base glass and also the doped samples is a directly allowed transition which we selected to obtain E_g values by using the Tauc relation.

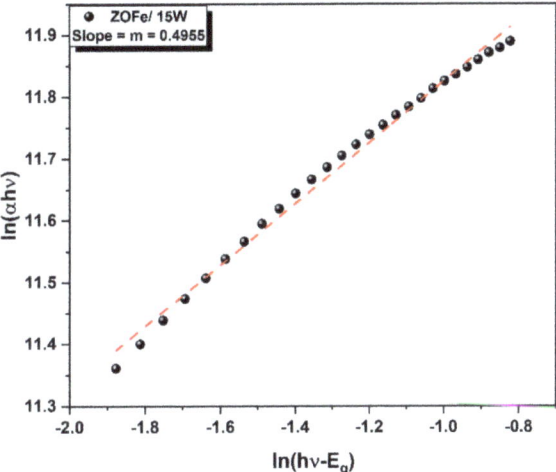

Figure 12. Type of the optical transition used in the undoped and Fe doped ZO thin films.

Thus, the estimated values of E_g for ZO doped and undoped by Fe are listed in Table 5 to compare with the other values obtained from different approaches.

One can see that in Table 5, the obtained values of E_g determined from the Tauc method are close to those determined by dT/dE and $dT/d\lambda$ and LD methods. Furthermore, the average values of the E_g are 3.26, 3.27, 3.29, 3.33 and 3.39 eV for the undoped and 3 W, 5 W, 10 W and 15 W of Fe-doped ZO thin films, respectively.

It is worthwhile to notice that two points may be inferred from these energy gap values. The first is that from the undoped to doped thin film, the E_g values increased. Furthermore, as the doping concentration of Fe increases, the E_g value increases.

The first point can be interpreted as follows. As we know, the incidence of a photon with energy of $h\nu$ on a semiconductor leads to a transition between the highest occupied state of valence band and the lowest unoccupied state of the conduction band. This phenomenon is applied to calculate the optical band gap energy of ZO thin films as we referred before. In Fe-doped ZO thin films, the optical band gap energy mainly depends on the valence state of Fe ions [38,39]. With Fe-doped ZO thin films, there is a large variability in the optical behavior which leads to inconsistent conclusions. The valence state of Fe is an important factor in the optical properties of ZO thin films. Fe dopants in ZO thin films can be in the two forms of Fe^{2+} or Fe^{3+}, or even coexistence of these two forms is seen. By addition of Fe into ZO films, the maximum of the valence band is increased and the minimum of the conduction band is decreased which leads to the reduction of band gap. However, substituting Fe^{3+} ions into Zn^{2+} leads to providing extra free carrier concentrations. As a result, the Fermi level moves toward the conduction band and the band gap increases. This makes the optical band gap of Fe-doped ZO thin films tunable.

Moreover, it is known that doping significantly alters the absorption spectrum of the pure semiconductor, resulting in degenerate energy levels which cause the Fermi level to push above the conduction band edge. This shift is known as doping induced band-filling called a Moss–Burstein shift and leads to an increase in the band gap from the undoped to doped thin film [40]. In addition, with Fe incorporation, there are initially sp-d exchange interactions that do not strongly affect the band gap. After a further doping percentage, the Moss–Burstein effect dominates the normal sp—d exchange interactions. In addition, ZO is normally n-type material stable at room temperature in which the Fermi level lies in the conduction band when doped with greater Fe [41]; as a result, there is an increase in the band gap with more Fe doping. Since no empty state is available inside the conduction band, so the absorption edge shifts towards the higher energy region, thereby giving rise to an increase in the band gap. With increasing Fe doping the d—d transition of Fe ions also increases and leads to increase in the band gap. According to Parra-Palomino et al. [38], increasing Fe doping concentration to ZO nanocrystals increased the band gap significantly. Chen et al. [42] and Wang et al. [43] found that raising the Fe doping concentration in ZO films can minimize the band gap.

3.4.3. Urbach Energy

One of the main properties which measure the degree of structural disorder in the films is the Urbach energy (E_U). The imperfection in the structurally disordered film leads to broadening the bands of localized states. As we mentioned, near the band edge, the absorption coefficient has an exponential dependence on photon energy. This dependence is given as follows:

$$\alpha = \alpha_o exp\left(\frac{h\nu}{E_U}\right) \quad (20)$$

where α_o is the band tailing parameter that can be obtained by:

$$\alpha_o = \sqrt{\frac{\sigma_o\left(\frac{4\pi}{c}\right)}{x\Delta E}} \quad (21)$$

where σ_0, c and ΔE are the electrical conductivity at absolute zero, the velocity of light and the width of the tail of localized states in the energy gap, respectively, and $x = 0.5$ for the directly allowed transitions [44]. Then the Urbach energy can emerge from Equation (22) as:

$$E_U = \left[\frac{d\,(ln(\alpha))}{d\,h\nu}\right]^{-1} \qquad (22)$$

As shown in Figure 13, the values of the E_U were obtained from the slopes of the linear fitting of the plots of ln (α) versus ($h\nu$) and are listed in Table 6. E_U increases from 225 to 403 meV as the Fe concentration increased. The dopant may contribute significantly to the width of localized states within the ZO's optical band. In other words, the E_U is responsible for the valence and conduction bands' tails [45]. Moreover, E_U values confirm the GIXRD analysis which shows that Fe doping is followed by an increase of the strain which can also create structural disorder.

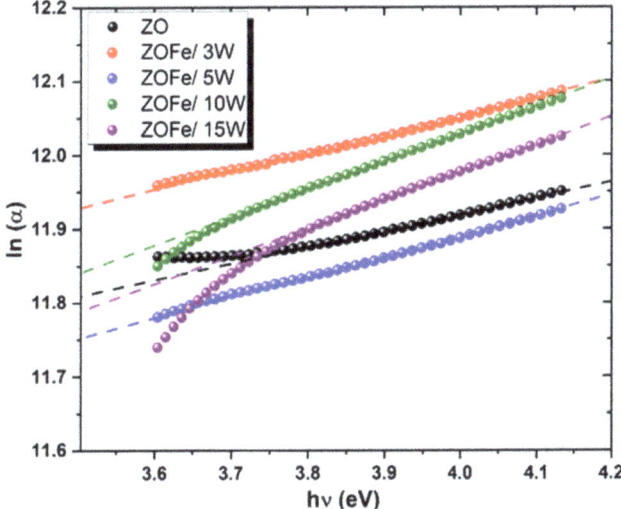

Figure 13. ln (α) vs. ($h\nu$) to determine the Urbach energy of ZO and Fe doped ZO thin films.

Table 6. Optical band gap and refractive index values for undoped ZO and Fe doped ZO thin films.

Film	Optical Band Gap, E_g (eV)	E_u (meV)	n_o [Moss]	n_o [Ravindra and Srivastava]	n_o [Dimitrov and Sakka]
ZO	3.26	255	2.323	2.399	2.330
ZOFe/3 W	3.27	236	2.321	2.397	2.327
ZOFe/5 W	3.29	263	2.318	2.393	2.323
ZOFe/10 W	3.33	400	2.311	2.386	2.313
ZOFe/15 W	3.35	403	2.307	2.382	2.308

3.4.4. Refractive Index

The two most interesting optical properties of a semiconductor are the absorption edge, or optical energy gap, and the refractive index. The relation between the optical band gap E_g and the refractive index n_o was proposed for first time by Moss in 1950 on the very general grounds that all energy levels in a solid are scaled down by a factor $1/\varepsilon_{opt}^2$, where $\varepsilon_{opt} = n^2$ is the optical dielectric constant; it was described in detail elsewhere [46,47]. Moss showed that the energy levels of the semiconducting materials can be scaled down by the term $1/n^4$ as follows [46]:

$$n_o^4 E_g = 9 \text{ eV} \qquad (23)$$

Then Ravindra and Srivastava [48] modified Equation (25) to be:

$$n_o^4 E_g = 108 \text{ eV} \tag{24}$$

The main difference between the Moss and Ravindra relations originated from the estimating of the reflection loss by Ravindra. Moreover, Equation (26) covers most of the semiconducting materials used for solar cell, photocatalytic and sensing applications. Dimitrov and Sakka [49] have deduced the correlation between the refractive index and the optical band gap from the Lorentz- Lorentz (L.L.) Equation:

$$\frac{n_o^2 - 1}{n_o^2 + 2} = 1 - \sqrt{\frac{E_g}{20}} \tag{25}$$

The estimated refractive index based on the optical band gap is shown in Table 6. The refractive index of ZO was compared with the optical data from references [22,49]. As we can see, by increasing the Fe content, the refractive index decreased and as we know the refractive index is correlated to the optical band gap. Subsequently, from Equations (25) and (26), the relation between them is inverse.

3.4.5. The Extinction Coefficient

The extinction coefficient (k) reflects the absorption of electromagnetic waves in the semiconductor due to inelastic scattering events and it can be calculated from the transmittance spectra using the relation

$$k = \frac{\alpha \lambda}{4\pi} \tag{26}$$

The dependence of k on the wavelength for all investigated thin films is shown in Figure 14. In the UV region ($\lambda < 400$ nm), the k value increases from 0.36 to 0.42 with increasing Fe concentration. Furthermore, the k values are very small and relatively constant in most of the visible region. This indicates the smoothness on the surface and homogeneity of the films and enhances its uses in blocking UV radiation [40]. As well, increase in k value by doping Fe can be ascribed to increasing topical donor levels formed within the energy levels that lead to an increased extinction coefficient, showing that the electronic transitions occur directly.

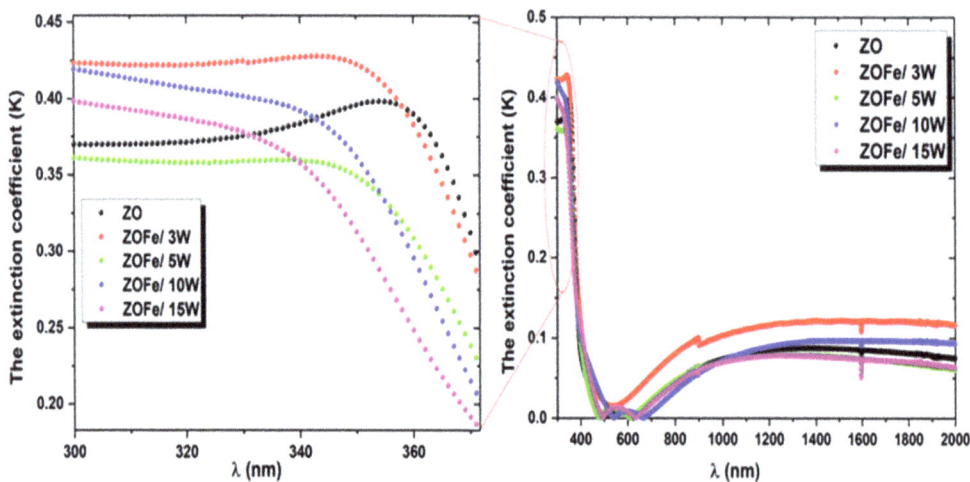

Figure 14. The extinction coefficient (K) as a function of the wavelength (λ) for the undoped and Fe doped ZO films.

3.5. Magnetic Properties

3.5.1. Magnetic Behavior of the Films

The hysteresis loops of ZO and Fe doped ZO thin films in Figure 15 show clear room temperature ferromagnetism (RTFM) for all the doping concentrations of Fe; they were obtained from room temperature vibrating sample magnetometer (VSM) measurements. It is clear from Figure 15 that undoped ZO and Fe doped ZO films show well-defined hysteresis loops with noticeable coercivity. The loops are S-shaped, which indicates the ferromagnetic nature of the samples. One can notice that there is non-saturation of the M-H loop at higher applied fields, which is a commonly observed phenomenon in semiconductor/metal nanocomposites. Rumpf et al. also observed such non-saturation behavior for ferromagnetic semiconductors and attributed it to the lattice defects and magnetic anisotropy [50].

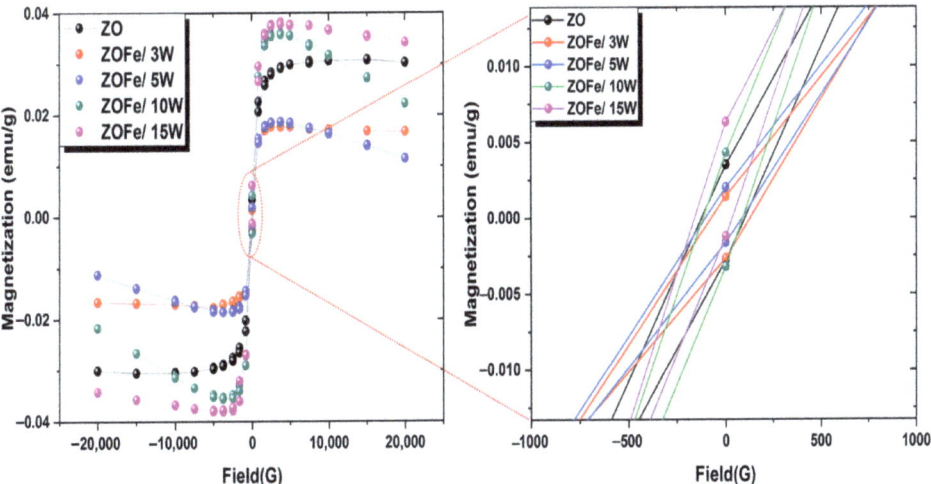

Figure 15. M-H curve showing ferromagnetic loops for the undoped and Fe doped ZO thin films.

The mechanism of RTFM in the ZO based system was still unclear for researchers. The dopant related secondary phase, magnetic cluster, metal precipitation formation will lead to magnetic behavior in the doped ZO. In a doped ZO system, the carrier exchange interactions such as RKKY, indirect double exchange interactions and super exchange interactions induced a magnetic nature [51–53]. In addition, the doped and co-doped ZO samples reveal ferromagnetism at RT due to the presence of intrinsic lattice defects. The presence of lattice defects as strain and texture, and concentration of Fe doping ions may also have a major impact on the room temperature magnetic properties [54,55]. Generally, RT ferromagnetism in TM doped ZO films could be explained on the basis of intrinsic and extrinsic effects associated with the local magnetic moments. If the spin of Fe ions has an exchange interaction with local magnetic moments of ZO then it is intrinsic effect. When there is a direct interaction between the local magnetic moments of the Fe clusters, it is known to be an extrinsic effect.

Accordingly, in Figure 15, the M-H curves of the pure ZO exhibit ferromagnetic (FM) behavior and Fe doped ZO thin films, with Fe concentration 2.4, 4.8 and 5 at. % of Fe dopants, exhibit diamagnetic behavior at a high magnetic field and FM behavior in the lowest field regime. There are several literature reports that pure ZO exhibited ferromagnetic nature at room temperature due to the exchange interaction between localized electron spin moments resulting from oxygen vacancies [56].

For more details about the diamagnetic behavior shown in the pure ZO films, a fitting with Langevin curve was performed, as shown in Figure 16; this type of analysis has been conducted on ZnO doped with transition metals [9].

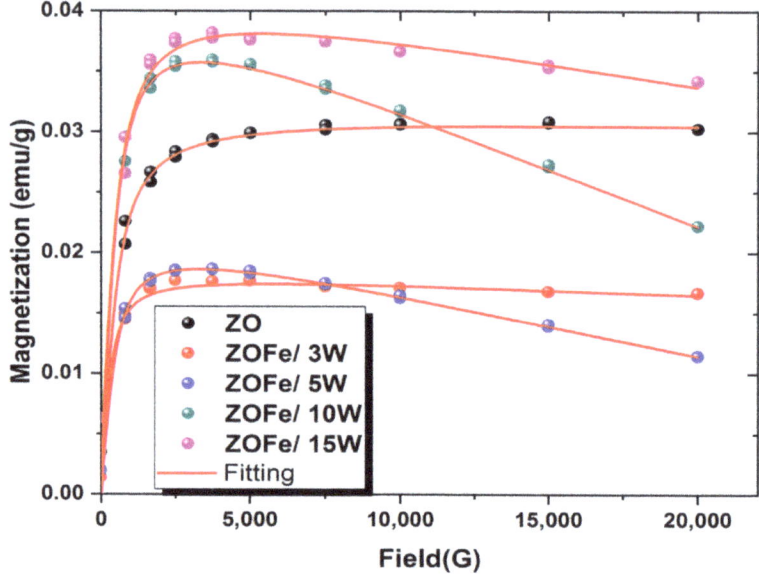

Figure 16. Fitting of the Langevin function with experimental M (H) curves.

Further, the ferromagnetic property for DMSs can be contributed from the secondary phase related to metal oxides or metal clusters with a ferromagnetic property. DMSs could have an intrinsic effect as well [57]. However, our GIXRD patterns clearly indicate the absence of clustering of metallic iron phase or iron oxide cluster in the films. Moreover, the possibility of attributing the observed ferromagnetic order of Fe doped ZO thin films to the carrier-mediated mechanism or RKKY exchange interaction is also ruled out because the DMSs usually show very high resistance. Therefore, RTFM of our Fe doped ZO films are not due to the carrier-mediated mechanism. A pertinent and appropriate explanation for the observed RTFM [58] in our case is the ferromagnetic exchange mechanism involving oxygen vacancies model. This could be explained in terms of two reasons: (1) the number of oxygen vacancies (O_V) and zinc interstitials (Zn_i) and (2) the exchange interaction between doped transition metal ion and O ion spins.

Further, Kittilstved et al. [59] stated that the ferromagnetism in TM doped ZO is due to the effective dopant defect hybridization arising from the energetic alignment of the shallow donors, Zn interstitial (Zn_i) and oxygen vacancy (O_V), with $TM^{+/2+}$ level. Thus, the magnetic coupling between Fe ions with ZO is FM mediated by (Zn_i) and (O_V) and this may account for the observed RTFM and indicates that defects play a role to gain a FM coupling. In addition, in our case, for low magnetic fields, the ferromagnetic behavior can be attributed to the presence of small magnetic dipoles located at the surface of thin film nanocrystals, which interact with their nearest neighbors inside the crystal involving oxygen vacancies [53].

3.5.2. The Magnetic Parameters

Table 7 shows the saturation magnetization M_s of undoped ZO~9×10^{-4} emu/g at 2000 G. Vettumperumal et al. [60] have observed RTFM of ZO thin film with M_s value ~ 6.9×10^{-5} emu/g emu/g at 300 K while Ariyakkani et al. [61] have obtained M_s value ~ 5.6×10^{-5} emu/g at 300 K for ZO thin film. Therefore, the magnetization value obtained in our work is higher as compared to previous reported values of undoped ZO. Furthermore, it is obvious, from Table 7, that the saturation magnetization of the films is increased by increasing the Fe concentration. As we mentioned before, the origin of ferromagnetism in transition metal doped ZO until now is still not clear. A number of studies indicate that ferromagnetism in transition metal doped ZO may come from secondary magnetic phases [62]. However, others reported that the RTFM in Fe-doped ZO could be due to the oxidation of Fe in mixed valence (Fe^{2+} and Fe^{3+}) states [63]. The exchange interaction between conductive electrons and local spin polarized electrons is responsible for the magnetic behavior of Fe-doped ZO. Likewise, the occurrence of Fe-Fe super exchange interaction could be one reasons for increased M_s values with increasing Fe concentration; it is found to be dominant at higher Fe doping concentrations [64,65].

Table 7. Magnetization, coercivity and remanence values of undoped and Fe doped ZO thin films.

Film	Magnetization $\times 10^{-4}$ emu/g	Coercivity (G)	Remanence $\times 10^{-3}$ emu/g
ZO	3.16 ± 0.04	110.7 ± 2	3.12 ± 0.2
ZOFe/3 W	1.84 ± 0.03	100.3 ± 1	2.05 ± 0.1
ZOFe/5 W	2.20 ± 0.02	94.3 ± 1	1.82 ± 0.1
ZOFe/10 W	4.24 ± 0.05	101.3 ± 1	3.76 ± 0.2
ZOFe/15 W	4.26 ± 0.06	95.7 ± 1	3.77 ± 0.2

The coercivity of Fe doped films is lower than the undoped film at any particular Fe concentration, as shown in Table 7. There are several factors influencing the reduction in the coercivity such as magnetic domain size, micro-strain, magnetocrystalline anisotropy, and shape anisotropy. However, such decrease in coercivity could be related to a decrease in the anisotropy field, which reduces the thickness of the domain wall [66]. However, the coercivity (H_c) is low (~95 Oe) for the doped films and it indicates that all Fe doped ZO thin film samples are ferromagnetically soft compared to ZO. The magnetization measurements by Kumar et al. [63] on 1% Fe doped ZO samples shows soft ferromagnetic behavior at room temperature with coercivity (H_c) about 27 Oe. Further, low coercivity and high saturation magnetization are required for spintronics memory devices and this is obtained for the ZO: Fe (5% at 15 W) thin films compared to the other thin films. It can be seen that the coercivities and saturation magnetizations were enhanced with the increase of Fe content. This phenomenon should be investigated further to understand it. Therefore, to enhance ferromagnetic order, the appropriate doping concentration of Fe in ZO wurtzite structure plays a vital role in obtaining an optimum number of charge carriers and oxygen vacancies, which may be useful for spintronics based solid state devices.

Thus, the remanence (M_r) is found to slightly increase with the increase in the high Fe concentration which is related to the saturation magnetization M_s. Table 7 shows the magnetic saturation (M_s), remanence magnetization (M_r) and coercivity (H_c) for all the thin films.

3.6. Electrical Properties

As already noted, ZO is a large band gap semiconductor and its electric conductivity is very low. Low electric conductivity is one of the most important reasons for limited performance in the devices based on ZO film. Therefore, to improve electron transport rates in ZO film would be a significant way to achieve high performance devices. It has been suggested by many researchers to substitute ZO films with TM [67].

The electrical resistivity of ZO thin films undoped and doped by Fe with different concentrations was investigated by a two probe method. Figure 17a shows the I-V characteristics of the thin film samples having Ohmic behavior. The lowest resistivity of these films (4.6×10^7 Ω·cm) was measured for the films doped with 5% Fe corresponding to the highest DC power of the thin films. Otherwise, the resistivity of the co-sputtered films showed an initial increase for a DC power of 3 W and also for 5 W. This was followed by a continuous reduction in resistivity as the DC power increased beyond 10 W, reaching the lowest value of (4.6 × 10⁷ Ω·cm) in the films deposited under a DC power of 15 W, i.e., two orders of magnitude lower than the lowest resistivity obtained in the films prepared from the doped targets. These results manifest the pivotal role played by the method of incorporation of the dopant.

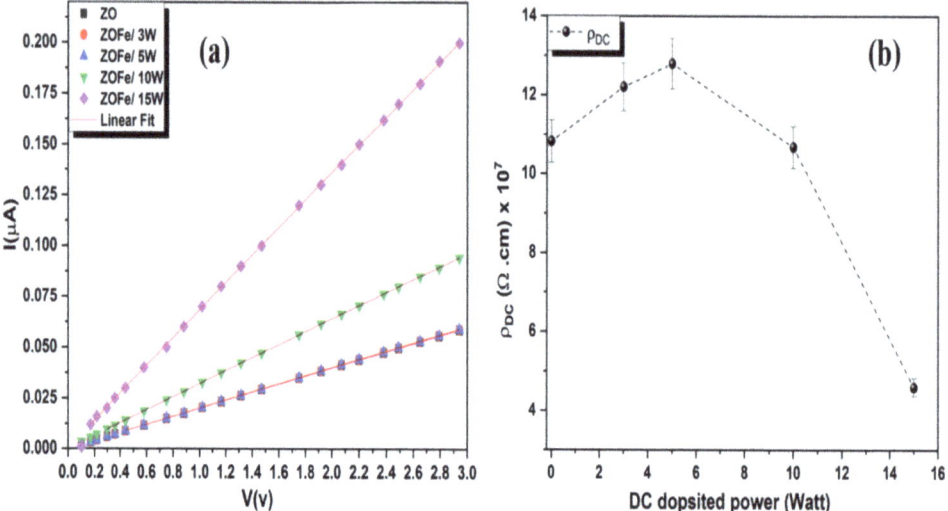

Figure 17. (a) I-V characteristics of undoped and Fe doped ZO thin films and (b) variations of the electrical resistivity ρ_{DC} with Fe concentration of all investigated thin films.

The electrical resistivity of undoped ZO thin films prepared by RF sputtering has been reported to be in the range of (10^4–10^8 Ω.cm), depending on the substrate temperature [68]. Further, a previous study on co-sputtered Fe—doped ZO thin films reported a decrease of resistivity by four orders of magnitude compared to the undoped film [16].

As we pointed above, the resistances of the films increase with increase in low dopant concentration of Fe then decrease with higher concentrations, as shown in Figure 17b. The increasing trend of resistivity at low dopant concentration was reported by Cherif et al. [69]. The probable causes of increase in resistivity from undoped to Fe dopants in low concentration could be due to carrier traps formation at the film surface causing impediment to charge carriers and due to Fe incorporation, with the formation of interstitial metal atoms causing a decrease in the oxygen vacancies [70]. Meanwhile, increasing the concentration of Fe led to an increase the number of surface defects in the ZO matrix. Hence, the electrical resistivity of ZO is reduced as the concentration of Fe in ZO is increased slightly or, in other words, the conductivity increased. This makes our samples of thin films a promising material in very wide range of applications.

4. Conclusions

ZO and ZOFe films are deposited using RF magnetron sputtering with thicknesses 150 nm. The effect of Fe dopant on the structure, morphology, optical, magnetic and electric properties in Fe-doped ZO films was investigated. Curiously, we found that, both ZO and ZOFe films show a high crystalline quality with sharp band edge emission. As an interesting point, the energy band gap (E_g) was obtained by different numerical methods based on the experimental data in order to determine the most precise values. In addition, it has been pointed out that the refractive index (n) decreased with increasing Fe content which correlated to the optical band gap behavior.

Likewise, an increasing of the saturation magnetization (M_s) values was noted with increasing Fe doped concentration while the coercivity (H_c) decreased. Finally, the resistivity of ZO is reduced as the concentration of Fe in ZO is increased slightly or, in other words, the conductivity is increased. The high saturation magnetization and conductivity values as well as the tunable optical properties of the investigated doped thin films make them promising multifunctional nanomaterials for a wide range of applications as spintronics memory and magneto-optoelectronic devices.

Author Contributions: Conceptualization, A.F. and G.M.; Data curation, H.E. and C.S.; Formal analysis, A.F. and C.S.; Investigation, A.F. and C.S.; Methodology, A.F. and C.S.; Resources, G.M.; Supervision, G.M.; Validation, A.F., H.E. and G.M.; Visualization, A.F., H.E. and G.M.; Writing—original draft, A.F.; Writing—review and editing, A.F., H.E. and G.M. All authors have read and agreed to the published version of the manuscript.

Funding: This research received no external funding.

Institutional Review Board Statement: Not applicable.

Informed Consent Statement: Not applicable.

Data Availability Statement: The data presented in this study are available on request from the corresponding author.

Conflicts of Interest: The authors declare no conflict of interest.

References

1. Theerthagiri, J.; Salla, S.; Senthil, R.A.; Nithyadharseni, P.; Madankumar, A.; Arunachalam, P.; Maiyalagan, T.; Kim, H.S. A review on ZnO nanostructured materials: Energy, environmental and biological applications. *Nanotechnology* **2019**, *30*, 392001. [CrossRef] [PubMed]
2. Tereshchenko, A.; Bechelany, M.; Viter, R.; Khranovskyy, V.; Smyntyna, V.; Starodub, N.; Yakimova, R. Optical biosensors based on ZnO nanostructures: Advantages and perspectives. A review. *Sens. Actuators B Chem.* **2016**, *229*, 664–677. [CrossRef]
3. Hwang, D.K.; Oh, M.S.; Lim, J.H.; Park, S.J. ZnO thin films and light-emitting diodes. *J. Phys. D Appl. Phys.* **2007**, *40*, R387–R412. [CrossRef]
4. Kumar, R.; Kumar, G.; Al-Dossary, O.; Umar, A. ZnO nanostructured thin films: Depositions, properties and applications—A review. *Mater. Express* **2015**, *5*, 3–23. [CrossRef]
5. Hamid, S.B.A.; Teh, S.J.; Lai, C.W. Photocatalytic Water Oxidation on ZnO: A Review. *Catalysts* **2017**, *7*, 93. [CrossRef]
6. Prellier, W.; Fouchet, A.; Mercey, B. Oxide-diluted magnetic semiconductors: A review of the experimental status. *J. Physics Condens. Matter* **2003**, *15*, R1583–R1601. [CrossRef]
7. Rong, P.; Ren, S.; Yu, Q. Fabrications and Applications of ZnO Nanomaterials in Flexible Functional Devices-A Review. *Crit. Rev. Anal. Chem.* **2018**, *49*, 336–349. [CrossRef]
8. Dietl, T. Zener Model Description of Ferromagnetism in Zinc-Blende Magnetic Semiconductors. *Science* **2000**, *287*, 1019–1022. [CrossRef]
9. Castro-Lopes, S.; Guerra, Y.; Silva-Sousa, A.; Oliveira, D.M.; Gonçalves, L.A.P.; Franco, A.; Padrón-Hernández, E.; Peña-Garcia, R. Influence of pH on the structural and magnetic properties of Fe-doped ZnO nanoparticles synthesized by sol gel method. *Solid State Sci.* **2020**, *109*, 106438. [CrossRef]
10. Soares, A.S.; Castro-Lopes, S.; Cabrera-Baez, M.; Milani, R.; Padrón-Hernández, E.; Farias, B.V.; Soares, J.M.; Gusmão, S.S.; Viana, B.C.; Guerra, Y.; et al. The role of pH on the vibrational, optical and electronic properties of the Zn1-xFexO compound synthesized via sol gel method. *Solid State Sci.* **2022**, *128*, 106880. [CrossRef]
11. Liu, C.; Yun, F.; Morkoç, H. Ferromagnetism of ZnO and GaN: A Review. *J. Mater. Sci. Mater. Electron.* **2005**, *16*, 555–597. [CrossRef]
12. Wang, Y.S.; Thomas, P.J.; O'Brien, P. Optical properties of ZnO nanocrystals doped with Cd, Mg, Mn, and Fe ions. *J. Phys. Chem. B* **2006**, *110*, 21412–21415. [CrossRef]
13. Aghgonbad, M.M.; Sedghi, H. Optical and Electronic Analysis of Pure and Fe-Doped ZnO Thin Films Using Spectroscopic Ellipsometry and Kramers–Kronig Method. *Int. J. Nanosci.* **2019**, *18*, 1850013. [CrossRef]
14. Bousslama, W.; Elhouichet, H.; Férid, M. Enhanced photocatalytic activity of Fe doped ZnO nanocrystals under sunlight irradiation. *Optik Stuttg.* **2017**, *134*, 88–98. [CrossRef]
15. Wang, L.; Meng, L.; Teixeira, V.; Song, S.; Xu, Z.; Xu, X. Structure and optical properties of ZnO:V thin films with different doping concentrations. *Thin Solid Films* **2009**, *517*, 3721–3725. [CrossRef]
16. Kamalianfar, A.; Halim, S.A.; Behzad, K.; Naseri, M.G.; Navasery, M.; Din, F.U.; Zahedi, J.A.M.; Lim, K.P.; Chen, S.K.; Sidek, H.A.A. Effect of thickness on structural, optical and magnetic properties of Co doped ZnO thin film by pulsed laser deposition. *J. Optoelectron. Adv. Mater.* **2013**, *15*, 239–243.

17. Wang, L.M.; Liao, J.-W.; Peng, Z.-A.; Lai, J.-H. Doping Effects on the Characteristics of Fe:ZnO Films: Valence Transition and Hopping Transport. *J. Electrochem. Soc.* **2009**, *156*, H138. [CrossRef]
18. Al-Kuhaili, M.F.; Durrani, S.M.A.; El-Said, A.S.; Heller, R. Influence of iron doping on the structural, chemical, and optoelectronic properties of sputtered zinc oxide thin films. *J. Mater. Res.* **2016**, *31*, 3230–3239. [CrossRef]
19. Wang, X.B.; Song, C.; Li, D.M.; Geng, K.W.; Zeng, F.; Pan, F. The influence of different doping elements on microstructure, piezoelectric coefficient and resistivity of sputtered ZnO film. *Appl. Surf. Sci.* **2006**, *253*, 1639–1643. [CrossRef]
20. Chu, W.-T.; Mayer, J.W.; Nicolet, M. *Backscattering Spectrometry*; Acadamic Press, INC.: London, UK, 1978. [CrossRef]
21. Fujimura, N.; Nishihara, T.; Goto, S.; Xu, J.; Ito, T. Control of preferred orientation for ZnOx films: Control of self-texture. *J. Cryst. Growth* **1993**, *130*, 269–279. [CrossRef]
22. Hammad, A.H.; Abdel-wahab, M.S.; Vattamkandathil, S.; Ansari, A.R. Structural and optical properties of ZnO thin films prepared by RF sputtering at different thicknesses. *Phys. B Condens. Matter.* **2018**, *540*, 1–8. [CrossRef]
23. Akl, A.A.; Mahmoud, S.A.; L-Shomar, S.M.A.; Hassanien, A.S. Improving microstructural properties and minimizing crystal imperfections of nanocrystalline Cu2O thin films of different solution molarities for solar cell applications. *Mater. Sci. Semicond. Process.* **2018**, *74*, 183–192. [CrossRef]
24. Debye, P.; Scherrer, P. Interferenz an regellos orientierten Teilchen im Röntgenlicht I. *Phys. Z.* **1916**, *17*, 277.
25. Köseoğlu, Y. Rapid synthesis of room temperature ferromagnetic Fe and Co co-doped ZnO DMS nanoparticles. *Ceram. Int.* **2015**, *41*, 11655–11661. [CrossRef]
26. Allen, J.A.; Murugesan, D.; Viswanathan, C. Circumferential growth of zinc oxide nanostructure anchored over carbon fabric and its photocatalytic performance towards p-nitrophenol. *Superlattices Microstruct.* **2019**, *125*, 159–167. [CrossRef]
27. Shaban, M.; El Sayed, A.M. Influences of lead and magnesium co-doping on the nanostructural, optical properties and wettability of spin coated zinc oxide films. *Mater. Sci. Semicond. Process.* **2015**, *39*, 136–147. [CrossRef]
28. Maniv, S.; Westwood, W.D.; Colombini, E. Pressure and Angle of Incidence Effects in Reactive Planar Magnetron Sputtered ZnO layers. *J. Vac. Sci. Technol.* **1982**, *20*, 162–170. [CrossRef]
29. Hur, T.B.; Hwang, Y.H.; Kim, H.K.; Lee, I.J.J. Strain effects in ZnO thin films and nanoparticles. *J. Appl. Phys.* **2006**, *99*, 64308. [CrossRef]
30. Patsalas, P.; Gravalidis, C.; Logothetidis, S. Surface kinetics and subplantation phenomena affecting the texture, morphology, stress, and growth evolution of titanium nitride films. *J. Appl. Phys.* **2004**, *96*, 6234–6235. [CrossRef]
31. Ulutas, C.; Gunes, M.; Gumus, C. Influence of post-deposition annealing on the structural and optical properties of γ-MnS thin film. *Optik Stuttg.* **2018**, *164*, 78–83. [CrossRef]
32. Yusuf, Y.; Azis, R.S.; Mustaffa, M.S. Spin-Coating Technique for Fabricating Nickel Zinc Nanoferrite ($Ni_{0.3}Zn_{0.7}Fe_2O_4$) Thin Films. In *Coatings and Thin-Film Technologies*; IntechOpen: London, UK, 2019. [CrossRef]
33. Tauc, J.; Grigorovici, R.; Vancu, A. Optical Properties and Electronic Structure of Amorphous Germanium. *Phys. Status Solidi* **1966**, *15*, 627–637. [CrossRef]
34. Che, H.; Huso, J.; Morrison, J.L.; Thapa, D.; Huso, M.; Yeh, W.J.; Tarun, M.C.; McCluskey, M.D.; Bergman, L. Optical properties of ZnO-alloyed nanocrystalline films. *J. Nanomater.* **2012**, *2012*, 963485. [CrossRef]
35. Marotti, R. Bandgap energy tuning of electrochemically grown ZnO thin films by thickness and electrodeposition potential. *Sol. Energy Mater. Sol. Cells* **2004**, *82*, 85–103. [CrossRef]
36. McLean, T.P. The absorption edge spectrum of semiconductors. *Prog. Semicond.* **1960**, *5*, 53–102.
37. Hammad, A.H.; Abdelghany, A.M.; Okasha, A.; Marzouk, S.Y. The influence of fluorine and nickel ions on the structural, spectroscopic, and optical properties of $(100 - x)[15NaF–5CaF_2–80B_2O_3]$-xNiO glasses. *J. Mater. Sci. Mater. Electron.* **2017**, *28*, 8662–8668. [CrossRef]
38. Parra-Palomino, A.; Perales–Perez, O.; Singhal, R.K.; Tomar, M.S.; Hwang, J.; Voyles, P. Structural, optical, and magnetic characterization of monodisperse Fe-doped ZnO nanocrystals. *J. Appl. Phys.* **2008**, *103*, 07D121. [CrossRef]
39. Salem, M.; Akir, S.; Ghrib, T.; Daoudi, K.; Gaidi, M. Fe-doping effect on the photoelectrochemical properties enhancement of ZnO films. *J. Alloys Compd.* **2016**, *685*, 107–113. [CrossRef]
40. Shanmuganathan, G.; Banu, I.B.S. Influence of Codoping on the Optical Properties of ZnO Thin Films Synthesized on Glass Substrate by Chemical Bath Deposition Method. *Adv. Condens. Matter Phys.* **2014**, *2014*, 761960. [CrossRef]
41. Dhiman, P.; Chand, J.; Kumar, A.; Kotnala, R.K.; Batoo, K.M.; Singh, M. Synthesis and characterization of novel Fe@ZnO nanosystem. *J. Alloys Compd.* **2013**, *578*, 235–241. [CrossRef]
42. Chen, Z.C.; Zhuge, L.J.; Wu, X.M.; Meng, Y.D. Initial study on the structure and optical properties of $Zn_{1-x}Fe_xO$ films. *Thin Solid Films* **2007**, *515*, 5462–5465. [CrossRef]
43. Wang, C.; Chen, Z.; He, Y.; Li, L.; Zhang, D. Structure, morphology and properties of Fe-doped ZnO films prepared by facing-target magnetron sputtering system. *Appl. Surf. Sci.* **2009**, *255*, 6881–6887. [CrossRef]
44. Shaban, M.; El Sayed, A.M. Effects of lanthanum and sodium on the structural, optical and hydrophilic properties of sol-gel derived ZnO films: A comparative study. *Mater. Sci. Semicond. Process.* **2016**, *41*, 323–334. [CrossRef]
45. Shaban, M.; Zayed, M.; Hamdy, H. Nanostructured ZnO thin films for self-cleaning applications. *RSC Adv.* **2017**, *7*, 617–631. [CrossRef]
46. Moss, T.S. Relations between the Refractive Index and Energy Gap of Semiconductors. *Phys. Status Solidi.* **1985**, *131*, 415–427. [CrossRef]

47. Kumar, V.; Singh, J.K. Model for calculating the refractive index of different materials. *Indian J. Pure Appl. Phys.* **2010**, *48*, 571–574.
48. Ravindra, N.M.; Srivastava, V.K. Variation of refractive index with energy gap in semiconductors. *Infrared Phys.* **1979**, *19*, 603–604. [CrossRef]
49. Dimitrov, V.; Sakka, S. Linear and nonlinear optical properties of simple oxides. II. *J. Appl. Phys.* **1996**, *79*, 1741–1745. [CrossRef]
50. Rumpf, K.; Granitzer, P.; Krenn, H. Porous Silicon/Metal Hybrid System with Ferro and Paramagnetic Behavior. *IEEE Trans. Magn.* **2008**, *44*, 2753–2755. [CrossRef]
51. Zener, C. Interaction between the d-shells in the transition metals. II. Ferromagnetic compounds of manganese with Perovskite structure. *Phys. Rev.* **1951**, *82*, 403–405. [CrossRef]
52. Anderson, P.W. New approach to the theory of superexchange interactions. *Phys. Rev.* **1959**, *115*, 2–13. [CrossRef]
53. Katayama-Yoshida, H.; Sato, K.; Fukushima, T.; Toyoda, M.; Kizaki, H.; Dinh, V.A.; Dederichs, P.H. Theory of ferromagnetic semiconductors. *Phys. Status Solidi Appl. Mater. Sci.* **2007**, *204*, 15–32. [CrossRef]
54. Pan, H.; Yi, J.B.; Shen, L.; Wu, R.Q.; Yang, J.H.; Lin, J.Y.; Feng, Y.P.; Ding, J.; Van, L.H.; Yin, J.H. Room-Temperature Ferromagnetism in Carbon-Doped ZnO. *Phys. Rev. Lett.* **2007**, *99*, 127201. [CrossRef] [PubMed]
55. Özgür, Ü.; Alivov, Y.I.; Liu, C.; Teke, A.; Reshchikov, M.A.; Doğan, S.; Avrutin, V.; Cho, S.J.; Morkoç, H. A comprehensive review of ZnO materials and devices. *J. Appl. Phys.* **2005**, *98*, 041301. [CrossRef]
56. Sundaresan, A.; Bhargavi, R.; Rangarajan, N.; Siddesh, U.; Rao, C.N.R. Ferromagnetism as a universal feature of nanoparticles of the otherwise nonmagnetic oxides. *Phys. Rev. B-Condens. Matter Mater. Phys.* **2006**, *74*, 161306. [CrossRef]
57. Zhang, W.; Zhao, J.; Liu, Z.Z.; Liu, Z.Z. Structural, optical and magnetic properties of $Zn_{1-x}Fe_xO$ powders by sol-gel method. *Appl. Surf. Sci.* **2013**, *284*, 49–52. [CrossRef]
58. Murugaraj, R. Room temperature ferromagnetism in Ist group elements co-doped ZnO:Fe nanoparticles by co-precipitation method. *Phys. B Phys. Condens. Matter.* **2016**, *487*, 102–108. [CrossRef]
59. Kittilstved, K.R.; Schwartz, D.A.; Tuan, A.C.; Heald, S.M.; Chambers, S.A.; Gamelin, D.R. Direct kinetic correlation of carriers and ferromagnetism in Co^{2+}:ZnO. *Phys. Rev. Lett.* **2006**, *97*, 2–5. [CrossRef] [PubMed]
60. Vettumperumal, R.; Kalyanaraman, S.; Santoshkumar, B.; Thangavel, R. Magnetic properties of high Li doped ZnO sol-gel thin films. *Mater. Res. Bull.* **2014**, *50*, 7–11. [CrossRef]
61. Ariyakkani, P.; Suganya, L.; Sundaresan, B. Investigation of the structural, optical and magnetic properties of Fe doped ZnO thin films coated on glass by sol-gel spin coating method. *J. Alloys Compd.* **2017**, *695*, 3467–3475. [CrossRef]
62. Wang, D.; Chen, Z.Q.; Wang, D.D.; Gong, J.; Cao, C.Y.; Tang, Z.; Huang, L.R. Effect of thermal annealing on the structure and magnetism of Fe-doped ZnO nanocrystals synthesized by solid state reaction. *J. Magn. Magn. Mater.* **2010**, *322*, 3642–3647. [CrossRef]
63. Kumar, S.; Kim, Y.J.; Koo, B.H.; Sharma, S.K.; Vargas, J.M.; Knobel, M.; Gautam, S.; Chae, K.H.; Kim, D.K.; Kim, Y.K.; et al. Structural and magnetic properties of chemically synthesized Fe doped ZnO. *J. Appl. Phys.* **2009**, *105*, 07C520. [CrossRef]
64. Sattar, A.A.; Elsayed, H.M.; Faramawy, A.M. Comparative study of structure and magnetic properties of micro- and nano-sized $Gd_xY_{3-x}Fe_5O_{12}$ garnet. *J. Magn. Magn. Mater.* **2016**, *412*, 172–180. [CrossRef]
65. Limaye, M.V.; Singh, S.B.; Das, R.; Poddar, P.; Kulkarni, S.K. Room temperature ferromagnetism in undoped and Fe doped ZnO nanorods: Microwave-assisted synthesis. *J. Solid State Chem.* **2011**, *184*, 391–400. [CrossRef]
66. Faramawy, A.M.; Mattei, G.; Scian, C.; Elsayed, H.; Ismail, M.I.M. Cr^{3+} substituted aluminum cobalt ferrite nanoparticles: Influence of cation distribution on structural and magnetic properties. *Phys. Scr.* **2021**, *96*, 125849. [CrossRef]
67. Ueda, K.; Tabata, H.; Kawai, T. Magnetic and electric properties of transition-metal-doped ZnO films. *Appl. Phys. Lett.* **2001**, *79*, 988–990. [CrossRef]
68. Chaabouni, F.; Abaab, M.; Rezig, B. Effect of the substrate temperature on the properties of ZnO films grown by RF magnetron sputtering. *Mater. Sci. Eng. B Solid-State Mater. Adv. Technol.* **2004**, *109*, 236–240. [CrossRef]
69. Cherifi, Y.; Chaouchi, A.; Lorgoilloux, Y.; Rguiti, M.; Kadri, A.; Courtois, C. Electrical, dielectric and photocatalytic properties of Fe-doped ZnO nanomaterials synthesized by sol gel method. *Process. Appl. Ceram.* **2016**, *10*, 125–135. [CrossRef]
70. Cheng, W.; Ma, X. Structural, optical and magnetic properties of Fe-doped ZnO. *J. Phys. Conf. Ser.* **2009**, *152*, 012039. [CrossRef]

Article

On the Statistics of Mechanical Failure in Flame-Sprayed Self-Supporting Components

Florian Kerber [1,*], Magda Hollenbach [1], Marc Neumann [1], Tony Wetzig [1], Thomas Schemmel [2], Helge Jansen [2] and Christos G. Aneziris [1]

[1] TU Bergakademie Freiberg, Institute of Ceramics, Refractories and Composite Materials, Agricolastraße 17, 09599 Freiberg, Germany
[2] Refratechnik Steel GmbH, Research and Development, Am Seestern 5, 40547 Düsseldorf, Germany
* Correspondence: florian.kerber@ikfvw.tu-freiberg.de

Abstract: The objective of this study was to investigate the variability of flexural strength for flame-sprayed ceramic components and to determine which two-parametric distribution function was best suited to represent the experimental data. Moreover, the influence of the number of tested specimens was addressed. The stochastic nature of the flame-spraying process causes a pronounced variation in the properties of potential components, making it crucial to characterise the fracture statistics. To achieve this, this study used two large data sets consisting of 1000 flame-sprayed specimens each. In addition to the standard Weibull approach, the study examined the quality of representing the experimental data using other two-parametric distribution functions (Normal, Log-Normal, and Gamma). To evaluate the accuracy of the distribution functions and their characteristic parameters, random subsamples were generated by resampling of the experimental data, and the results were assessed based on the sampling size. It was found that the experimental data were best represented by either the Weibull or Gamma distribution, and the quality of the fit was correlated with the number of positive and negative outliers. The Weibull fit was more sensitive to positive outliers, whereas the Gamma fit was more sensitive to negative outliers.

Citation: Kerber, F.; Hollenbach, M.; Neumann, M.; Wetzig, T.; Schemmel, T.; Jansen, H.; Aneziris, C.G. On the Statistics of Mechanical Failure in Flame-Sprayed Self-Supporting Components. *Ceramics* **2023**, *6*, 1050–1066. https://doi.org/10.3390/ceramics6020062

Academic Editors: Amirhossein Pakseresht and Kamalan Kirubaharan Amirtharaj Mosas

Received: 21 March 2023
Revised: 18 April 2023
Accepted: 23 April 2023
Published: 25 April 2023

Copyright: © 2023 by the authors. Licensee MDPI, Basel, Switzerland. This article is an open access article distributed under the terms and conditions of the Creative Commons Attribution (CC BY) license (https://creativecommons.org/licenses/by/4.0/).

Keywords: flame-spraying; fracture statistics; QQ analysis; Weibull distribution; Gamma distribution

1. Introduction

The flame-spraying of ceramic materials is a highly versatile technique [1,2]. The unique microstructure of the flame-sprayed coatings, including cracks and pores, results in exceptional thermal shock resistance [3,4]. That offers numerous possibilities for the preparation and modification of various substrate materials for refractories in high-temperature applications. One motivation was the functionalisation of refractory hollowware based on uniaxially pressed rice husk ash [5] via flame-sprayed alumina potentially used in steel ingot casting (prototype, see Figure 1). These components must provide excellent resistance against thermal shock and erosion caused by the flow of the molten steel, as well as high chemical inertness against the steel melt itself [6,7]. However, materials traditionally used for this purpose, such as refractory clay, are limited in terms of these properties [8]. With the flame-spraying technique, a wide range of materials can be utilised [1,2,9]. Accordingly, their mechanical characteristics are of particular interest when considering the service life time of such materials in both room and high-temperature applications.

Figure 2 shows the cross-section obtained via scanning electron microscopy of a material composite based on rice husk ash and a flame-sprayed alumina coating. The nature of the rice husk ash and the shaping process results in a rough surface of the bulk material. This causes an interlocking with the flame-sprayed alumina coating on the coating/substrate interface. This interlocking must be taken into account with regard to the mechanical properties of the composite material, which is non-trivial. Consequently, the contributing materials has to be characterised separately in a first step. In this regard, the flame-sprayed ceramic coating is of particular interest.

Figure 1. Prototype of refractory hollowware based on biogenic silica with flame-sprayed alumina coating.

Recent studies provided an overview regarding the mechanical characteristics of flame-sprayed components based on different ceramic materials (Al_2O_3 and Al_2O_3 with TiO_2 and ZrO_2 additions) [3,4]. The stochastic nature of the flame-spraying process can result in a pronounced variability in the mechanical properties of the flame-sprayed components, particularly in terms of their fracture statistics [4,10]. Thus, a comprehensive characterisation of these properties is essential in order to predict their mechanical behaviour and ensure their successful use in high-temperature applications [11]. Hence, the sampling size required to obtain statistically reliable data has to be questioned.

Following the suggestion of Danzer et al. [12], a minimum of 30 specimens is required for determining strength data for ceramic components. According to the law of large numbers, the probability to obtain the true value for a variable from an experiment approaches one for the number of performed experiments approaching infinity [13]. However, in practice, the number of tested specimens needs to be limited and minimised as much as possible. Therefore, selecting an appropriate sampling size for determining strength data involves balancing the cost of sample preparation with the desired level of experimental accuracy. Danzer et al. [12,14] assessed the scatter of strength data for ceramic parts in dependency of the sampling size. To achieve this, strength data were randomly generated from an idealised theoretical Weibull distribution. Obviously, the scatter of the Weibull modulus determined for the generated sample around the true value of the distribution increases as the sampling size decreases. However, the analysis revealed an asymmetric deviation with a larger positive difference, leading to an overestimation of the true value. Thus, the standard suggests a correction factor for the Weibull modulus dependent on the sampling size. Conventionally, flexural strength data of ceramic components are represented using the Weibull distribution [15]. However, recent studies showed that the Weibull modulus of flame-sprayed ceramic components is low ($m = 3$ to 5.5 [3,4]) compared to densely sintered ceramic components fabricated by injection moulding ($m = 11.8$ [16]). Because of the unique microstructure of flame-sprayed ceramic components (see Figure 2), it is uncertain whether the assumption of a Weibull distribution, i.e., failure because of the weakest link, is met in the first place [15,17–19]. Thus, it is also uncertain, whether a similar behaviour as a function of the sampling size can be expected.

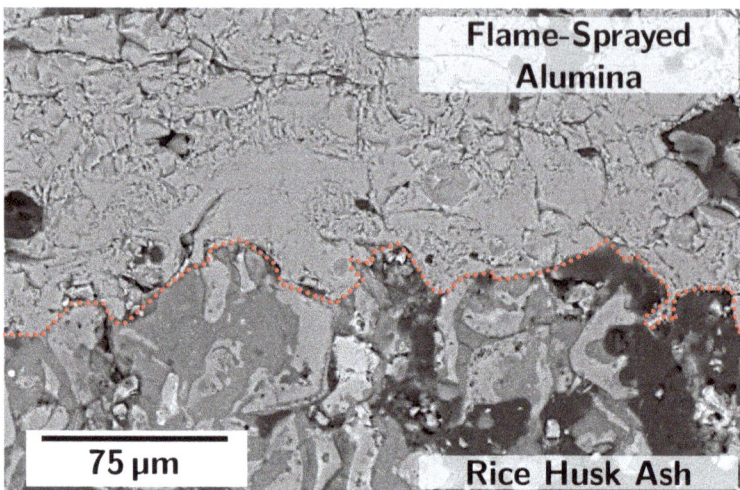

Figure 2. Microstructure of a composite material based on rice husk ash (**lower part**) and a flame-sprayed alumina coating (**upper part**) obtained by scanning electron microscopy; interface (dotted red line).

Therefore, this work statistically investigated the scatter of flexural strength for flame-sprayed ceramic components on two large data sets including 1000 specimens each. Recently, Gorjan and Ambrožič [20] investigated a large experimental data set of flexural strength data for low-pressure injection-moulded ceramic parts with regard to its representation by different two-parametric distribution functions. It was found that the Weibull distribution describes the experimental data best, while the Normal, Log-Normal, and Gamma distributions yielded lower fitting accuracy. However, Fedorov and Gulyaeva [21] pointed out that the obtained strength data of porous alumina pellets were significantly better represented by the Gamma distribution compared to the Weibull distribution.

Thus, the current study aims to investigate the suitability of the two-parametric distribution functions Normal, Log-Normal, and Gamma distribution, in addition to the standard Weibull approach, for representing flexural strength data of flame-sprayed components. Furthermore, this study presents a simulation based on resampling of experimentally determined strength data to investigate the influence of the sampling size on the robustness of different distribution functions and their characteristic parameters. This complements the theoretical simulation presented by Danzer et al. [12,14]. The study examines to what extent the flexural strength data of flame-sprayed components follows the theoretically expected fracture statistics of ceramic components.

2. Materials and Methods

2.1. Fabrication and Testing of Specimens

For flexural strength testing, disc-shaped Al_2O_3 specimens were fabricated using the flame-spraying technique (Flexicord feedstock, Saint-Gobain Coating Soluations S.A.S., Avignon, France) on a graphite substrate (Graphite Materials GmbH, Zirndorf, Germany) with pins of diameter $D = 8$ mm and $D = 15$ mm, respectively. The spray distance between the flame-spraying gun and the graphite substrate was 150 mm. Table 1 summarises the parameters of the flame-spraying process. Details of the experimental setup were described by Neumann et al. [4]. After cooling, the specimens were removed from the graphite pins, resulting in self-supporting components. To avoid pre-selection of specimens for B3B testing, the proportion of rejects r, i.e., specimens, which were not removed from the substrate without rupture, was determined by $r = 1 - q\,q_a^{-1}$, where q is the number of usable specimens and q_a the number of spraying attempts [4]. This ratio depends on the specimens' geometry and thickness and the substrate properties. For each disc geometry, a total of 1000 specimens were fabricated with $r < 0.01$. Following the suggestion by Neumann et al. [4], such geometries were suitable for estimating Weibull parameters. The two populations, designated as D08 and D15 according to the specimen diameter, each consisting of 1000 specimens, were tested for their flexural strength.

Table 1. Parameters of the flame-spraying process (m^3 represents a standard cubic metre).

Parameter	Dimension	Value
flow rate of O_2	$m^3 \cdot h^{-1}$	3.2
flow rate of C_2H_2	$m^3 \cdot h^{-1}$	1.3
flow rate of pressurised air	$m^3 \cdot h^{-1}$	38
feed rate	$m \cdot min^{-1}$	0.28–0.30

The flexural strength was determined using the ball on three ball test (B3B) [22,23]. For both populations, a similar experimental setup was used. The diameter of the used alumina balls was 5 mm and 10 mm for D08 and D15, respectively. All specimens were tested with a loading rate of $0.5 \, mm \, min^{-1}$. The flexural strength was calculated according to Equation (1) using the maximum load at rupture F_c, the specimens' thickness t, and the dimensionless factor f^*. The latter is dependent on Poisson's ratio ν and the ratio of both the specimens' thickness t and the support balls' diameter D_s to the specimens' diameter D and was calculated by an empirical formula presented by Börger et al. [23]. The Poisson's ratio ν of the flame-sprayed material was approximated by $\nu = 0.2$. A thickness measurement of the flame-sprayed components was carried out using a digital caliper (resolution 0.01 mm).

$$\sigma_{c,B3B} = f^* \left(\frac{2t}{D}, \frac{D_s}{D}, \nu \right) \frac{F_c}{t^2} \quad (1)$$

Figure 3 shows exemplary load-displacement curves of one tested specimen each of the D08 and D15 populations. For both curves, deviation from linearity at peak load was observed.

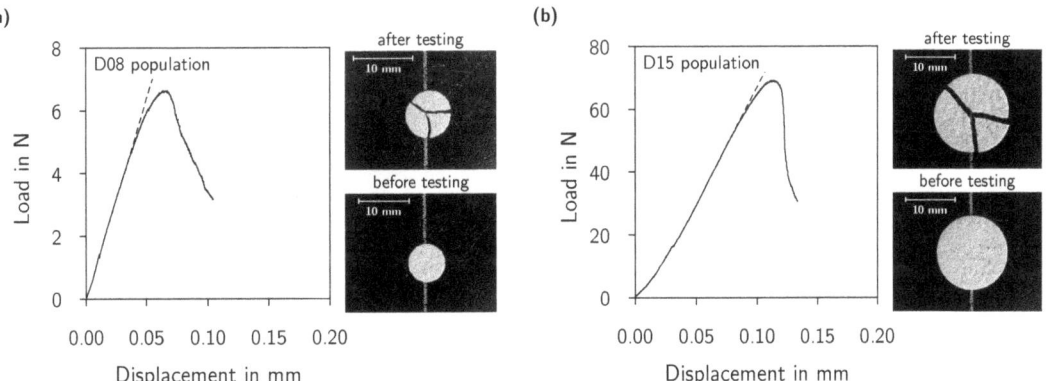

Figure 3. Exemplary Load-Displacement graphs for a tested specimen and specimen photographs before and after testing from the D08 population (**a**) and the D15 population (**b**).

2.2. Statistical Analyses

All calculations described in this work were performed using the statistical software package R [24]. Whenever required, a local weighted regression (LWR) was applied using the *loess* function from base R [25] to smooth certain curves for clarity and readability. The figure captions indicate the use of LWR. A span width of 0.05 was used, corresponding to 5% of the total data points used as nearest surrounding neighbours for the local regression at each point.

The compatibility of different two-parametric distribution functions, namely Weibull (W), Normal (N), Log-normal (LN), and Gamma (G) (Equations (2)–(5)) with the experimental data was investigated in the following analysis. The statistical variable was the flexural strength obtained by B3B designated as σ.

$$p_W(\sigma) = \frac{m}{\sigma_{0W}} \left(\frac{\sigma}{\sigma_{0W}}\right)^{m-1} \exp\left(-\left(\frac{\sigma}{\sigma_{0W}}\right)^m\right) \tag{2}$$

$$p_N(\sigma) = \frac{1}{\delta\sqrt{2\pi}} \exp\left(-\frac{1}{2}\left(\frac{\sigma - \sigma_{0N}}{\delta}\right)^2\right) \tag{3}$$

$$p_{LN}(\sigma) = \frac{1}{\sigma} \frac{1}{\omega\sqrt{2\pi}} \exp\left(-\frac{1}{2}\frac{ln(\sigma) - ln(\sigma_{0LN})}{\omega}^2\right) \tag{4}$$

$$p_G(\sigma) = \frac{1}{\sigma_{0G}^k \Gamma(k)} \sigma^{k-1} \exp\left(-\frac{\sigma}{\sigma_{0G}}\right) \tag{5}$$

The compatibility was evaluated using quantile-quantile probability plots (Q-Q plots) constituting the experimental data against the values calculated with a theoretical distribution. Moreover, the visual examination of the fit quality was complemented by calculating the coefficient of determination R^2 for each distribution according to Equation (6). Therefore, the fitting parameters were determined using the maximum likelihood method (for details, see [20]), while the probability of failure (P_i) was determined for the ascending ranked experimental data according to Equation (7). Note that quantities determined for the experimental data receive the index 'ex'.

$$R^2 = 1 - \frac{\sum_{i=0}^{N_j}(\sigma_i - \sigma_{i,th})^2}{\sum_{i=0}^{N_j}(\sigma_i - \langle\sigma_i\rangle)^2} \tag{6}$$

$$P_i = (i - 0.5)N_j^{-1} \tag{7}$$

The experimentally determined strength data were assumed to be a representative distribution of the true strength data of the flame-sprayed components. This assumption allowed for the application of the random resampling technique to investigate the influence of the sampling size on the robustness of different fitted distributions and their corresponding shape and scale parameters.

Random resampling was used to generate sub-populations from each experimentally determined population. The sub-populations were of size N_j, with $N_j = \{30; 31; \ldots; 1000\}$. For each value of N_j, 1000 sub-populations were generated by drawing N_j strength values with replacement from the data sets. This resulted in a total of 971,000 distinct sub-populations being generated from each of the populations D08 and D15. For each sub-population, the coefficient of determination, shape, and scale parameters for the Gamma and Weibull distributions were determined.

Whenever required, percentiles were calculated using the *quantile* function from R *stats* package [24]. The uth percentile of variable X is denoted as $X|^u$. The median of X would thus follow the denotation $X|^{50}$. The $u - v$th interpercentile range $X|_v^u$ for the variable X is calculated by subtracting the vth percentile $X|^v$ from the uth percentile $X|^u$.

3. Results and Discussion

In Figure 4, the complete experimental data are presented, i.e., the flexural strength over the specimen thickness. As presented and irrespective of the test series D08 or D15, for both the specimen thickness and the determined flexural strength, a certain scattering was observed. As outlined, the basic flame spray settings were kept consistent for both test series (number of oversprays, etc.). Consequently, from the processing point of view, the test series themselves can be considered consistent, meaning a sufficient resemblance of all individual specimens per series.

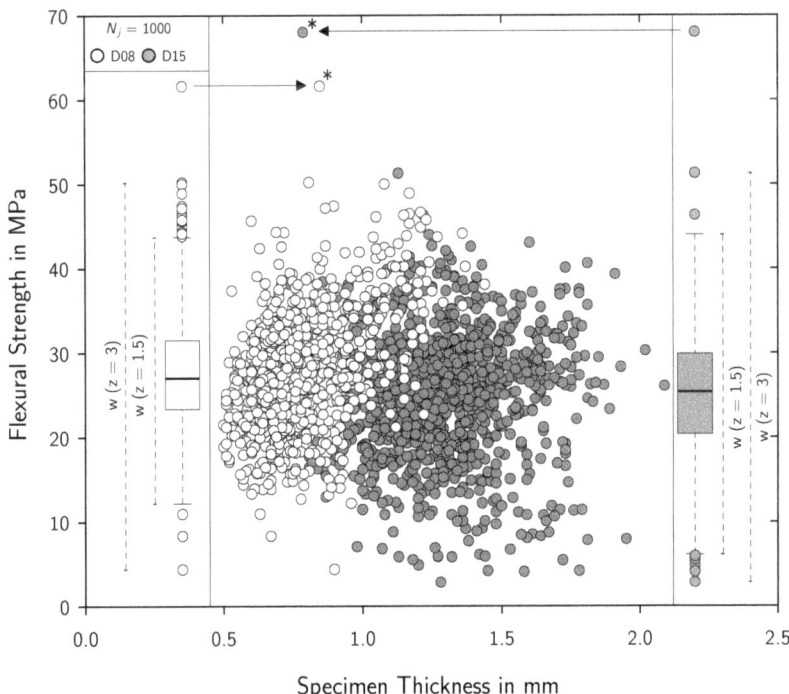

Figure 4. Complete experimental data: flexural strength over the specimen thickness and basic boxplots of the strength data ($w(z) = \{\sigma_c \in (Q_1 - zIQR, Q_3 + zIQR)\}$, IQR = inter quartile range, Q_1 = 25% quantile, Q_3 = 75% quantile). Asterisks * indicate extreme positive outliers in the scatter plot, derived from the boxplot representation, i.e., outliers above $w(z=3)$.

Following Equation (1), the influence of the specimen thickness should already be accounted for and thus the flexural strength should be independent of the specimen thickness. Referring to the scatter plots from Figure 4; however, a correlation between the specimen thickness and the flexural strength cannot be rejected from visual inspection alone. For that reason, the data were tested for correlation by the non-parametric Spearman rank sum correlation test. Basically, such test results in a correlation factor $\rho \in [-1, 1]$. For $\rho \in [-0.1, 0.1]$, no correlation would be given, for $\rho \in \pm[0.1, 0.5]$, a mild correlation can be assumed, and for $\rho \in \pm[0.5, 1.0]$, a strong or very strong correlation is present [26]. In the present study, mild correlation is considered the maximum tolerated correlation in order to hold the basic premise of one consistent data set. Mild correlation may trace back to the influence of the human operator and could be arbitrary. Known from the literature, several flame-spray process parameters can affect the final coating properties, such as the spray angle, distance between the spray unit and the substrate, or the horizontal travel- or rotation-speed of the spray unit, to name a few [1,2,27]. As an operator, a human would be unable to reproduce those settings exactly for each specimen, for example, due to human fatigue. Eventually, that may result in an inherent scatter of the specimen thickness and the derived flexural strength, as it was observed (cf. Figure 4). That is why mild correlation between the specimen thickness and the flexural strength is accepted (per test series). In contrast, a strong or very strong correlation would mean that the specimens are not similar due to strong variation in the spray process parameters. If so, the specimens could differ too much in terms of their microstructure, modulus of elasticity, local phase composition, etc. In the present study, the resulting ρ of D08 and D15 amounted to 0.39 and 0.10, respectively. Henceforth, for both series, only a mild correlation is present from the data and both series can be considered as one entire population each.

The boxplot representation reveals a rather symmetric distribution of the strength data around its median for both test series. Equal for both, a certain amount of outliers can be derived from the standard boxplots, i.e., for $w\,(z = 1.5)$. Also equal for both, an extreme positive outlier was found, i.e., for $w\,(z = 3)$ (cf. Figure 4). For the present study, extreme positive outliers are excluded from the experimental data (in Figure 4 indicated by asterisks). Extreme positive outliers oppose the conservative point of view insofar as they would lead to an overestimation of the flexural strength limits even for robust statistical measures (median, upper and lower quartile, etc.). Overestimation is considered the more harmful way of misestimation. Excluding such positive outliers can be understood as a 'concept of maximum doubt'. Extreme negative outliers on the contrary (whenever present), should not be excluded at any rate. Hence, for the following resampling procedures, the experimental data pool comprises $N_{ex} = 999$ strength values per series.

3.1. Fitting of Two-Parametric Distributions for Experimental Data

Figure 5 presents Q-Q plots for D08 and D15 assuming either the Weibull, Normal, Log-Normal, or Gamma distribution for the experimental data. The corresponding coefficients of determination R^2_{ex} as well as shape and scale parameter for each fit are summarised in Table 2.

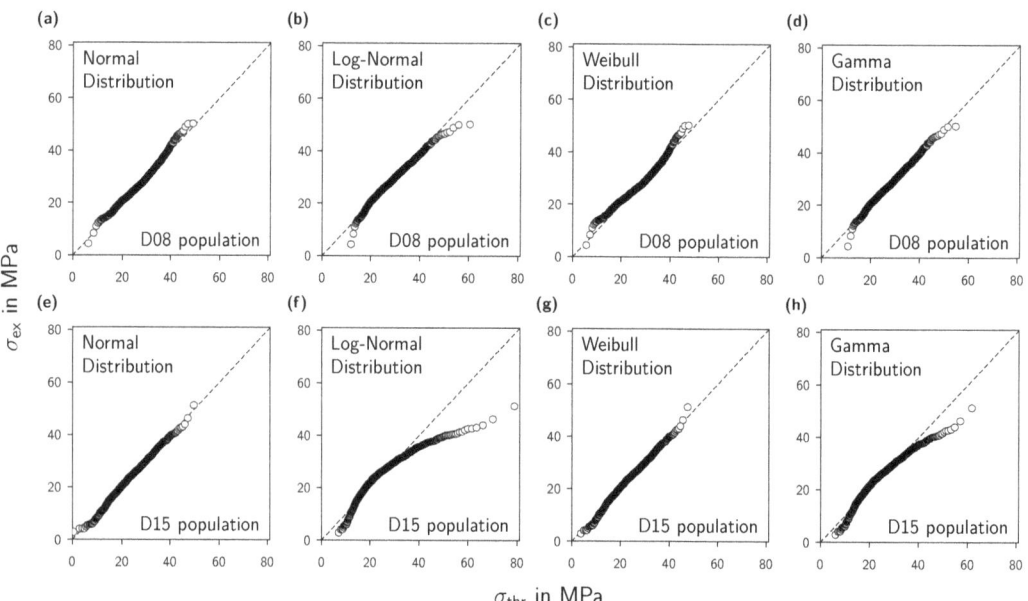

Figure 5. Q-Q plots assuming Weibull, Normal, Log-normal, and Gamma distribution for D08 population (**a**–**d**) and D15 population (**e**–**h**).

Visually, both populations fit well to a straight line for the Weibull or the Normal distribution. However, the fit for the Weibull distribution was slightly worse for D08 compared to D15. Accordingly, the Normal or the Weibull fit gave the highest R^2_{ex} values for the D15 population and slightly lower values for the D08 population (Table 2). In contrast, the data points deviate from a straight line when assuming Log-Normal distribution, resulting in lower R^2_{ex} values, particularly for the D15 population. This indicates a poor representation of the data points, leading to the rejection of the Log-Normal distribution for flexural strength data of flame-sprayed materials.

The Gamma distribution gave an excellent fit to the D08 population with an $R^2_{G,ex} = 0.9954$. However, the accuracy of the fit was worse for the D15 population. Nevertheless, a high $R^2_{G,ex} > 0.94$ was observed for both populations. Therefore, the Gamma distribution was considered for further investigation.

Figure 6 shows the histogram of the experimental data for both populations with the fitted lines for the Weibull and the Gamma distributions. It can be seen that the right tail of the Gamma distribution is larger compared to the Weibull distribution. This means that positive outliers are less detrimental to the Gamma distribution fit. On the other hand, when fitting the Weibull distribution, the left tail was found to be larger. This suggests a greater tolerance for negative outliers. The box plots show that the outliers of the D08 population were mainly positive outliers (Figure 6a), while those of the D15 population were mainly negative outliers (Figure 6b). Because of that, the fit quality of the D08 population was better with the Gamma distribution than with the Weibull distribution ($R^2_{G,ex} > R^2_{W,ex}$). Conversely, the fit quality of the D15 population was better with the Weibull distribution than with the Gamma distribution ($R^2_{W,ex} > R^2_{G,ex}$), see Table 2.

Table 2. Parameters of fitted distributions for experimental data of D08 and D15 populations (shape parameter of Normal distribution in MPa).

Distribution	Coeff. of Det.		Shape Par.		Scale Par. in MPa	
	D08	D15	D08	D15	D08	D15
Normal	0.9892	0.9959	6.481	7.460	27.640	25.003
Log-Normal	0.9847	0.8057	0.246	0.366	26.857	23.631
Weibull	0.9691	0.9957	4.478	3.753	30.189	27.650
Gamma	0.9954	0.9410	17.556	9.025	1.574	2.770

The Q-Q analysis showed that both D08 and D15 were equally well represented by the Normal distribution. However, Neumann et al. [4] reported that the Normal distribution underestimates the loss of strength in flame-sprayed components due to thermal shock. This is critical for components subjected to substantial thermal shock such as refractory hollowware [6,7]. Therefore, the Normal distribution was rejected.

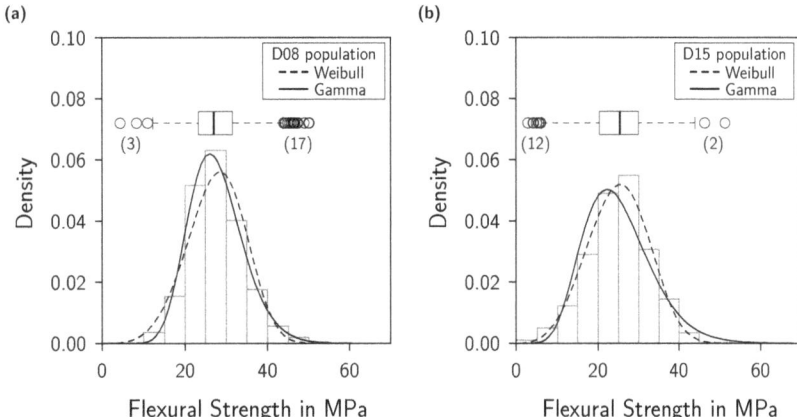

Figure 6. Histogram and fitted Weibull and Gamma distribution for experimental data of the D08 population (**a**) and the D15 population (**b**).

3.2. Random Sub-Population Analysis–Random Resampling

Based on the Q-Q analyses, the Weibull and the Gamma distribution functions were found to be suitable for representing the experimental strength data. Consequently, these distributions were used to fit the sub-populations generated by random resampling of the experimental strength data.

3.2.1. Coefficient of Determination

Figure 7 shows the relationship between R^2 and the sampling size (N_j) based on random resampling of the strength data. With the increasing sampling size, the median fit quality (represented by $R^2|^{50}$) improved and approached the fit quality of the underlying experimental data set (R^2_{ex}, Table 2). Accordingly, the interpercentile range between the 5th and the 95th percentiles ($R^2|^{95}_5$) decreased with the increasing N_j. This trend is in line with the law of large numbers [13].

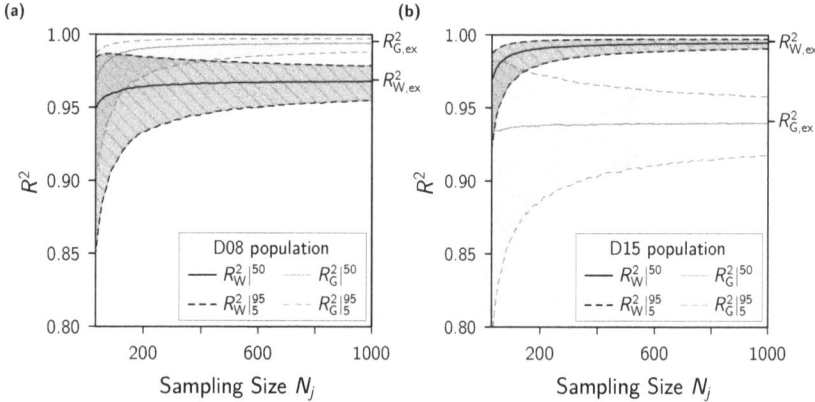

Figure 7. Fit quality R^2 dependent on sampling size N_j for assuming either the Weibull (W) or the Gamma (G) distribution for the sub-populations from D08 (**a**) and D15 (**b**); curves smoothed using LWR.

However, the span width of $R^2|^{95}_5$ for the D08 and D15 populations varied depending on the assumed distribution. Specifically, when the Weibull distribution was used, $R^2_W|^{95}_5$ was larger for the D08 population compared to D15, and the lower boundary reached smaller values for the decreasing N_j. On the other hand, when the Gamma distribution was used, $R^2_G|^{95}_5$ of the D15 population was larger than that of the D08 population, and the lower boundary reached smaller values for decreasing N_j.

The following has to be discussed:

- Among others, the fit quality of a data set depends on the number of its outliers, which also affects the fit quality of sub-populations derived from the data.
- A sub-population from a data set containing a high number of outliers may have a high or low fit quality, depending on chance.
- Generating a sub-population with a high fit quality is more likely if the number of outliers in the underlying experimental data is low. This also reduces the scatter in the fit quality of the generated sub-populations.

Figure 8 indicates that sub-populations from D08 had a larger median number of positive outliers than negative outliers, regardless of N_j. Moreover, the variability of the positive outliers was higher than that of the negative outliers (wider span of the IPR). Because the Weibull distribution is more sensitive to positive outliers than the Gamma distribution, this caused $R^2_W|^{50}$ to be lower than $R^2_G|^{50}$ for all N_j and a higher variability in R^2_W compared to R^2_G for the D08 sub-populations. The share of positive and negative outliers showed an opposite behaviour for sub-populations from D15, resulting in a reversal of the

R^2 values and their scatter for the Weibull and Gamma distributions. This was because the Gamma distribution is more sensitive to negative outliers than the Weibull distribution.

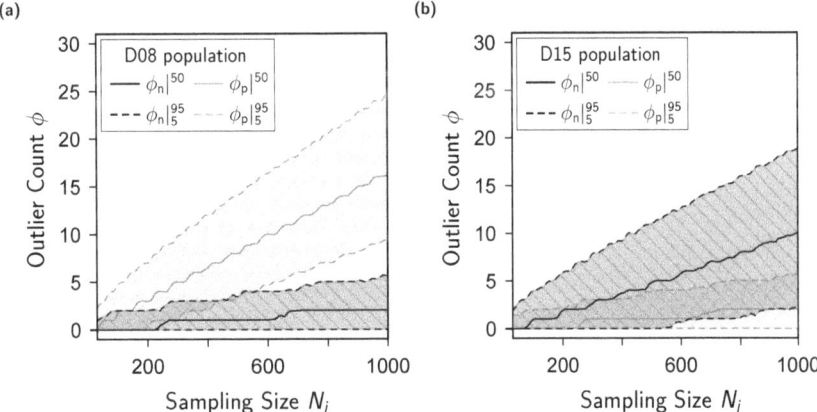

Figure 8. Number of positive outliers ϕ_p and negative outliers ϕ_n for the D08 sub-populations (**a**) and the D15 sub-populations (**b**) dependent on the sampling size N_j for 1000 resamplings with the median number of positive and negative outliers ($\phi_p|^{50}$ and $\phi_n|^{50}$, respectively) and the interpercentile range between the 5th and the 95th percentile of positive and negative outliers ($\phi_p|^{95}_5$ and $\phi_n|^{95}_5$, respectively); curves smoothed using LWR.

Nevertheless, the median fit quality $R^2|^{50}$ of the generated sub-populations was similar to the fit quality of the corresponding experimental data independently of N_j. Table 3 illustrates the required N_j to achieve R^2_{ex} within specified boundaries. This shows that $R^2|^{50}$ was robust even for small sampling sizes.

Table 3. Minimum sampling size $N_{j,\min}$ for which $R^2|^{50} \in R^2_{ex} \cdot (1 \pm i)$ with $i = \{0.05; 0.025; 0.01\}$.

i	0.05	0.025	0.01
$R^2_W|^{50}$-D08	30	30	74
$R^2_W|^{50}$-D15	30	34	93
$R^2_G|^{50}$-D08	30	35	98
$R^2_G|^{50}$-D15	30	30	30

3.2.2. Distribution Parameters

In the following, the shape and scale parameters of the Weibull distribution (m and σ_{0W}) and the Gamma distribution (k and σ_{0G}) were investigated as a function of the sampling size N_j for the resampled sub-populations.

Figure 9 presents the shape parameters m and k for assuming the Weibull and Gamma distributions, respectively, for the D08 and D15 sub-populations as a function of the sampling size (N_j). As N_j increases, the IPRs ($m|^{95}_5$ and $k|^{95}_5$) and the median shape parameters ($m|^{50}$ and $k|^{50}$) of the resampled populations approach the experimentally determined values (m_{ex} and k_{ex}) due to the law of large numbers [13]. However, the IPRs were found to span asymmetrically around m_{ex}, with a larger upper range. As presented, $m|^{50} > m_{ex}$ was valid independent of the investigated range of N_j. However, the difference $m|^{50} - m_{ex}$ decreased as N_j increased, which is consistent with previous studies [12,14]. This suggests that m_{ex}, which in this regard was considered as the true m of the resampled distribution, would be likely overestimated, particularly for small sampling sizes. Similarly, the median shape parameter for the Gamma distribution $k|^{50}$ was found to positively deviate from k_{ex} for small sampling sizes (Figure 9b,c).

Therefore, the standard EN 843 5 recommends a correction of the Weibull modulus m to m_{cor} using Equation (8), while b can be approximated with Equation (9) where $s = 1.593$ and $h = 1.047$ (determined based on a Monte-Carlo-Simulation).

$$m_{\text{cor}} = b \cdot m \quad (8)$$
$$b = 1 - s \cdot N_j^{-h} \quad (9)$$
$$k_{\text{cor}} = c \cdot k \quad (10)$$
$$c = 1 - s \cdot N_j^{-h} \quad (11)$$

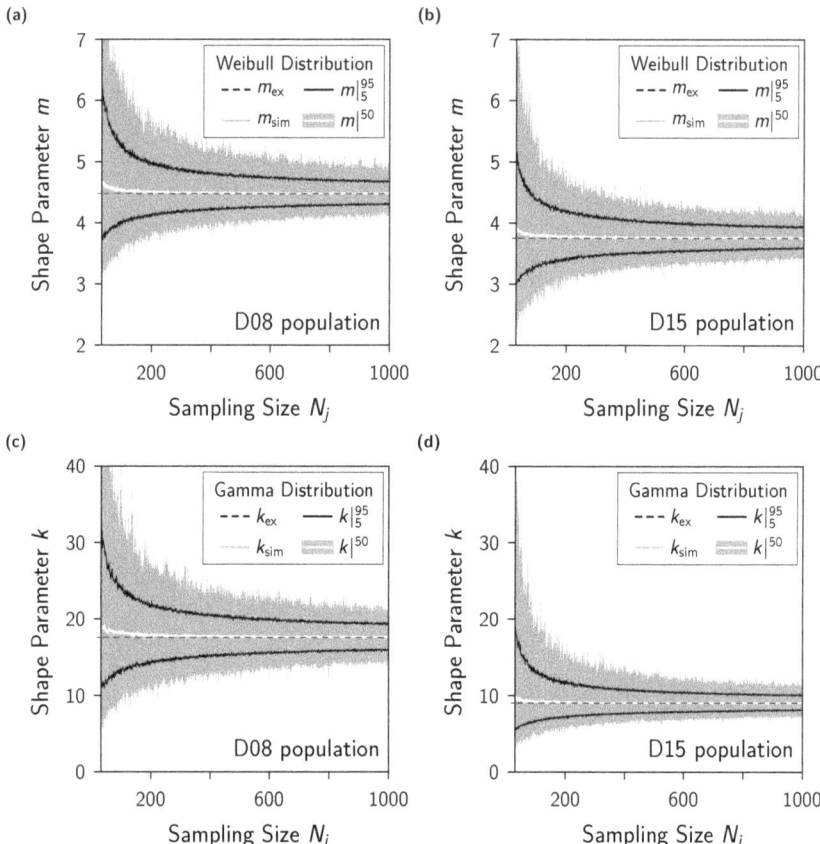

Figure 9. Shape parameter of the Weibull distribution (**a**,**b**) and the Gamma distribution (**c**,**d**) in dependence on the sampling size N_j for D08 and D15 sub-populations.

Similarly, an empirical correction factor was estimated based on the data obtained from the random resampling. The b_{adj} was fitted with $m_{\text{cor}} = m_{\text{ex}}$ and $m = m|^{50}(N_j)$ using Equation (8) and a non-linear least-squares method. Using the same approach and Equations (10) and (11) with $k_{\text{cor}} = k_{\text{ex}}$ and $k = k|^{50}(N_j)$, a correction factor $c_{\text{adj,D08}}$ and $c_{\text{adj,D15}}$ was determined for the Gamma distribution based on the data of D08 and D15, respectively. As mentioned above, extreme positive outliers were removed from both populations, which was a single data point in each case. At this point, the influence of a single extreme outlier shall be investigated. Therefore, the correction factors b^*_{adj} and c^*_{adj} were estimated for an identical random resampling of the distribution of strength data including the extreme outliers. Table 4 summarises the obtained coefficients s and h.

Table 4. Empirical determined parameters s and h for different correction factors.

Correction Factor	s	h
Weibull Distribution:		
b_{std}	1.5931	1.0470
$b_{adj,D08}$	1.4361	1.0328
$b_{adj,D15}$	0.9905	0.9694
$b^*_{adj,D08}$	0.9070	0.7458
$b^*_{adj,D15}$	0.6821	0.9694
Gamma Distribution:		
$c_{adj,D08}$	1.6765	0.8711
$c_{adj,D15}$	2.7654	1.0399
$c^*_{adj,D08}$	1.2296	0.8085
$c^*_{adj,D15}$	1.8863	0.9664

Figure 10 shows the adjusted correction factors as a function of N_j for the D08 and D15 populations. It can be seen that the adjusted correction factors $b_{adj,D08}$ and $b_{adj,D15}$ were in excellent agreement with b_{std}. Thus, using b_{std} as a correction factor for a Weibull analysis of flexural strength data for flame-sprayed components is suitable. However, for the Gamma distribution, a stronger correction factor than b_{std} was required for both populations. A stronger correction b^*_{adj} for the determined Weibull Modulus was required for both populations, if extreme outliers were included in the generation of sub-populations. Conversely, the correction factor c^*_{adj} differed only negligibly from c_{adj} of the corresponding distribution. This again emphasises the robustness of the Gamma distribution against positive outliers, while the Weibull distribution was susceptible in this regard.

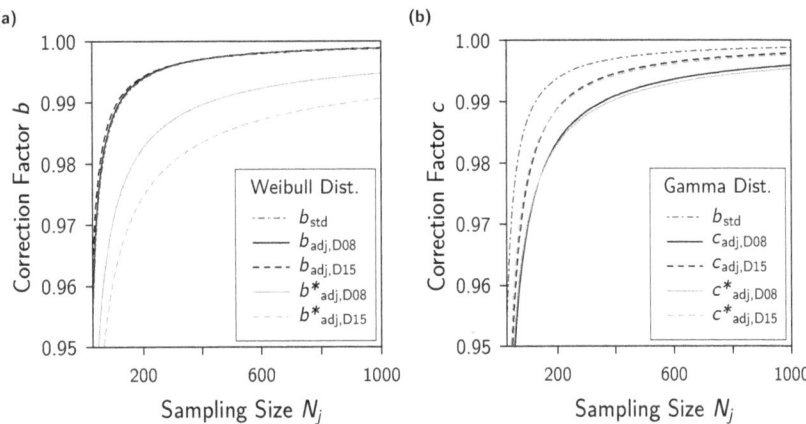

Figure 10. Empirical determined correction factors for the shape parameter of the Weibull distribution (**a**) and the Gamma distribution (**b**).

Figure 11 shows that the 5th to 95th interpercentile range of the characteristic strength $\sigma_{0W}|_0^{95}$ was symmetrical around $\sigma_{0W,ex}$ independently of N_j and became smaller as N_j increases. This was in excellent agreement with the simulation based on a theoretical Weibull distribution presented by Danzer et al. [12,14]. Moreover, the obtained median $\sigma_{0W}|^{50}$ of the simulated data deviated negatively from $\sigma_{0W,ex}$ for small values of N_j. Similarly, $\sigma_{0G}|_5^{95}$ spans symmetrically around $\sigma_{0G,ex}$. The median scale parameter $\sigma_{0G}|^{50}$ determined for the Gamma distribution deviated negatively from $\sigma_{0G,ex}$ as well. This means an underestimation of the experimentally determined scale parameter of the underlying distribution. While an overestimation would have more adverse effects and presents a higher degree of risk, an underestimation is a more conservative approach and therefore acceptable.

Therefore, no correction factor was required, which is in accordance with the EN 843-5 standard. In the case of the Gamma distribution, the characteristic strength is obtained by multiplication of shape and scale parameter. Thus, the underestimation of the scale parameter compensates for the overestimation of the shape parameter k to a certain extent.

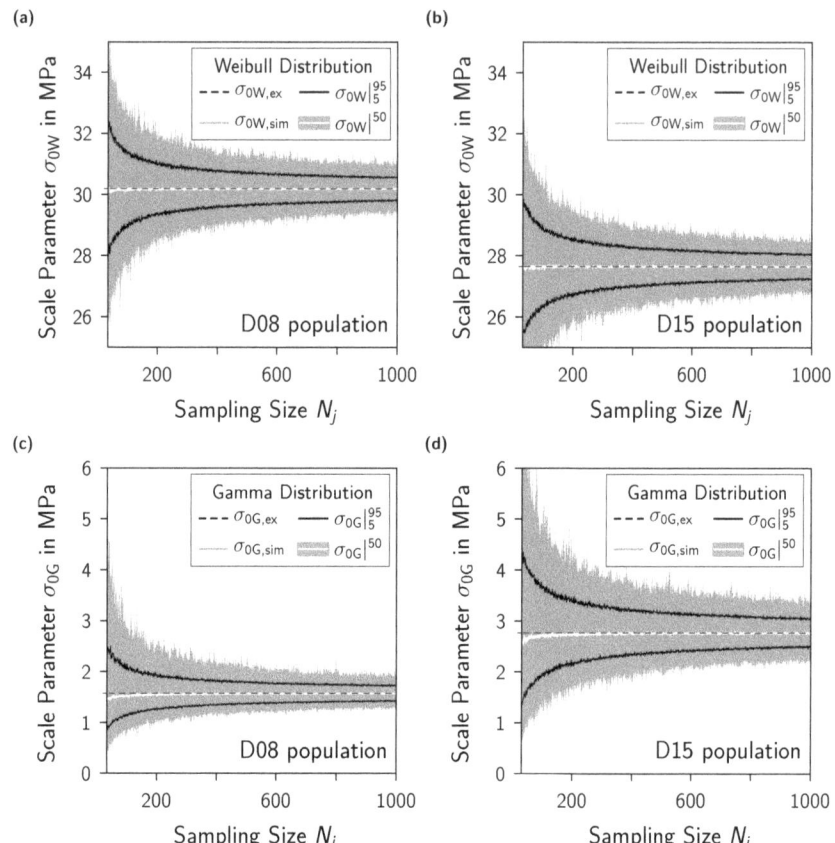

Figure 11. Scale parameter of the Weibull distribution (**a**,**b**) and the Gamma distribution (**c**,**d**) in dependency on the sampling size N_j for D08 and D15 sub-populations.

Figure 11 shows that the median characteristic strength $\sigma_{0W}|^{50}$ of the Weibull distribution was higher for the D08 population compared to the D15 population. This was in excellent agreement with the size effect theory, which predicts a decreasing characteristic strength with increasing volume under load [11]. Taking into account the multiplication of shape and scale parameter from the Gamma distribution, the characteristic strength of 27.6 MPa and 25.0 MPa was determined based on the experimental data from the D08 and D15 population, respectively. Thus, an identical trend of the size effect was observed.

3.3. Conditional Sub-Population Analysis–Conditional Data Cropping

A final note shall be made on the correlation between the specimen thickness and the flexural strength, exemplary for D08. Among the two test series, despite being of an overall mild nature, a positive correlation was present for D08. However, to isolate an origin of this relation remains a difficult question to be answered. A likely reason was already discussed, i.e., the fact of relying on a human operator. Despite this uncertainty, a resampling on the basis of the variation in the specimen thickness as a condition allows for some insight, taking the scattering of specimen thickness as an inherent characteristic of the manual flame-spray process. First, the quantiles of the specimen thicknesses were determined. Then, starting at the median of the specimen thickness $t|^{50}$, interpercentile ranges $t|^{50+x}_{50-x}$ around this median can be defined (abbreviated by IPR; including the borders). Referring to the Figures 9 and 11, this principle of interpercentile ranges was applied before in this study. In consequence, the higher the interpercentile range of the specimen thickness around the median is set, the more variability in the specimen thickness is accounted for. By cropping the experimental D08 data set on the basis of this premise, conditioned sub-populations were built, which were analysed for their statistics as performed before, including the correction of the determined Weibull modulus.

From that conditioned data cropping, two boundaries can be derived: allowing for no variation in the specimen thickness (lower boundary) or allowing for the maximum scatter of the specimen thickness (upper boundary). For the lower boundary, only the specimens with a thickness equal to the median of all thicknesses were taken into account and for the upper boundary, all specimens were taken into consideration (minus the extreme positive outlier, excluded before). Consequentially, the conditioned data cropping results in sub-populations of different sampling sizes N and thus, for comparability, it is necessary to account for that in regard of the Weibull modulus. At this point, the measurement precision of the specimen thickness became relevant. In the present study, the thickness was measured on two decimal digits. Due to the mere number of experimentally tested specimens and this measuring precision, the original experimental data sets contain multiple specimens of similar thickness. As a consequence, it was possible to form a data subset at the lower boundary (only specimens of median thickness), which comprises a statistically reasonable number of 30 individual specimens. Note that the following statements exclusively apply to the D08 series. Their transfer to other test series, materials, or testing methods has to be considered separately.

Starting at the lower boundary, no correlation between the specimen thickness and the flexural strength can be postulated, i.e., an ideal Spearman correlation factor of $\rho = 0$. The respective statistical parameters of the Weibull and Gamma fitting are listed in Table 5, relative to the corresponding experimental parameters from Table 2. For the upper boundary, i.e., the full data set, the Spearman correlation factor amounts to $\rho = 0.39$ (as presented before). Regarding the Weibull fitting (actually the less suited distribution for D08) it occurred that after reducing the specimen thickness variation to zero, the Weibull fitting was improved ($R^2_W/R^2_{W,ex} > 1$). Simultaneously, the Weibull modulus was increased ($m/m_{ex} > 1$), while the characteristic strength was lower ($\sigma_{0W}/\sigma_{0W,ex} < 1$). In comparison, the difference in the characteristic strength by 1% (only median thickness vs. the full data set) is lower than the difference in the Weibull modulus by about 30%. As further presented in Figure 12, a difference in either the Weibull modulus (positive difference up to 18%) or the characteristic strength (negative difference of about 2%) was observed nearly for the complete range of IPR. For the Gamma distribution fitting, it was observed that the fitting improved at a higher variation in the specimen thickness ($R^2_G/R^2_{G,ex} < 1$). Moreover, both distribution parameters k and σ_{0G} lay above the experimental data for $0 < \text{IPR} < 100$ (cf. Figure 12). The strictest conclusion that can be drawn from this is that the experimental scatter in the flexural strength of the D08 series was to some extent affected by the variation in specimen thickness, i.e., it was increased. Here, a trade off has to be found: stating a more pronounced scattering in the flexural strength would again be the more cautious/conservative point of view and prevent an overestimation of the material. In the present case of flame-sprayed alumina, it was thus indeed permitted to neglect the mild correlation between the flexural strength and the specimen thickness.

Table 5. Relative Weibull and Gamma fit parameters for specimens of only the median thickness $t|^{50}$ resulting in $N = 30$ considered specimens (R^2—coefficient of determination of the QQ analysis, m—Weibull modulus, σ_{0W}—characteristic strength, k—Gamma distribution shape parameter, σ_{0G}—Gamma distribution scale parameter; extension 'ex' indicates experimental supported data of the D08 test series).

$R^2_W / R^2_{W,ex}$	1.012	$R^2_W / R^2_{G,ex}$	0.955
m / m_{ex}	1.329	k / k_{ex}	1.539
$\sigma_{0W} / \sigma_{0W,ex}$	0.990	$\sigma_{0G} / \sigma_{0G,ex}$	1.531

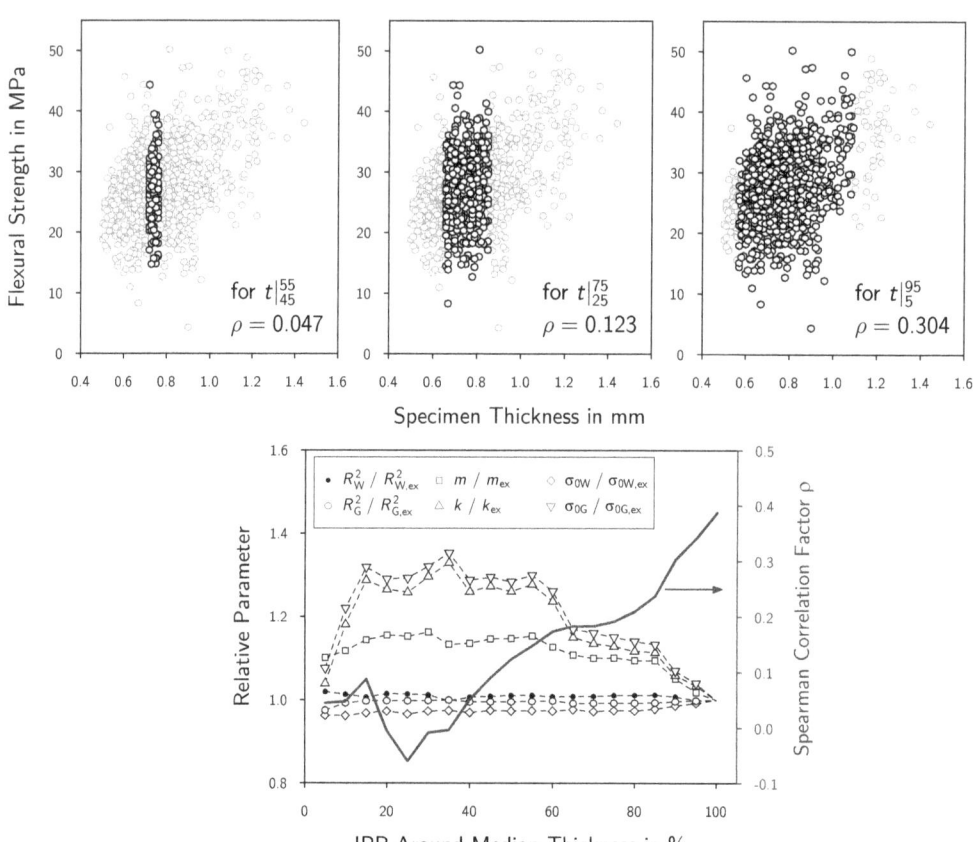

Figure 12. Exemplary sub-populations of D08 for different interpercentile ranges around the median specimen thickness (IPR, for $t|^{45}_{55}$, $t|^{25}_{75}$, and $t|^{95}_{5}$), generated by thickness-conditioned data cropping and the resulting relative Weibull and Gamma fit parameters and the Spearman correlation factor as function of the IPR (R^2—coefficient of determination of the QQ analysis, m—Weibull modulus, σ_{0W}—characteristic strength, k—Gamma distribution shape parameter, σ_{0G}—Gamma distribution scale parameter; extension 'ex' indicates experimental supported data of the D08 test series).

4. Conclusions

According to textbook standards, the variability of failure, i.e., the scattering of the mechanical strength, poses a fundamental and design-relevant material-specific feature to be introduced into the lifetime prediction of ceramic components, next to the fracture resistance and the sub-critical cracking behaviour. In this work, the flexural strength of two populations of flame-sprayed self-supporting ceramic components based on alumina including 1000 disc-shaped specimens each, were tested using the ball on three ball test. The suitability of the two-parametric distribution functions, i.e., the Normal, Log-Normal, Weibull, and Gamma distribution to describe the data were investigated using Q-Q analysis. Based on random resampling of the flexural strength data, the suitability of different distribution functions in dependency of the sampling size in the range from 30 to 1000 was investigated. Moreover, the distribution parameters, i.e., the shape and the scale parameters were determined and their dependence on the sampling size was evaluated.

The Q-Q analysis revealed the Weibull and Gamma distribution to be most suitable to represent flexural strength data of flame-sprayed self-supporting components. Moreover, it was shown that because of their skew, the Gamma distribution is more robust against positive outliers in a data set, while the Weibull distribution is more robust against negative outliers. The random resampling of the strength data showed that the fitting accuracy of the distribution was robust independently of the sampling size. Even for the smallest sample size investigated (30), the median fit quality was in a range of ±5% of the fit quality from the underlying experimental data set. An analysis of the distribution parameters revealed that for both the Weibull and the Gamma distribution, the shape parameter of the underlying distribution would be overestimated for small sampling sizes. Nevertheless, the study also showed that the suggested correction factor for the Weibull modulus was in excellent agreement with the simulated data (as long as extreme outliers were excluded from the analysis). Based on these results, a correction factor for the shape parameter of the Gamma distribution was suggested. Conversely, the characteristic strength (scale parameter) was slightly underestimated by both distribution functions for small sampling sizes and requires no correction.

Author Contributions: Conceptualization, F.K., M.N., T.W. and C.G.A.; methodology, F.K. and M.N.; software, M.N. and F.K.; validation, M.H., T.W. and C.G.A.; formal analysis, F.K., M.N. and M.H.; investigation, M.H., F.K. and M.N.; resources, T.S. and C.G.A.; data curation, F.K. and M.N.; writing—original draft preparation, F.K.; writing—review and editing, F.K., M.N. and T.W.; visualization, F.K. and M.N.; supervision, T.S., H.J. and C.G.A.; project administration, H.J. and C.G.A.; funding acquisition, T.S., H.J. and C.G.A. All authors have read and agreed to the published version of the manuscript.

Funding: This research was funded by the German Research Foundation (DFG) as part of the transfer project AN 322 39-1.

Institutional Review Board Statement: Not applicable.

Informed Consent Statement: Not applicable.

Data Availability Statement: Data available upon request.

Acknowledgments: The authors would like to thank N. Brachhold for reviewing the manuscript.

Conflicts of Interest: The authors declare no conflict of interest. The funders had no role in the design of the study; in the collection, analyses, or interpretation of data; in the writing of the manuscript; or in the decision to publish the results.

References

1. Fauchais, P.L.; Heberlein, J.V.; Boulos, M.I. *Thermal Spray Fundamentals: From Powder to Part*; Springer Science & Business Media: New York, NY, USA, 2014.
2. Vardelle, A.; Moreau, C.; Akedo, J.; Ashrafizadeh, H.; Berndt, C.C.; Berghaus, J.O.; Boulos, M.; Brogan, J.; Bourtsalas, A.C.; Dolatabadi, A.; et al. The 2016 thermal spray roadmap. *J. Therm. Spray Technol.* **2016**, *25*, 1376–1440. [CrossRef]
3. Aneziris, C.G.; Gehre, P.; Kratschmer, T.; Berek, H. Thermal shock behavior of flame-sprayed free-standing coatings based on Al_2O_3 with TiO_2- and ZrO_2-additions. *Int. J. Appl. Ceram. Technol.* **2011**, *8*, 953–964. [CrossRef]
4. Neumann, M.; Gehre, P.; Hubálková, J.; Zielke, H.; Abendroth, M.; Aneziris, C.G. Statistical Analysis of the Flexural Strength of Free-Standing Flame-Sprayed Alumina Coatings Prior and After Thermal Shock. *J. Therm. Spray Technol.* **2020**, *29*, 2026–2032. [CrossRef]
5. Stein, V.; Schemmel, T. Sustainable rice husk ash-based high-temperature insulating materials. *Interceram-Int. Ceram. Rev.* **2020**, *69*, 30–37. [CrossRef]

6. Wetzig, T.; Neumann, M.; Schwarz, M.; Schöttler, L.; Abendroth, M.; Aneziris, C.G. Rapid Prototyping of Carbon-Bonded Alumina Filters with Flame-Sprayed Alumina Coating for Bottom-Teeming Steel Ingot Casting. *Adv. Eng. Mater.* **2022**, *24*, 2100777. [CrossRef]
7. Zhang, L.; Thomas, B.G. State of the art in the control of inclusions during steel ingot casting. *Metall. Mater. Trans. B* **2006**, *37*, 733–761. [CrossRef]
8. Schönwelski, W.; Ruwier, K.; Föllbach, S.; Sperber, J. High-quality refractories for high quality steel. In Proceedings of the ICRF 2014, 2nd International Conference on Ingot Casting, Rolling and Forging, Milan, Italy, 7–9 May 2014.
9. Chagnon, P.; Fauchais, P. Thermal spraying of ceramics. *Ceram. Int.* **1984**, *10*, 119–131. [CrossRef]
10. Ostojic, P.; Berndt, C. The variability in strength of thermally sprayed coatings. *Surf. Coat. Technol.* **1988**, *34*, 43–50. [CrossRef]
11. Munz, D.; Fett, T. *Ceramics, Mechanical Properties, Failure Behaviour, Materials Selection*; Springer: Berlin, Germany, 2001; Volume 36, pp. 25–26.
12. Danzer, R.; Lube, T.; Supancic, P.; Damani, R. Fracture of ceramics. *Adv. Eng. Mater.* **2008**, *10*, 275–298. [CrossRef]
13. Schervish, M.J.; DeGroot, M.H. *Probability and Statistics*; Pearson Education: London, UK, 2012.
14. Danzer, R.; Lube, T.; Supancic, P. Monte Carlo simulations of strength distributions of brittle materials—Type of distribution, specimen and sample size. *Int. J. Mater. Res.* **2001**, *92*, 773–783. [CrossRef]
15. Weibull, W. A statistical distribution function of wide applicability. *J. Appl. Mech.* **1951**. [CrossRef]
16. Michálek, M.; Michálková, M.; Blugan, G.; Kuebler, J. Strength of pure alumina ceramics above 1 GPa. *Ceram. Int.* **2018**, *44*, 3255–3260. [CrossRef]
17. Danzer, R. A general strength distribution function for brittle materials. *J. Eur. Ceram. Soc.* **1992**, *10*, 461–472. [CrossRef]
18. Danzer, R. Some notes on the correlation between fracture and defect statistics: Are Weibull statistics valid for very small specimens? *J. Eur. Ceram. Soc.* **2006**, *26*, 3043–3049. [CrossRef]
19. Keleş, Ö.; García, R.E.; Bowman, K.J. Deviations from Weibull statistics in brittle porous materials. *Acta Mater.* **2013**, *61*, 7207–7215. [CrossRef]
20. Gorjan, L.; Ambrožič, M. Bend strength of alumina ceramics: A comparison of Weibull statistics with other statistics based on very large experimental data set. *J. Eur. Ceram. Soc.* **2012**, *32*, 1221–1227. [CrossRef]
21. Fedorov, A.; Gulyaeva, Y. Strength statistics for porous alumina. *Powder Technol.* **2019**, *343*, 783–791. [CrossRef]
22. Danzer, R.; Harrer, W.; Supancic, P.; Lube, T.; Wang, Z.; Börger, A. The ball on three balls test—Strength and failure analysis of different materials. *J. Eur. Ceram. Soc.* **2007**, *27*, 1481–1485. [CrossRef]
23. Börger, A.; Supancic, P.; Danzer, R. The ball on three balls test for strength testing of brittle discs: Stress distribution in the disc. *J. Eur. Ceram. Soc.* **2002**, *22*, 1425–1436. [CrossRef]
24. R Core Team. *R: A Language and Environment for Statistical Computing*; R Foundation for Statistical Computing: Vienna, Austria, 2022.
25. Cleveland, W.S. Robust locally weighted regression and smoothing scatterplots. *J. Am. Stat. Assoc.* **1979**, *74*, 829–836. [CrossRef]
26. Schober, P.; Boer, C.; Schwarte, L.A. Correlation coefficients: Appropriate use and interpretation. *Anesth. Analg.* **2018**, *126*, 1763–1768. [CrossRef] [PubMed]
27. Gadelmoula, A.; Al-Athel, K.; Akhtar, S.; Arif, A. A Stochastically Generated Geometrical Finite Element Model for Predicting the Residual Stresses of Thermally Sprayed Coatings Under Different Process Parameters. *J. Therm. Spray Technol.* **2020**, *29*, 1256–1267. [CrossRef]

Disclaimer/Publisher's Note: The statements, opinions and data contained in all publications are solely those of the individual author(s) and contributor(s) and not of MDPI and/or the editor(s). MDPI and/or the editor(s) disclaim responsibility for any injury to people or property resulting from any ideas, methods, instructions or products referred to in the content.

Article

Tribological Study of Simply and Duplex-Coated CrN-X42Cr13 Tribosystems under Dry Sliding Wear and Progressive Loading Scratching

Maria Berkes Maros *[ ] and Shiraz Ahmed Siddiqui [†]

TriboTeam Research Group, Department of Structural Integrity, Institute of Materials Science and Technology, Faculty of Mechanical Engineering and Informatics, University of Miskolc (UM), 3515 Miskolc, Hungary
* Correspondence: maria.maros@uni-miskolc.hu
† Current address: C-117, Kalyanpur, Lucknow 226010, India.

Citation: Maros, B.M.; Siddiqui, S.A. Tribological Study of Simply and Duplex-Coated CrN-X42Cr13 Tribosystems under Dry Sliding Wear and Progressive Loading Scratching. *Ceramics* **2022**, *5*, 1084–1101. https://doi.org/10.3390/ceramics5040077

Academic Editors: Amirhossein Pakseresht and Kamalan Kirubaharan Amirtharaj Mosas

Received: 30 September 2022
Accepted: 15 November 2022
Published: 24 November 2022

Publisher's Note: MDPI stays neutral with regard to jurisdictional claims in published maps and institutional affiliations.

Copyright: © 2022 by the authors. Licensee MDPI, Basel, Switzerland. This article is an open access article distributed under the terms and conditions of the Creative Commons Attribution (CC BY) license (https://creativecommons.org/licenses/by/4.0/).

Abstract: CrN coatings are widely used in the industry due to their excellent mechanical features and outstanding wear and corrosion resistance. Using scratch and ball-on-disk wear tests, the current study deals with the tribological characterisation of CrN coatings deposited onto an X42Cr13 plastic mould tool steel. Two surface conditions of the secondary-hardened substrate are compared—the plasma nitrided (duplex treated) and the un-nitrided (simply coated) states. The appropriate combination of secondary hardening providing the maximum toughness and the high-temperature nitriding of this high Cr steel is a great challenge due to the nitrogen-diffusion-inhibiting effect of Cr. The beneficial effect of the applied duplex treatment is proven by the 34% improvement of the adhesion strength and the 43% lower wear rate of the investigated duplex coatings. Detailed morphological analyses give insight into the characteristic damage mechanisms controlling the coating failure processes during scratching and wearing. For the simply CrN-coated sample, a new type of scratch damage mechanism, named "SAS-wings", is identified, providing useful information in predicting the final failure of the coating. The tribological results obtained on tribosystems with the investigated high Cr steel/CrN constituents represent a novelty in the given field.

Keywords: CrN coating; X42Cr13 tool steel; duplex treatment; high-temperature nitriding; scratch test; wear test; damage mechanism

1. Introduction

The class of transition metal nitride coatings has proved to be beneficial in improving the useful service life of the components in the metal tool industry. The performance of the coated components in various engineering applications is largely dependent on not only the tribological characteristics of the coatings but the bonding (adhesion) strength between the coating and the substrate. The adhesion can be improved by adequately controlling the coating process parameters and modifying the substrate prior to the deposition of the coating.

Protective surface coatings prepared by physical vapour deposition (PVD) have been effectively used in mechanical components to increase their lifetime. CrN coatings possess wide application due to their excellent mechanical properties [1,2], low friction coefficient, high hardness, wear resistance [3,4], chemical inertness [5,6], superior oxidation resistance at high temperatures up to ~700 °C, or corrosion resistance under severe operating conditions. These coatings provide excellent abrasion resistance [7]; thus, they are used as cutting, milling, and screw threading tools. Using CrN coatings, a considerable reduction of ejection forces or cavity pressure is achievable in injection moulding; therefore, they are preferred as protective coatings for tool cores, punches, or dies [8] and in various other tribological applications [9]. TiN coatings having large-scale application in the 1980s–1990s were replaced by CrN coatings in the 1990s–2000s due to their better wear performance in dry and lubricated conditions [9–11], higher oxidation resistance at elevated temperatures, better thermal stability, corrosion resistance, and three times higher achievable deposition rate [12] compared with that of TiN or Ti-C-N coatings [13].

The unique combination of mechanical and physical characteristics can significantly impact the tribological performance of a coated system, that is, a more favourable combined response of the substrate and the surface layer under a wear type of loading. The performance of the coated systems can be significantly improved by duplex treatment [14], which combines two or more surface technologies to produce a composite surface on the component with improved loadability and durability, not achievable by the individual processes alone. This improvement indirectly leads to a considerable cost reduction and increases the productivity of the manufacturing technologies applying tools with duplex-treated surfaces [15].

Plasma nitriding is an advanced and versatile tool for the thermochemical treatment of steel, widely used in the industry. A duplex treatment consisting of plasma nitriding the substrate followed by ceramic coating deposition represents an effective solution to improve the wear resistance, hardness, and durability of a coating/steel substrate material system [16,17]. The diffusion of nascent nitrogen into the surface layer results in a nitrided layer consisting of two structural zones. The compound layer consists of $\varepsilon(Fe_{2-3}N)$ and $\gamma'(Fe_4N)$ nitrides formed by the alloying elements with the basic constituents, while the diffusion layer contains interstitially solute N atoms along with fine and coherent precipitates of nitrides, where the solubility limit is exceeded. The nitrided layer provides several advantages, such as increased surface hardness, better corrosion and fatigue resistance, or reduced wear and friction [18,19], which is utilised as a sublayer in duplex-treated coatings.

The base material investigated in the current research is a high-alloy plastic mould tool steel designed to cater to the requirements of die casting and plastic injection moulding dies in polymer processing. It requires good machinability, high strength, toughness, corrosion resistance, high wear resistance, and good photoetching characteristics. The increasing demand for this steel in the automobile, mechanical, and structural component industries triggered intensive research and development of these steels. The material properties desired in these applications can be efficiently improved by plasma nitriding; however, the material's response to nitriding depends on several parameters, such as substrate composition, treatment time, temperature, and nitrogen potential.

Alloying elements such as Al, Ti, Cr, V, and Mo can interact with nitrogen to form nitrides, which affect surface hardening. The investigated X42Cr13 steel has a high chromium content. It is known that a Cr content below 5.6 wt% has an intermediate effect on N diffusion during nitriding. At the same time, above this value, a strong interaction between Cr and N occurs [20], resulting in the formation of CrN precipitates that limit the nitrogen diffusion into the surface layer and significantly reduce the achievable hardness, besides causing high tensile residual stresses.

This nitriding issue has been a critical obstacle to the broader-range application of such high Cr steels. Several researchers have dealt with this problem [21,22], and they suggested the reduction of the nitriding temperature to below 450 °C [23] and applying a long-term nitriding [24] to avoid overaging during nitriding of the precipitation-hardened steels of high hardness (470–550 HV) obtained by a tempering temperature below 500 °C. In contrast, when lower hardness/higher toughness is the purpose, realised by a tempering temperature greater than 500 °C, better wear resistance can be reached by applying a nitriding temperature above 520 °C and shorter (~4 h) holding times. In this case, there is no risk of overaging [20], but other undesirable consequences of the treatment—delayed cracking [25], increased sensitivity to corrosion [24], steep hardness profile, or lower case-depth—may occur. Studies on this nitriding problem are available mainly for the steels of AISI 420 type, mainly focusing on the corrosion resistance of low-temperature nitrided martensitic stainless steels [26–30]. A minimal amount of studies on wear resistance of high Cr steel treated by high-temperature nitriding [27,28] is found, and to the best of the authors' knowledge, no results on the nitriding problem and wear resistance of the X42Cr13 steel have been reported up to now.

Adhesion of the coating to the substrate plays a vital role in the successful application and durability. There are several techniques to evaluate the adhesion strength of the coating. The scratch test is one of the most reliable and efficient standardised methods for determining the L_C critical load, causing the delamination of the coating from the substrate [31]. This critical load is a function of several factors, such as the mechanical properties of the coating and the substrate, the adhesion strength between them, the coating thickness, the internal stresses in the coating or the subsurface region, or the flaw size distribution at the substrate–coating interface [32]. Furthermore, the measured critical load depends on the loading rate, scratching speed, load gradient, stylus tip radius, or friction between the stylus and the coating [33]. However, the critical load is constant if the load gradient is constant [34], regardless of the loading rate and scratching speed, which allows for a comparison of scratch test results carried out with different test parameters but identical load gradients. In coated systems, a decisive role is played by the composite hardness of the coated material system influenced by the hardness of both the coating and the substrate, as well as the hardness gradient between them. The critical force derived from the scratch test can also be correlated directly with this composite hardness [35,36].

Wear is a damaging process involving progressive material loss that takes place due to the interaction of surfaces being in relative motion. During wear, within the area of contact, several physical and chemical elementary processes between the sliding pairs occur, leading to changes in the material structure and the shape of the frictional counterparts. Under a given loading, usually, more wear mechanisms operate simultaneously, and the damage process is controlled by the dominant one, which is possible to identify by analysing the wear track morphology using optical or electron microscopy. For quantitative characterisation of the wear resistance, several wear tests are available. The ball-on-disk method is one of the most frequently used standardised test procedures [37]. Similar to the case of scratching, the wear damage process, the controlling damage mechanisms, and the wear resistance of the tribosystem are also influenced by several extrinsic and intrinsic factors [38–41].

The current research focuses on the effect of the applied duplex treatment on the tribological performance of CrN coatings deposited on the investigated high Cr content tool steel (X42Cr13).

Tribological analyses—either wear or scratch tests—accomplished on tribosystems constituted by the combination of the studied substrate and coating materials have not been reported in the professional literature. Some tribological studies on coated systems with AISI 420, that is a similar substrate material, are available [29,30,42,43]. However, AISI 420 and X42Cr13 have compositional differences in terms of C and other alloying elements, such as V, Mo, Ni, Cu, and Al, which influence both the results of the surface modification (nitriding) technology and the tribological behaviour of the coated system. It should also be noted here that the tribological results for AISI 420 steel found in the literature refer to AISI 420 steel grades with a very diverse chemical composition in terms of the listed constituents, so in some cases, they are difficult to compare with each other and with the results presented in this research.

This study on the influence of the duplex treatment on the scratch and wear resistance of the applied CrN coating deposited on a high Cr steel substrate is a novel contribution to the topic and presents the first-ever results on trybosystems with high Cr, secondary-hardened, martensitic stainless-steel substrates, therefore may attract the attention of tribologists and material scientists.

2. Materials and Methods

Test samples were made of X42Cr13 plastic mould tool steel of the following chemical composition (wt%): C = 0.40, Si = 0.42, Mn = 0.38, P = 0.021, S = 0.002, Cr = 13.1, V = 0.03, Ni = 0.17, Cu = 0.12, Al = 0.019, and remaining is Fe. Disc-shaped samples of

$\varphi 30$ mm × 10 mm were cut from a commercially available annealed steel road and sectioned using a universal lathe machine. It was followed by grinding to refine the dimensions and geometric shape, for example, the parallelism of the surfaces. Then samples were sonicated with acetone to remove dirt and grease; then holes of $\varphi 3$ mm in diameter were created by drilling to hang the samples during nitriding. Finally, samples were given an identification mark.

Samples were subjected to bulk heat treatment of secondary hardening, consisting of austenitising and oil quenching, followed by a high-temperature tempering. Austenitisation was accomplished at a temperature of $T_{aust} = 1020\ °C$ for a time of $t_{aust} = 20$ min, and double tempering at $T_{temp} = 580\ °C$ for the duration of $t_{temp} = 2$ h. After bulk heat treatment, the samples were divided into two categories. One set of the samples, denoted by (S), was not nitrided, only ground and polished, then coated by a hard CrN layer deposited by PVD. During grinding, SiC papers with a grit size of P400, P1000, and P1500 were used, and for polishing, Al_2O_3 suspension of a grain size of 0.3 µm was applied. The other set of samples, denoted by (D) and called duplex treated, was nitrided in a plasma nitriding equipment (type Nitrion 10, from SC PLASMATERM SA, Targu Mures, Romania) at a temperature of $T_{nitr} = 520\ °C$, for a holding time of $t_{nitr} = 8$ h. The primary voltage was 600 V, and the pressure was 200 Pa. The source of nitrogen was decomposed ammonia ($N_2:H_2 = 1:3$). Thus, the S (simply coated) samples are secondary-hardened + PVD coated specimens, while the D (duplex-coated) samples are secondary-hardened + nitrided + PVD coated test materials.

CrN coating is deposited using PVD magnetron sputtering, assisted by the microwave technique. Microwave increases the quantity and the kinetic energy of Ar ions. The equipment used was TSD 550 (from HEF Durferrit M&S SAS, Andrézieux-Bouthéon, France). The steps involved are as follows: First, a Cr adhesion layer was deposited using a Cr target (99.99%) in Ar (60 cm^3/min). The nitrogen gas flow was controlled by light measurement (luminosity), a chamber working pressure of 0.3–0.5 Pa using a pulsed DC sputtering power source, held at 6 kW. A bias voltage of 100 V was applied to the substrates. The thickness of the CrN layers was controlled by adjusting the deposition time. The substrate temperature was 400 K, resulting from the bombarding species only. The deposition rate for the CrN films was 1.5 µm/h.

The elemental distribution in the near-surface layer was defined by glow discharge optical emission spectroscopy (GDOES) using the equipment of GD-Profiler 2 (from HORIBA Jobin Yvon, Montpellier Cedex 4, France). Test parameters were as follows: RF generator, flushing time: 5 s, preintegration time: 60 s, depth: 66.966 µm, pressure: 500 Pa, power: 25 W, module: 6 V, phases: 5 V.

The hardness of the samples with different treatment conditions and the composite hardness of the coated systems, furthermore the near-surface hardness profile of the nitrided layers, were determined by the standard micro-Vickers test (with Mitutoyo HVK-H1 hardness tester from Mitutoyo Corp., Kanagawa, Japan) with normal loads of $F = 1$ N and 0.5 N. The thickness of the coatings was defined by the ball cratering method using a Compact CAT^{2}c Calotester from Anton Paar GmbH, Graz, Austria [44]). The tests were performed with a $\varphi 30$ mm quenched steel ball and a rotational speed of n = 3000 1/min for t = 3 min.

Optical microscopy (OM) using a Zeiss Axio Observer D1m type inverse microscope from Zeiss, Jena, Germany, and scanning electron microscopy (SEM) using a Zeiss Evo MA10 machine equipped with energy dispersive X-ray spectroscopy (Zeiss, Jena, Germany) was used to reveal the microstructure of the samples of different—secondary-hardened, nitrided, and coated—conditions. Secondary electron (SE) and backscattered electron (BSE) imaging during the SEM were used for detailed morphological analyses of the scratched and worn surfaces and to visualise the microstructural features of the nitrided surface.

For evaluating the adherence of the coatings, an instrumented scratch tester (type SP15, from Sunplant, Miskolc, Hungary [45]) was used by applying a progressive loading scratch test, widely used for rapid assessment and quality assurance of surface coatings [31].

During this test, a standard HRC diamond cone, as a stylus, is moved over the coated surface with a linearly increasing load, which varied from 2 to 150 N in the current study. The velocity of the sample holder table, moving below the stylus, was 5 mm/min, and the gradient of the normal force was 10 N/mm. The critical force at which the coating failure, that is, delamination, occurs can be found from the friction coefficient-loading force versus the scratching distance diagram recorded by the test machine. Failure modes (damage mechanisms), occurring at and below the critical forces, were identified by morphological analyses of the scratch grooves using OM and SEM.

The wear resistance of the coatings was investigated using a ball-on-disk test (equipment: UNMT-1, manufacturer: Center for Tribology, Inc., CETR, Campbell, CA, USA). The friction counterpart was a SiC ball of $\varphi 6$ mm in diameter, the normal load was 20 N, the total sliding distance was 108 m, the radius of the wear track was 3 mm, and the sliding rate was 30 mm/s. The integrated software program recorded the values of the coefficient of friction as a function of sliding distance. The specific wear rate (k) was calculated according to Archard's equation [46]:

$$k = (A \times C)/(F \times L_t), \qquad (1)$$

where A (mm^2) is the worn cross section obtained from profilometry (with Altisurf 520 surface tester using a confocal head from Altimet, Thonon-les-Bains, France), F (N) is the normal load, L_t (m) is the total sliding distance, and C (mm) is the length of the wear track centreline.

During the hardness test, the scratch tests, the coating thickness measurements, and in the case of determining the hardness profiles of the nitrided layer, three valid measurements on two samples of both the simply and duplex-coated test groups were accomplished. Wear tests with the given test condition were performed on two samples accomplishing two measurements per sample. During profilometry, the wear tracks were scanned along four mutually perpendicular directions of the circumferential worn track. The average of these four data for each worn track is represented by A_t, while the A value is the average of the A_t values obtained on two worn tracks created on one sample. The ball cratering thickness tests were performed three times on two samples.

3. Results and Discussion

3.1. Microstructural and Compositional Characterisation and Hardness

After the bulk heat treatment, the secondary-hardened microstructure consists of a martensitic matrix containing high Cr content precipitates, which are usually $Cr_{23}C_6$- and $(Cr, Fe)_{23}C_6$-type carbides [47]. The dimensional stability is provided by the low amount of residual austenite, characteristic of the given microstructure in the case of the applied tempering temperature [48]. A SEM image of the microstructure combined with EDX compositional analyses in two spots is shown in Figure 1.

Spot 1 is indicative of the tempered martensitic matrix, with an amount of alloying elements corresponding to the composition of the raw material. Spot 2 shows a Cr-rich precipitation with higher C content, suggested by the C peak appearing in the related EDX spectrum that is not observed in Spot 1. Compared with the maximum (540 HV) hardness achievable for this steel, the applied bulk heat treatment results in a lower—HV_{PH} = 290 $HV_{0.1}$ ± 4.7—substrate hardness, providing higher toughness and better dimensional stability. It is applied primarily in structural components, as well as in the case of tools, where good toughness is more important than high hardness. In the current study, the bulk heat treatment parameters were chosen to avoid the significant hardness reduction, dimensional inhomogeneity, and high residual stresses that are possible to occur during subsequent nitriding.

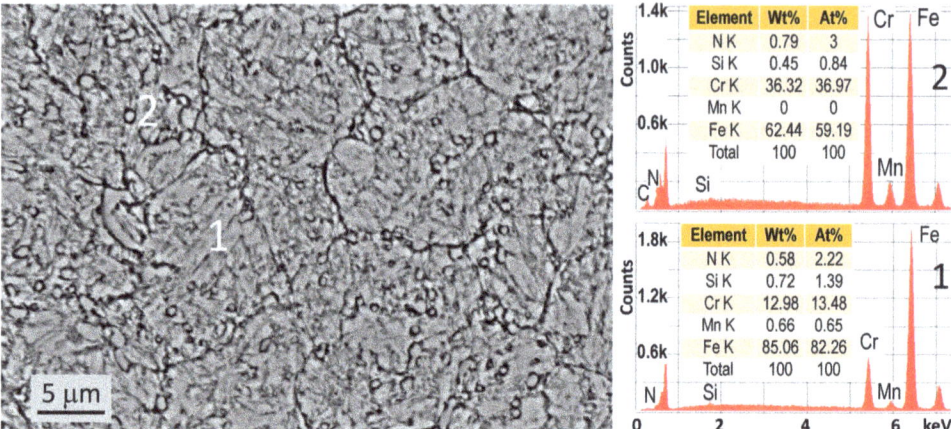

Figure 1. The SEM image of the secondary-hardened (quenched and highly tempered) substrate along with the EDX spectra in the indicated two spots (etchant: Nital 5%).

A SEM cross-sectional image and the hardness profile of the nitrided layer (520 °C, 8 h) are shown in Figure 2. The microstructure of the nitrided layer contains a large number of precipitates along the prior austenite grain boundaries and in the bulk, which is characteristic of high Cr steels. The most common type of these precipitates is CrN, hindering the N diffusion strongly and resulting in a flat hardness profile shown in Figure 2, and may cause high tensile stresses [26,49]. The surface hardness measured after removing the white layer—and before the coating process—was 360 $HV_{0.05}$ ± 37, which dropped sharply below the surface towards the core. The nitrided layer thickness, $h_N = 40 \pm 5.6$ µm, was small.

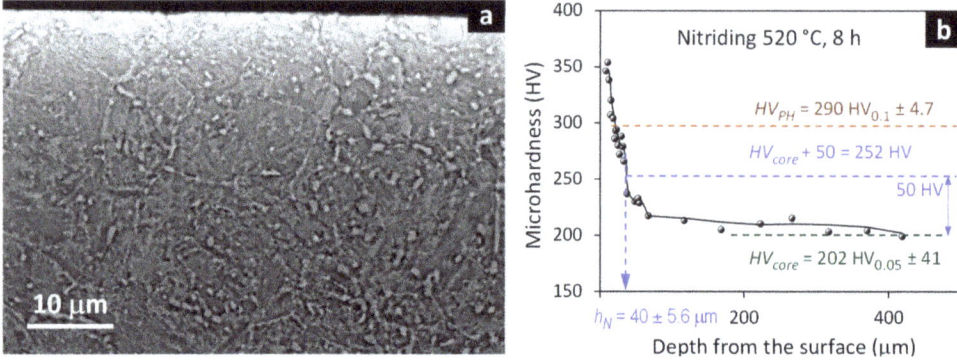

Figure 2. SEM image of the cross section of the nitrided surface layer in the quenched and highly tempered (secondary-hardened) substrate (**a**); defining the nitrided layer depth (h_N) based on the hardness profile (**b**).

The nitrogen and chromium distribution obtained by GDOES in the near-surface layer (Figure 3) explains the hardness profile shown in Figure 2.

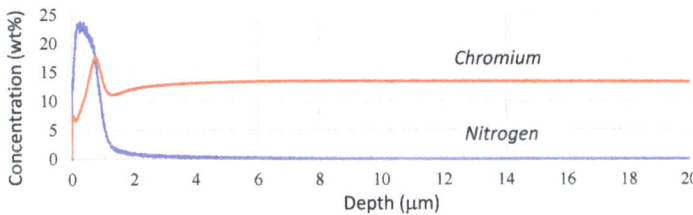

Figure 3. Distribution of the nitrogen and chromium in the nitrided sample (obtained by GDOES).

The nitrogen content decreases sharply in the 0.5–1 μm depth region and approaches 0% at the deeper locations. It is reasoned by the intermediate nitriding temperature, resulting in a low penetration depth of the nitrogen atoms responsible for the moderate surface hardness after nitriding. Obtaining a thicker and harder layer would require an increase in the nitriding temperature and holding time. The Cr peak, observed at a 0.8 μm depth, appears supposedly due to the strong gradient of the N distribution in this region. Since there is no significant amount of N at higher depths that would attract the Cr atoms, Cr is accumulated in a very thin, high-N content subsurface region.

SEM images of the coatings are seen in Figure 4. The coating thicknesses on the simple and duplex-treated samples were 3.8 ± 0.23 and 4.0 ± 0.27 μm, respectively.

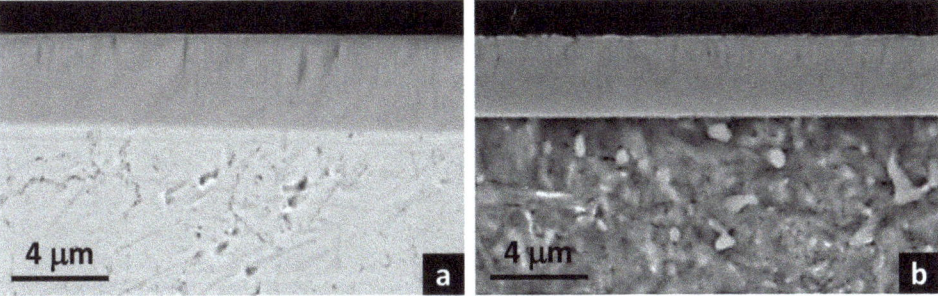

Figure 4. SEM image of the CrN coatings on the simply coated (**a**) and duplex-treated (**b**) samples.

The composite hardness of the coated systems was 632 ± 61 $HV_{0.5}$ for the simply coated and 1250 ± 204 $HV_{0.5}$ for the duplex-treated samples.

3.2. Scratch Resistance of the Simply and Duplex-Coated Systems

During the progressive loading scratch test, the friction coefficient/loading force vs. scratch length curves for the simply and duplex-coated specimens was recorded. Scratch diagrams, characteristic of both coating systems, are illustrated in Figure 5.

The L_{C3} critical force causing delamination is easy and quick to read directly from the scratch diagrams at the location of the abrupt change in the slope of the friction coefficient curve. These values are denoted by $L_{C3(S)}$ and $L_{C3(D)}$ for the simply and duplex-coated samples, respectively, and indicate the better adhesion strength of the CrN coating for the duplex-treated system.

On the contrary, identification of the subcritical loading forces—i.e., L_{C1} and L_{C2} indicated in the diagram—that initiate distinct characteristic damage mechanisms at the lower loading regions requires morphological analyses using optical and scanning electron microscopy.

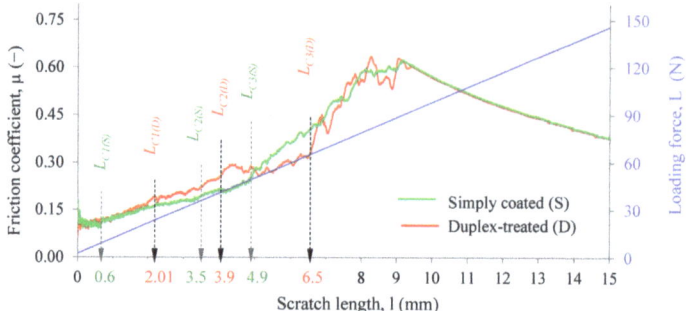

Figure 5. Friction coefficient/loading force vs. scratch length diagrams for simply coated (S) and duplex-treated (D) samples.

Figure 6 illustrates the damage mechanisms for the simply coated and duplex-treated samples at these characteristic subcritical and critical loadings.

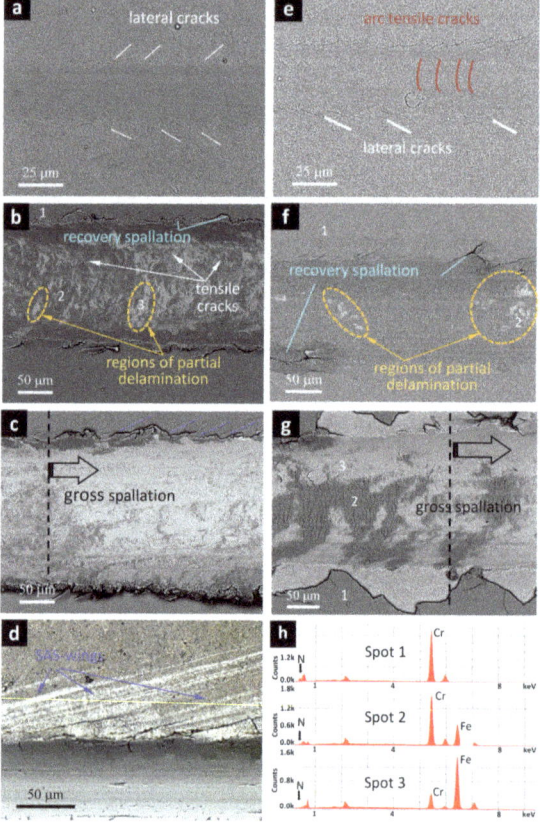

Figure 6. (**a**–**c**) SEM images of the characteristic damage mechanisms at the L_{C1}, L_{C2}, and L_{C3} loadings for the simply coated sample, respectively; (**d**) magnified OM image of the SAS-wing region; (**e**–**g**) SEM images of the damage mechanisms at the L_{C1}, L_{C2}, and L_{C3} loadings, respectively, for the duplex-coated specimen; (**h**) EDS spectra in some characteristic points of the scratched and intact regions, denoted by 1, 2, and 3 in the (**b**,**f**,**g**) insets of the figure.

For the simply coated sample, the initial cracking appears in the form of *lateral cracking* at $l_{1(S)} = 0.6$ mm scratch length (Figure 6a) at a subcritical load of $L_{C1(S)} = 9$ N. As the force increases, *recovery spallation* can be observed on both sides of the groove (Figure 6b) at $l_{2(S)} = 3.5$ mm scratch length, belonging to the $L_{C2(S)} = 36.6$ N subcritical load and the friction coefficient of $\mu = 0.18$. The *arc tensile cracks* become more expressed, and *partial delamination* also begins in this region. At the critical force of $L_{C3(S)} = 49.7$ N, ($l_{3(S)} = 4.9$ mm, $\mu = 0.23$), the coating starts to delaminate by *gross spallation* (Figure 6c), a damage mechanism, characteristic of coatings with low adhesion strength. A continuous increase in the corresponding friction coefficient curve, even at subcritical loads, can also be observed (Figure 5, red curve) because small particles are torn off the coating and agglomerate in the vicinity of the stylus. The debris particles create an additional obstacle to the stylus' motion, thereby increasing the μ value. In Figure 6d, presenting a magnified OM image of the region indicated with blue dashed lines in Figure 6c, a new type of damage mechanism is identified, named "*SAS-wings*" by the authors. It is supposed that a special plastic deformation mechanism, that is, grain boundary sliding of the columnar grains, characteristic of CrN coatings at high compressional stresses [50], occurs due to the plastic shear deformation of the underlying soft substrate. Thus, "SAS-wings" may be considered a forecast of the critical condition belonging to the exhaustion of the load-bearing capacity of the substrate. SAS-wings were observed exclusively in the simple CrN coating and were not seen in duplex versions at the same loading level. The visibility of the SAS-wings is characteristically different in OM and SEM images. They appear as a set of white parallel lines on OM pictures, while they look like barely visible, dark shadow lines on SEM images.

For the duplex-treated sample, the damage mechanisms are similar to those found in the case of the simply coated sample, but the subcritical and critical forces at which the coating starts to fail, or delaminate from the substrate, are significantly higher. At the subcritical force of $L_{C1(D)} = 22.5$ N, which is 2.5 times greater than that of the simply coated sample, the initial cohesive failure of the coating appears in the form of *lateral cracking* and *arc tensile cracking* (Figure 6e). At $L_{C2(D)} = 40.8$ N, the adhesive failure starts in the form of *recovery spallation*, followed by partial delamination. Additionally, a lateral and very dense tensile crack formation is continued in this region (Figure 5f). At the critical load of $L_{C3(D)} = 66.5$ N, gross spallation starts, like in the case of the simply coated sample (Figure 5g). The 34% higher critical load suggests a better adhesion and scratch resistance of the duplex-coated sample compared with the simply coated one.

In the numbered locations, as shown in the Figure 6b,f,g insets, we made EDS spot analyses and observed the following characteristic compositional differences. In the spot denoted by "1", representing an intact region of the coating, we found a high amount of Cr and N content at a ratio of about 1:4, which corresponds to the composition of the CrN chemical compound. In Spot "2", similarly, a high amount of Cr and N was seen, combined with Fe, alluding to the presence of a mixture of the coating and the substrate material. In contrast, In Spot "3", a very high amount of Fe and a significantly lower Cr and N content were obtained, corresponding to the composition of the substrate material due to the delamination of the coating.

Thus, nitriding the substrate prior to coating deposition proved to be an effective method to increase the critical loads, resulting in improved scratch resistance of the investigated CrN coating systems.

Here, it should be noted that scratch tests results for duplex tribosystems with CrN coating are extensively reported [51–55] and demonstrated that duplex treatment might result in a considerable improvement of the critical load. However, no publication was found by the authors for tribosystems with substrate material investigated in the current research. It is an essential distinguishing factor since substrate material plays a crucial role in a coated system's behaviour.

The direct comparison of the results obtained in the current study with those available in the professional literature is also difficult due to the differences in scratch test parameters, tribosystems, variations in the coating technology, and treatments of the substrate. Nevertheless, some further remarks may be useful.

A significant consideration in scratch tests is that the critical load depends on various factors, such as, the load gradient mentioned in the Introduction section. Scratch tests with the same tribosystems, applying different load gradients—20 N/mm for the simple and 40 N/mm for the duplex coating and a slightly higher scratching speed of 6 mm/min—have been carried out by the authors in the framework of a round-robin test [56]. The results revealed a 53% improvement in the critical force in the case of the duplex-treated system, which is attributed partially to the different and considerably higher load gradients [31,32]. The 34% improvement obtained at a significantly lower (10 mm/min) load gradient in this study proves unambiguously the favourable and remarkable effect of the duplex treatment on the scratch resistance and the role of the substrate's loadability in the performance of the composite system formed by the substrate and coating together. Additionally, some conclusions can be derived regarding the non-negligible effect of the load gradient. Considering the simply coated case, the load gradient in the referred work was twice higher, resulting in a 7% higher critical load, while for the duplex-coated system, the load gradient was four times higher, leading to a critical load higher by 22%. It draws attention to the importance of identical test conditions when comparing the results.

During scratch analyses, several damage modes—depending on the substrate/coating combination—can be recognised using OM analysis of the scratch groove. An efficient guide in this respect is represented by the ASTM C1624 standard [31], which notes that in the case of new types of coatings or substrate/material combinations, novel damage mechanisms are expected to be identified; therefore, describing them as detailed as possible and their systematisation together with the previously known ones may be helpful for researchers. Identifying the SAS-wing-type scratch damage mechanism can contribute to a better understanding and a more efficient description of the scratching behaviour of the given coated system.

3.3. Wear Behaviour of the Simply and Duplex-Coated Tribosystems

The wear resistance of the coatings was compared based on the friction coefficient curves, compositional and phase analysis of the worn surface, wear track morphology, cross-sectional profiles of the wear tracks, and specific wear rate, k.

For the simply coated sample, the running-in period of the friction coefficient vs. the sliding distance curve is somewhat longer with a higher peak of μ, suggesting more intensive damage of the coating on the softer and less stiff substrate compared with that of the duplex-treated sample (Figure 7). As the ball contacts the sample, adhesion occurs at the contact area, leading to sticking and increasing the frictional force, consequently, μ. This well-known transient behaviour is responsible for the first peaks at the very beginning of both curves; however, it does not explain the peaks appearing at 10 and 5 m for the simply and duplex-coated systems, respectively. The harsh acoustic noise experienced during the whole wear test indicated that the load-bearing capacity of the CrN/substrate systems had already been exceeded at the beginning of the test for both types of tribosystems. With the premature loss of the integrity of the coating, debris is formed, and a three-body abrasive wear mechanism occurs in the contact zone. As a result, the friction coefficient increases to a large extent, which is more pronounced in the case of the simply coated sample, where the coating could not adapt to the larger plastic deformation of the weaker substrate. In the case of the duplex-treated tribosystem, although the coating is damaged early too, the process is presumably associated with less amount of particle detachment, so the increase in the coefficient of friction is less. With continuous debris fragmentation, the obstacle to the movement of the counterpart (ball) also decreases, and the coefficient of friction begins to decrease. This phenomenon can be observed at a sliding distance of 10–40 m in the simply coated system, while it occurs at about 5–10 m in the duplex-treated sample.

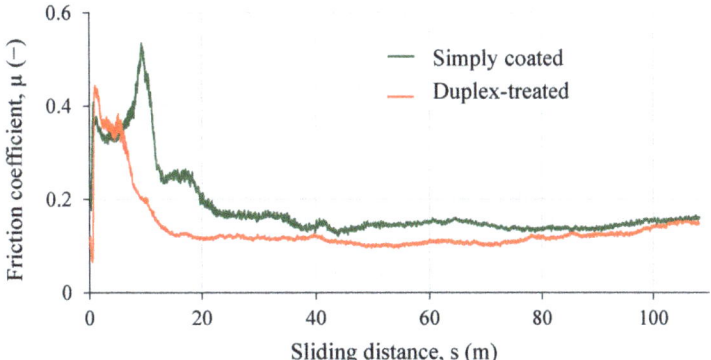

Figure 7. Friction coefficient curves obtained for the simply coated and duplex-treated samples during the ball-on-disk wear test.

Following this period, the friction coefficient curves show a stabilising character with an increasing sliding distance that suggests a tribofilm formation to occur. The process behind this is that as the coating is damaged, the steel substrate that gradually becomes exposed also participates in the wear process by forming an oxide-type—presumably Fe_3O_4—tribofilm, a characteristic of steels. With an increasing sliding distance, the amount of tribofilm increases, providing a continuous layer to be formed, which ensures a gradual decrease in the friction coefficient.

The steady-state friction coefficient is 0.16 and 0.11 for the simply and duplex-coated samples. Here, it should be mentioned that preliminary measurements with an Al_2O_3 ball showed a similar tendency of the wear behaviour of the differently treated CrN coatings. However, considerably higher friction coefficients were obtained, namely, 0.94 and 0.88, respectively, for the simply and duplex-coated systems.

In the last quarter of the sliding distance, the thickening tribofilm becomes increasingly brittle and starts to break up. Debris particles from the tribofilm cause abrasive wear, and the friction coefficient increases again. In this stage, the two competing processes are tribofilm formation and delamination, and their ratio defines the controlling wear mechanism. Consequently, the worn surface will be partially covered by massive tribofilm and tribofilm debris and contain exposed steel substrate areas. Similar damage mechanisms are reported for CrN coatings against Si_3N_4 balls [57]. However, a strong chemical reaction between the ball material and the CrN coating occurred.

Analysing the dark regions in the OM image of the wear track, appearing on the worn surface of the simply CrN-coated specimen (Figure 8), the presence of Fe_3O_4 tribofilm was proven by laser-induced breakdown spectroscopy (LIBS) using a digital microscope (Keyence VHX-5000 series) equipped with a laser material inspection head (EA-300 series) from Keyence Corp., Osaka, Japan [58].

The compositional analysis was accomplished in two spots having different appearances inside the wear track. Spot 1 in Figure 8 is a highly worn region, while Spot 2 is an area covered by black smoke assumed to be the Fe_3O_4 tribofilm. Based on quantitative elemental analysis, the substrate material, that is, stainless steel, was identified with Cr content corresponding to the applied base material in Spot 1, while in the black area, the presence of Fe_3O_4 film was confirmed in Spot 2.

The OM pictures of the worn tracks reveal an identical wear mechanism for the two coated systems (Figure 9) controlled by tribo-oxidation of the steel substrate after the coatings have been damaged.

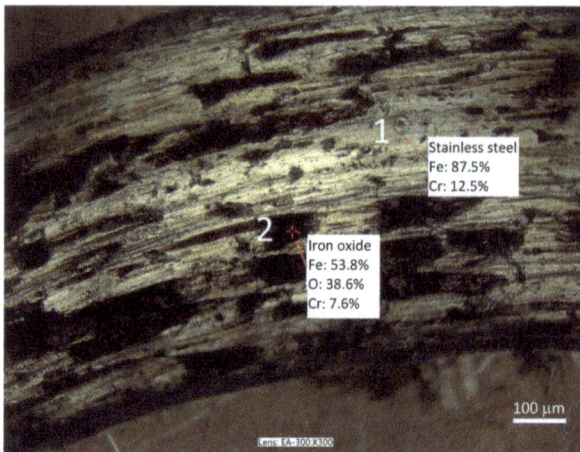

Figure 8. Compositional analysis and phase identification in two spots of the wear track of the simply CrN-coated sample. Spot 1: a highly stripped region, Spot 2: an area covered by black smoke assumed to be the Fe_3O_4 tribofilm.

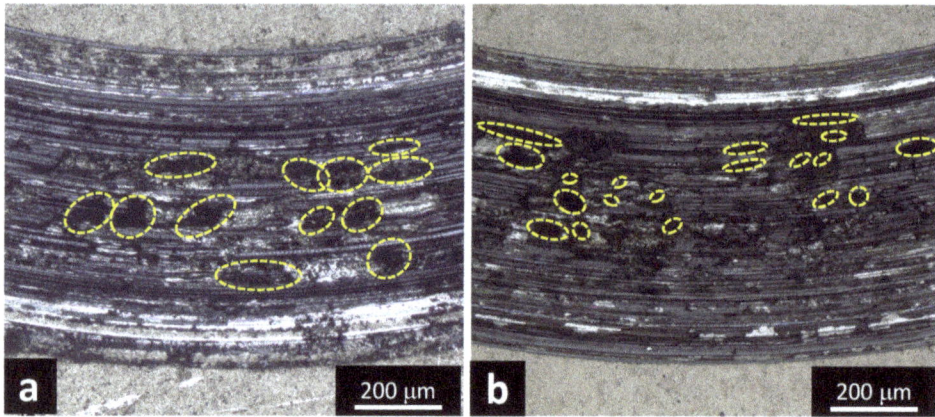

Figure 9. Morphology of the worn track on the simply (**a**) and the duplex-coated sample (**b**) with Fe_3O_4 islands on the damaged coatings.

The thinner wear track and the significantly lower amount of exposed substrate (Figure 9b) reflect the duplex-treated sample's more favourable wear behaviour. It can be attributed to the higher strength of the nitrided steel substrate, giving better support to the brittle ceramic coating and the formed tribofilm.

While the optical microscopy of the wear rings showed only a slight difference in the width of the wear tracks of the simply and duplex-treated samples, the profilometry measurements already revealed a significantly higher worn cross section. Consequently, there was a higher amount of material loss for the simply coated sample compared with the duplex-treated one (Figure 10).

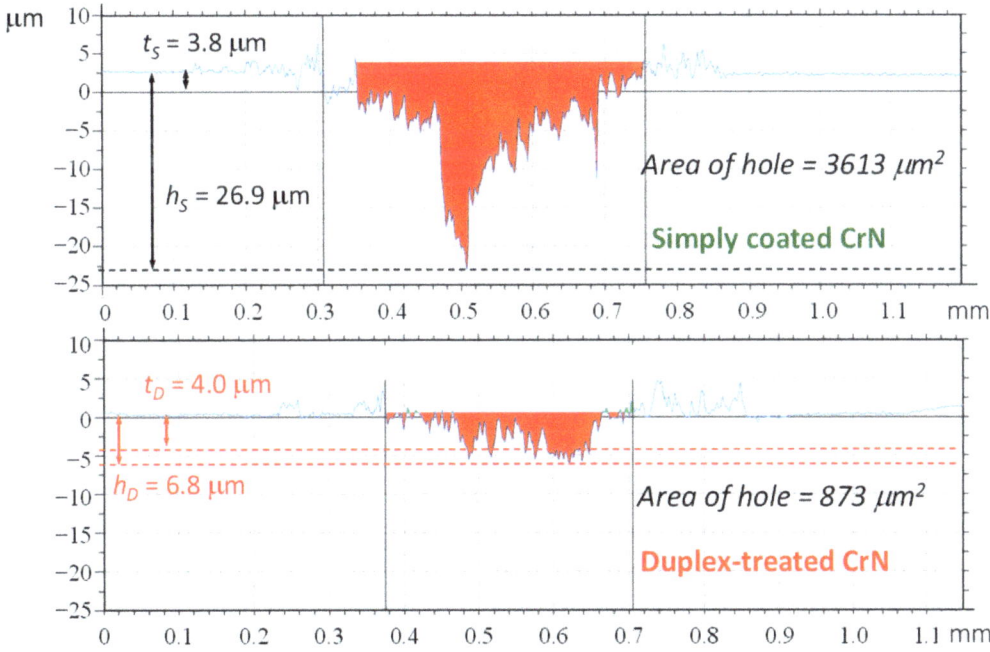

Figure 10. Worn track profiles of simply coated (S) and duplex-treated (D) samples (*t*: coating thickness; *h*: depth of wear track).

Besides, the ratio of the depth of the wear track (h) to the coating thickness (t) is approximately seven times higher for the simply coated sample, that is, h_S/t_S = 7.1, while it is h_D/t_D = 1.7 in the case of the duplex-treated coating system. This finding gives unambiguous evidence of the improvement of the wear resistance of the applied CrN coating deposited on the nitrided steel substrate.

In addition, the profile diagrams reveal that in the case of the simply coated specimen, the coating was entirely removed from the substrate, while the duplex-treated coating remained partially intact in the contact zone.

The specific wear rate values, calculated by Equation (1), were an average of (2.8 ± 0.11) × 10^{-5} mm^3/Nm and (1.6 ± 0.76) × 10^{-5} mm^3/Nm for the simply coated and duplex-treated samples, respectively. This change represents a 43% reduction in the wear rate due to the duplex treatment, as shown in Figure 11.

The applied duplex surface treatment resulted in a significant improvement of the wear resistance of the investigated CrN coatings in terms of the reduction of the wear rate.

Formerly, we established that nitriding the substrate preceding the coating deposition does not alter the wear mechanism of the duplex-coated system compared with that of the simply coated one for the investigated loading conditions (Figure 9). The measured wear rate values, which fall into the same order of magnitude (10^{-5}), are in harmony with this observation.

Usually, a significant change in the wear rate can be expected if the wear mechanism is modified. Here, the observed reduction of the wear rate cannot be attributed to such kind of phenomenon. The duplex treatment led to the observed favourable wear behaviour by altering the substrate microstructure, improving its loadability, and providing better support for the hard and brittle coating, significantly increasing the composite hardness and the joint performance of the substrate/coating composite system.

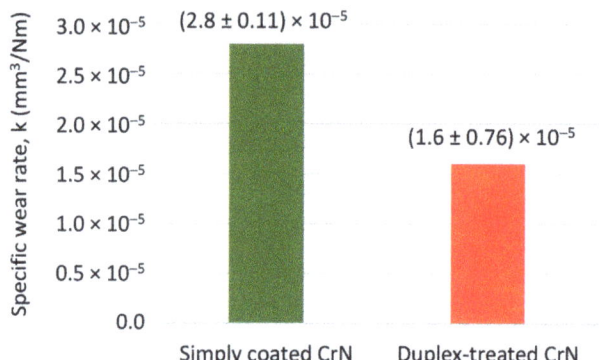

Figure 11. The specific wear rate for the simply and duplex-treated samples.

Here, it should also be noted that the applied F = 20 N normal load already causes failure of the CrN coating at the very early stage of the wear test. Research works on CrN coating reflect that the applied loading conditions represent a severe loading regime for this tribosystem [59,60].

The measured wear rate values, especially those obtained for the duplex-treated sample, fall close to the values of 10^{-6} mm^3/Nm representing the border between the mild and severe wear regimes. The enhancement of the substrate loadability through optimising the duplex treatment—particularly regarding the high-temperature nitriding—would provide a breakthrough to this border, changing the wear regime to mild wear.

Similarly, as mentioned in connection to scratch studies available in the literature, several reports on wear tests for CrN coatings can also be found [61–64]. However, the possibility for comparison with the current results is very limited due to the different substrate characteristics and differences in the applied test parameters that play a vital role in the wear mechanisms and wear resistance of the tribosystem.

4. Conclusions

The tribological performance of CrN coatings deposited by the PVD technique on a high Cr content X42Cr13 tool steel was analysed by comparing two different conditions of the substrate/coating system. On the one hand, the coating was deposited by PVD directly to the quenched and highly tempered (secondary-hardened) substrate. On the other hand, duplex treatment, consisting of nitriding and a subsequent CrN coating of the substrate material, was applied. Tribological behaviour was studied in scratch and wear-type loadings. The main findings of the research work can be summarised as follows.

The investigated high Cr content steel represents a key issue regarding the appropriate combination of nitriding parameters during duplex treatment. The applied high-temperature nitriding technology—characteristic of this steel in case of maximum toughness requirements—is less investigated and reported in the literature. The related problems were demonstrated during structural and compositional analyses of the obtained nitrided layer, demonstrating a steep hardness profile, low nitrided case depth, and characteristic high residual stresses due to the intense CrN precipitation. Despite these problems, the applied duplex treatment resulted in a considerable improvement of the tribological performance reflected by the 34% higher critical loads during scratching, that is, higher resistance to coating delamination and a 43% improvement of the wear resistance in terms of the calculated wear rate.

During morphological analyses of the scratch grooves, a new damage mechanism named "SAS-wings" was identified, which is a characteristic of the simple CrN tribosystems and a preliminary indicator of the critical failure of the ceramic coating not observed for the duplex-treated systems.

The research work confirmed that the duplex treatment is an effective and advantageous solution to improve the tribological behaviour of the tested CrN coating by increasing the load-bearing capacity of the applied X42Cr13 steel substrate. It allows for a significant decrease in material loss due to wearing, thereby increasing the lifetime of the tool.

One of the important application areas of X42Cr13 steel is the plastic mould tool industry. Due to dynamic loadings, the lifetime during plastic injection moulding or extrusion is often controlled by the toughness of the tool, and it is limited significantly by the abrasive wear caused by reinforcing particles, such as glass and carbon fibres of the processed plastic materials. The use of duplex technology in the case of high-toughness, low-hardness tool steels is significant because the hard CrN coating, applied for high wear resistance, can lead to a significant hardness gradient occurring at the interface between the substrate and the coating, limiting this way considerably the achievable improvement of the wear resistance. The duplex treatment may essentially reduce this hardness gradient and the accompanying weaker performance.

X42Cr13 steels are often replaced by prehardened steels—e.g., X30Cr13 (DIN) equivalent with SUS420J2 (JIS), AISI 420, or 420S45 (BS)—the heat treatment and the corresponding hardness of which are very similar to those of the tested steel. However, these types of steels are marketed in large blocks; therefore, manufacturing test specimens required for the wear test is rather complex, labour intensive, and expensive. The wear tests performed can also be considered model tests for these steels, and the obtained results can be effectively used for the wear problem solutions of such prehardened steels too.

Another crucial area of utilisation is represented by improving the mould release capability of the plastic forming tools and reducing the tool opening force. Using the CrN coating, the interfacial energy can be significantly reduced; consequently, the tool opening force can be reduced. This green technology solution can take the place of using release agents, which both pollute the environment and contaminate the finished product and the tool.

The presented tribological investigations with the novel combination of constituents applied in the tribosystem open many aspects where a significant study and further research are proposed. These novel findings provide a reasonable basis for future work to elaborate an optimised set of technological parameters for the duplex technology of the given tribosystem. Further improvements in the efficiency of the scratch and wear resistance can be achieved if a more diffused hardness profile with a higher case depth is obtained by the optimised parameters of the high-temperature nitriding.

A new series of wear tests with a decreasing normal load is suggested to determine the threshold of the load-bearing capacity of the investigated tribosystem. Such a programme would also allow for a more thorough examination of the wear resistance by avoiding the premature failure of the CrN coating observed in the current study. Reducing the normal load during the wear test allows for reaching the transition from severe to mild wear regime. This change in wear behaviour may be promoted by elaborating the appropriate high-temperature nitriding technology, less studied, reported, and applied currently, but promising regarding improving the wear resistance of the analysed tribosystem.

Author Contributions: Conceptualization, methodology, M.B.M.; test execution, data processing, S.A.S.; writing—original draft preparation, S.A.S.; writing—review and editing, M.B.M.; investigation, data evaluation and analysis, S.A.S. and M.B.M.; visualisation: S.A.S. and M.B.M.; supervision: M.B.M. All authors have read and agreed to the published version of the manuscript.

Funding: This research received no external funding.

Institutional Review Board Statement: Not applicable.

Informed Consent Statement: Not applicable.

Data Availability Statement: The data presented in this study are available on request from the corresponding author.

Acknowledgments: The authors thank the support of the Stipendium Hungaricum scholarship by the Tempus foundation, Hungary. The authors express their gratitude to Attila Széll and Gyula Juhász at T.S. Hungary; Andrea Szilágyiné Biró at Oerlikon Balzers GmbH Kft., Székesfehérvár, Hungary; and Norbert Luczai at Voestalpine High-Performance Metals Hungary Kft. for their assistance in heat treatment, coating, and valuable professional consultations. Thanks also go to the colleagues at the University of Miskolc, namely, Csaba Felhő for the profilometry analyses, Tibor Kulcsár for the GDOS tests, and Árpád Kovács for the SEM and EDX measurements, as well as to Zsombor Hegyi (Keyence International (Belgium) NV/SA) for the LIBS tests. The authors thank the Institute of Materials Science and Technology of UM for the financial support covering the costs of the mechanical, tribological, and microstructural testing.

Conflicts of Interest: The authors declare no conflict of interest.

References

1. Galindo, R.E.; van Veen, A.; Schut, H.; Janssen, G.; Hoy, R.; de Hosson, J. Adhesion behaviour of CrNx coatings on pre-treated metal substrates studied in situ by PBA and ESEM after annealing. *Surf. Coat. Technol.* **2005**, *199*, 57–65. [CrossRef]
2. Choi, E.Y.; Kang, M.C.; Kwon, D.H.; Shin, D.W.; Kim, K.H. Comparative studies on microstructure and mechanical properties of CrN, Cr–C–N and Cr–Mo–N coatings. *J. Mater. Process. Technol.* **2007**, *187–188*, 566–570. [CrossRef]
3. Dobrzański, L.; Lukaszkowicz, K. Erosion resistance and tribological properties of coatings deposited by reactive magnetron sputtering method onto the brass substrate. *J. Mater. Process. Technol.* **2004**, *157–158*, 317–323. [CrossRef]
4. Gåhlin, R.; Bromark, M.; Hedenqvist, P.; Hogmark, S.; Håkansson, G. Properties of TiN and CrN coatings deposited at low temperature using reactive arc-evaporation. *Surf. Coat. Technol.* **1995**, *76–77*, 174–180. [CrossRef]
5. Aouadi, S.; Schultze, D.; Rohde, S.; Wong, K.-C.; Mitchell, K. Growth and characterization of Cr_2N/CrN multilayer coatings. *Surf. Coat. Technol.* **2001**, *140*, 269–277. [CrossRef]
6. Warcholinski, B.; Gilewicz, A. Tribological properties of CrNx coatings. *J. Achiev. Mater. Manuf. Eng.* **2009**, *37*, 498–504.
7. Aouadi, K.; Tlili, B.; Nouveau, C.; Besnard, A.; Chafra, M.; Souli, R. Influence of Substrate Bias Voltage on Corrosion and Wear Behavior of Physical Vapor Deposition CrN Coatings. *J. Mater. Eng. Perform.* **2019**, *28*, 2881–2891. [CrossRef]
8. Tillmann, W.; Stangier, D.; Dias, N.F.L.; Gelinski, N.; Stanko, M.; Stommel, M.; Krebs, E.; Biermann, D. Reduction of Ejection Forces in Injection Molding by Applying Mechanically Post-Treated CrN and CrAlN PVD Films. *J. Manuf. Mater. Process.* **2019**, *3*, 88. [CrossRef]
9. Navinšek, B.; Panjan, P. Novel applications of CrN (PVD) coatings deposited at 200 °C. *Surf. Coat. Technol.* **1995**, *74–75*, 919–926. [CrossRef]
10. Huang, Z.; Sun, Y.; Bell, T. Friction behaviour of TiN, CrN and (TiAl)N coatings. *Wear* **1994**, *173*, 13–20. [CrossRef]
11. Warcholinski, B.; Gilewicz, A.; Ratajski, J. Cr_2N/CrN multilayer coatings for wood machining tools. *Tribol. Int.* **2011**, *44*, 1076–1082. [CrossRef]
12. Olaya, J.; Rodil, S.; Muhl, S.; Huerta, L. Influence of the energy parameter on the microstructure of chromium nitride coatings. *Surf. Coat. Technol.* **2006**, *200*, 5743–5750. [CrossRef]
13. Wieciński, P.; Smolik, J.; Garbacz, H.; Kurzydłowski, K.J. Thermal Stability and Corrosion Resistance of Cr/CrN Multilayer Coatings on Ti6Al4V Alloy. *Solid State Phenom.* **2015**, *237*, 47–53. [CrossRef]
14. Bell, T.; Dong, H.; Sun, Y. Realising the potential of duplex surface engineering. *Tribol. Int.* **1998**, *31*, 127–137. [CrossRef]
15. Vetter, J.; Barbezat, G.; Crummenauer, J.; Avissar, J. Surface treatment selections for automotive applications. *Surf. Coat. Technol.* **2005**, *200*, 1962–1968. [CrossRef]
16. Sun, Y.; Bell, T. Plasma surface engineering of low alloy steel. *Mater. Sci. Eng. A* **1991**, *140*, 419–434. [CrossRef]
17. Alsaran, A.; Karakan, M.; Çelik, A. The investigation of mechanical properties of ion-nitrided AISI 5140 low-alloy steel. *Mater. Charact.* **2002**, *48*, 323–327. [CrossRef]
18. Wen, D.-C. Plasma nitriding of plastic mold steel to increase wear- and corrosion properties. *Surf. Coat. Technol.* **2009**, *204*, 511–519. [CrossRef]
19. Sirin, S.Y.; Kaluc, E. Structural surface characterization of ion nitrided AISI 4340 steel. *Mater. Des.* **2012**, *36*, 741–747. [CrossRef]
20. Pinedo, C.E.; Monteiro, W.A. Surface hardening by plasma nitriding on high chromium alloy steel. *J. Mater. Sci. Lett.* **2001**, *20*, 147–150. [CrossRef]
21. Jasiński, J.J.; Frączek, T.; Kurpaska, Ł.; Lubas, M.; Sitarz, M. Effects of different nitriding methods on nitrided layer structure and morphology. *Arch. Metall. Mater.* **2018**, *63*, 337–345. [CrossRef]
22. Alphonsa, I.; Chainani, A.; Raole, P.; Ganguli, B.; John, P. A study of martensitic stainless steel AISI 420 modified using plasma nitriding. *Surf. Coat. Technol.* **2002**, *150*, 263–268. [CrossRef]
23. Borisyuk, Y.V.; Oreshnikova, N.M.; Pisarev, A.A. Low-Temperature Plasma Nitriding of Low- and High-Chromium Steels. *Bull. Russ. Acad. Sci. Phys.* **2020**, *84*, 736–741. [CrossRef]
24. Dalibon, E.; Charadia, R.; Cabo, A.; Brühl, S.P. Short Time Ion Nitriding of AISI 420 Martensitic Stainless Steel to Improve Wear and Corrosion Resistance. *Mater. Res.* **2019**, *22*, e20190415. [CrossRef]

25. Tuckart, W.; Forlerer, E.; Iurman, L. Delayed cracking in plasma nitriding of AISI 420 stainless steel. *Surf. Coat. Technol.* **2007**, *202*, 199–202. [CrossRef]
26. Pinedo, C.E.; Monteiro, W.A. Influence of heat treatment and plasma nitriding parameters on hardening martensitic stainless steel AISI 420. In Proceedings of the 18th IFHTSE Congress–International Federation for Heat Treatment and Surface Engineering, Rio de Janeiro, Brazil, 26–30 July 2010; pp. 4750–4757.
27. Li, Y.; He, Y.; Xiu, J.; Wang, W.; Zhu, Y.; Hu, B. Wear and corrosion properties of AISI 420 martensitic stainless steel treated by active screen plasma nitriding. *Surf. Coat. Technol.* **2017**, *329*, 184–192. [CrossRef]
28. Xi, Y.-T.; Liu, D.-X.; Han, D. Improvement of corrosion and wear resistances of AISI 420 martensitic stainless steel using plasma nitriding at low temperature. *Surf. Coat. Technol.* **2007**, *202*, 2577–2583. [CrossRef]
29. Dai, L.; Niu, G.; Ma, M. Microstructure Evolution and Nanotribological Properties of Different Heat-Treated AISI 420 Stainless Steels after Proton Irradiation. *Materials* **2019**, *12*, 1736. [CrossRef]
30. Dalibon, E.L.; Charadia, R.; Cabo, A.; Trava-Airoldi, V.; Brühl, S. Evaluation of the mechanical behaviour of a DLC film on plasma nitrided AISI 420 with different surface finishing. *Surf. Coat. Technol.* **2013**, *235*, 735–740. [CrossRef]
31. *ASTM C1624-05*; Standard Test Method for Adhesion Strength and Mechanical Failure Modes of Ceramic Coatings by Quantitative Single Point Scratch Testing. ASTM International: West Conshohocken, PA, USA, 2015. [CrossRef]
32. He, Y.; Wang, L.; Wu, T.; Wu, Z.; Chen, Y.; Yin, K. Facile fabrication of hierarchical textures for substrate-independent and durable superhydrophobic surfaces. *Nanoscale* **2022**, *14*, 9392–9400. [CrossRef]
33. Randall, N.; Favaro, G.; Frankel, C. The effect of intrinsic parameters on the critical load as measured with the scratch test method. *Surf. Coat. Technol.* **2001**, *137*, 146–151. [CrossRef]
34. Steinmann, P.; Tardy, Y.; Hintermann, H. Adhesion testing by the scratch test method: The influence of intrinsic and extrinsic parameters on the critical load. *Thin Solid Films* **1987**, *154*, 333–349. [CrossRef]
35. Ichimura, H.; Rodrigo, A. The correlation of scratch adhesion with composite hardness for TiN coatings. *Surf. Coat. Technol.* **2000**, *126*, 152–158. [CrossRef]
36. Puchi-Cabrera, E.S. A new model for the computation of the composite hardness of coated systems. *Surf. Coat. Technol.* **2002**, *160*, 177–186. [CrossRef]
37. *ASTM G99-95a (200)*; Standard Test Method for Wear Testing with a Pin on Disk Apparatus. ASTM International: West Conshohocken, PA, USA, 2010.
38. Maros, M.B.; Németh, A. Wear maps of HIP sintered Si_3N_4/MLG nanocomposites for unlike paired tribosystems under ball-on-disc dry sliding conditions. *J. Eur. Ceram. Soc.* **2017**, *37*, 4357–4369. [CrossRef]
39. Viáfara, C.; Sinatora, A. Influence of hardness of the harder body on wear regime transition in a sliding pair of steels. *Wear* **2009**, *267*, 425–432. [CrossRef]
40. Viáfara, C.; Sinatora, A. Unlubricated sliding friction and wear of steels: An evaluation of the mechanism responsible for the T1 wear regime transition. *Wear* **2011**, *271*, 1689–1700. [CrossRef]
41. Duarte, M.C.; Godoy, C.; Wilson, J.A.-B. Analysis of sliding wear tests of plasma processed AISI 316L steel. *Surf. Coat. Technol.* **2014**, *260*, 316–325. [CrossRef]
42. Dalibon, E.L.; Brühl, S.P.; Trava-Airoldi, V.J.; Escalada, L.; Simison, S.N. Hard DLC coating deposited over nitrided martensitic stainless steel: Analysis of adhesion and corrosion resistance. *J. Mater. Res.* **2016**, *31*, 3549–3556. [CrossRef]
43. Falsafein, M.; Ashrafizadeh, F.; Kheirandish, A. Influence of thickness on adhesion of nanostructured multilayer CrN/CrAlN coatings to stainless steel substrate. *Surf. Interfaces* **2018**, *13*, 178–185. [CrossRef]
44. *DIN EN 1071-2*; Advanced Technical Ceramics—Methods of Test for Ceramic Coatings—Part 2: Determination of Coating Thickness by the Crater Grinding Method. Deutsches Institut fur Normung E.V. (DIN): Berlin, Germany, 2003; pp. 1–17.
45. Czél, G.; Baán, M.; Makk, P.; Raffay, C.s.; Fancsali, J.; Janovszky, D. Research Scene to Examine Mechanical Strength of Coatings on Metallic Substrate. In Proceedings of the Fourth International Symposium on Measurement Technology and Intelligent Instruments, ISMT II '98, Miskolc-Lillafured, Hungary, 2–4 September 1998; pp. 335–338.
46. Archard, J.F. Contact and Rubbing of Flat Surfaces. *J. Appl. Phys.* **1953**, *24*, 981–988. [CrossRef]
47. Woodyatt, L.R.; Krauss, G. Iron-Chromium-Carbon System at 870 °C. *Metall. Trans. A* **1976**, *7*, 983–989. [CrossRef]
48. ASM International Handbook Committee. *ASM Handbook, Vol. 9: Metallography and Microstructures*; Vander Voort, G.F., Ed.; ASM International: Metals Park, OH, USA, 2004; pp. 644–669.
49. Brühl, S.P.; Charadia, R.; Sanchez, C.; Staia, M.H. Wear behavior of plasma nitrided AISI 420 stainless steel. *Int. J. Mater. Res.* **2008**, *99*, 779–786. [CrossRef]
50. Bobzin, K.; Brögelmann, T.; Kruppe, N.; Arghavani, M.; Mayer, J.; Weirich, T. Plastic deformation behavior of nanostructured CrN/AlN multilayer coatings deposited by hybrid dcMS/HPPMS. *Surf. Coat. Technol.* **2017**, *332*, 253–261. [CrossRef]
51. Zhang, X.; Tian, X.-B.; Zhao, Z.-W.; Gao, J.-B.; Zhou, Y.-W.; Gao, P.; Guo, Y.-Y.; Lv, Z. Evaluation of the adhesion and failure mechanism of the hard CrN coatings on different substrates. *Surf. Coat. Technol.* **2019**, *364*, 135–143. [CrossRef]
52. Montesano, L.; Pola, A.; Gelfi, M.; Brisotto, M.; Depero, L.E.; La Vecchia, G.M. Effect of microblasting on cathodic arc evaporation CrN coatings. *Surf. Eng.* **2013**, *29*, 683–688. [CrossRef]
53. Wang, Q.; Zhou, F.; Yan, J. Evaluating mechanical properties and crack resistance of CrN, CrTiN, CrAlN and CrTiAlN coatings by nanoindentation and scratch tests. *Surf. Coat. Technol.* **2016**, *285*, 203–213. [CrossRef]

54. Li, Z.; Guan, X.; Wang, Y.; Li, J.; Cheng, X.; Lu, X.; Wang, L.; Xue, Q. Comparative study on the load carrying capacities of DLC, GLC and CrN coatings under sliding-friction condition in different environments. *Surf. Coat. Technol.* **2017**, *321*, 350–357. [CrossRef]
55. Lee, J.M.; Ko, D.C.; Kim, B.M. Evaluation of Adhesive Properties of Arc PVD Coatings on Non-Nitrided and Nitrided Various Substrates. *Key Eng. Mater.* **2007**, *340–341*, 77–82. [CrossRef]
56. Siddiqui, S.A.; Favaro, G.; Maros, M.B. Investigation of the Damage Mechanism of CrN and Diamond-Like Carbon Coatings on Precipitation-Hardened and Duplex-Treated X42Cr13/W Tool Steel by 3D Scratch Testing. *J. Mater. Eng. Perform.* **2022**, *31*, 7830–7842. [CrossRef]
57. Polcar, T.; Parreira, N.; Novák, R. Friction and wear behaviour of CrN coating at temperatures up to 500 °C. *Surf. Coat. Technol.* **2007**, *201*, 5228–5235. [CrossRef]
58. Keyence. Available online: https://www.keyence.eu/products/microscope/elemental-analyzer/ea-300/ (accessed on 29 September 2022).
59. Singh, S.K.; Chattopadhyaya, S.; Pramanik, A.; Kumar, S. Experimental investigation of CrN coating deposited by PVD Process. In *IOP Conference Series: Materials Science and Engineering*; IOP Publishing: Bristol, UK, 2019; Volume 691, p. 12042.
60. Zhang, C.; Gu, L.; Tang, G.; Mao, Y. Wear transition of CrN coated M50 steel under high temperature and heavy load. *Coatings* **2017**, *7*, 202. [CrossRef]
61. Jerina, J.; Kalin, M. Aluminium-alloy transfer to a CrN coating and a hot-work tool steel at room and elevated temperatures. *Wear* **2015**, *340–341*, 82–89. [CrossRef]
62. Cai, F.; Huang, X.; Yang, Q.; Wei, R.; Nagy, D. Microstructure and tribological properties of CrN and CrSiCN coatings. *Surf. Coat. Technol.* **2010**, *205*, 182–188. [CrossRef]
63. Su, Y.; Yao, S.; Leu, Z.; Wei, C.; Wu, C. Comparison of tribological behavior of three films—TiN, TiCN and CrN—Grown by physical vapor deposition. *Wear* **1997**, *213*, 165–174. [CrossRef]
64. Mo, J.; Zhu, M. Sliding tribological behaviors of PVD CrN and AlCrN coatings against Si_3N_4 ceramic and pure titanium. *Wear* **2009**, *267*, 874–881. [CrossRef]

Article

Comparative Study on the Scratch and Wear Resistance of Diamond-like Carbon (DLC) Coatings Deposited on X42Cr13 Steel of Different Surface Conditions

Shiraz Ahmed Siddiqui * and Maria Berkes Maros

Tribo Team Research Group, Department of Structural Integrity, Institute of Materials Science and Technology, Faculty of Mechanical Engineering and Informatics, University of Miskolc (UM), Egyetemváros, H-3515 Miskolc, Hungary
* Correspondence: ahmedshiraz432@gmail.com

Citation: Siddiqui, S.A.; Maros, M.B. Comparative Study on the Scratch and Wear Resistance of Diamond-like Carbon (DLC) Coatings Deposited on X42Cr13 Steel of Different Surface Conditions. *Ceramics* **2022**, *5*, 1207–1224. https://doi.org/10.3390/ceramics5040086

Academic Editors: Amirhossein Pakseresht and Kamalan Kirubaharan Amirtharaj Mosas

Received: 16 October 2022
Accepted: 2 December 2022
Published: 8 December 2022

Publisher's Note: MDPI stays neutral with regard to jurisdictional claims in published maps and institutional affiliations.

Copyright: © 2022 by the authors. Licensee MDPI, Basel, Switzerland. This article is an open access article distributed under the terms and conditions of the Creative Commons Attribution (CC BY) license (https:// creativecommons.org/licenses/by/ 4.0/).

Abstract: Tribological investigations are of great importance, especially in the case of novel combinations of materials used for the tribosystem. In the current research, multilayer diamond-like carbon coating deposited by plasma enhanced chemical vapour deposition on an X42Cr13 plastic mould tool steel is studied with two different surface conditions of the substrate. On the one hand, it is secondary hardened; on the other hand, it is additively plasma nitrided preceding the diamond-like carbon coating. This latter combined treatment, called duplex treatment, has an increasingly wide range of applications today. However, its effectiveness largely depends on applying the appropriate nitriding technology. The tribological behaviour was characterised by an instrumented scratch test and a reciprocating ball-on-plate wear test. The results demonstrate better scratch resistance for the duplex-treated samples, while they show weaker performance in the applied wear type of loading. The current comparative study reveals the reason for the unexpected behaviour and highlights some critical aspects of the heat treatment procedure. The architecture of the tested multilayer DLC coating is unique, and no tribological results have yet been published on tribosystems combined with an X42Cr13 steel substrate. The presented results may particularly interest tribologists and the materials research community.

Keywords: multilayer DLC coating; CrN interlayer; X42Cr13 tool steel; duplex treatment; nitriding; scratch test; reciprocating wear test; damage mechanism

1. Introduction

Duplex surface treatment is a unique method combining two (or more) surface technologies to produce a composite surface having improved properties [1]. The significant efforts and advances in this field have led to considerable developments in reducing the wear, friction, and damage of the components, thereby increasing the durability of the components [2–5]. The duplex treatment combines surface modification and surface coating [6–8]. Surface modification refers to the changes in the microstructure, realised by thermo-chemical diffusional processes (such as nitriding, carburising), mechanical processes (such as forging, shot blasting, etc.), or ion implantation. Surface coating is a technique of adding a new layer onto the underlying substrate. The coating can be produced from the vapour state of the constituents, chemically (sol-gel, anodising, etc.) or by thermal spray (air plasma, vacuum plasma). The vapour state method is most widely used in mechanical engineering industries. One of the main objectives of surface coating is to increase wear resistance and reduce interfacial shear stresses and strains. This way, the material properties and performance of the surface layer can be significantly improved without altering the bulk materials [9–11].

Plasma nitriding is one of the most versatile nitriding techniques. It is a thermo-chemical process for diffusing nascent nitrogen onto the surface of steel or cast iron. It increases the surface hardness and improves the fatigue strength of alloy steels causing minimum distortion or damage to bulk materials. However, the application of the nitriding technology for high (more than 13%) Cr-content plastic mould tool steels may impose limitations on their wide range of applicability. The reason for it can be the formation of the CrN precipitates during the high temperature (>450 °C) nitriding, accompanied by the simultaneous reduction in the free Cr content in the steel matrix leading to low corrosion resistance. This unfavourable structural change represents a key issue in the metal-mechanical industry, especially in the plastic injection moulding sector, where high alloyed tool steels are widely used. The general solution for this problem is the application of an anti-wear and anti-corrosion coating to enhance the tool's steel performance and durability.

Carbon-based coatings such as diamond-like carbon (DLC) have proven highly advantageous in tribological applications [12–14]. They consist of hydrogen and carbon atoms with sp^2 and sp^3 hybridisation. The classification of DLC coatings has been extensively reported in the literature [15–17]. The two main classes are hydrogen-free and hydrogenated films. The properties of DLC are largely dependent on the hydrogen content. An increasing amount of hydrogen makes DLCs less rigid and less stressed.

DLC coatings possess high hardness, chemical inertness, wear resistance, corrosion resistance, excellent adhesion to several substrate materials [18], self-lubricating capability, and low friction coefficient [19]. Applications of such coatings include automotive [20], aerospace industry, protective layer for cutting, forming tools, and biomedical implants [21–23].

These coatings have found wide applications in plastic injection moulds owing to high durability and influencing favourably specific technological characteristics, such as ejection force, surface lubrication, cavitation pressure, etc. The key requirement for a mould component coating is to effectively increase the productive service life of the components while reducing operational and maintenance costs. DLC coating on a plastic injection mould proved to show a non-sticking behaviour, with close to zero ejection forces [24]. A comprehensive overview of the beneficial effects of DLC coating in plastic injection mould tooling is provided by Delaney et al. [25]. Among the others, DLC significantly decreases the ejection force when polypropylene (PP)], polyethene terephthalate (PET) [26], polylactic acid (PLA), polystyrene (PS) [27,28], polyamide (PA), or polyoxymethylene (POM) [28] parts are removed from the tool. PVD-deposited DLC films—characteristic thickness of 0.5–5 µm—effectively reduced the cavity pressure for molten PS and PET plastics [29] and reduced the filling flow resistance of PET in thin-wall injection moulding.

Various technological solutions are available for depositing DLC coatings, e.g., magnetron sputtering (MS) [30,31], high power impulse magnetron sputtering (HiPIMS) [32,33]. The group of the plasma enhanced chemical vapour deposition (PECVD) techniques involves several types of processes, such as the radio-frequency plasma enhanced chemical vapor deposition (RFPECVD) [34,35] and the low-frequency plasma-enhanced chemical vapor deposition LFPECVD [36–38]. For industrial purposes, RFPECVD and LFPECVD are the most widely used methods due to their versatility in scale [39] and coating structure, e.g., for doped or multilayered DLC systems.

PECVD techniques are characterised by a low deposition temperature, allowing them to obtain a uniform layer structure and size and making them suitable for coating tool steel substrates of different treatment conditions. The efficiency of duplex treatment involving plasma nitriding was emphasised in decreasing the wear rate for AISI 4140 steel with MS DLC coating [40] and improving the wear and scratch resistance of the K340 cold work tool steel with PACVD DLC coating used as a deep drawing die in sheet metal forming [41]. Combined mechanical and thermochemical treatment—involving mechanical turning and burnishing combined with plasma nitriding—improved the tribological performance of multilayer PECVD DLC-coated Sverker 21 (AISI D2) steel and Vanadis 8 powder metallurgical (P/M) steel [42]. The importance of the effect of the surface finish and removal of the compound layer on friction and wear behaviour was emphasised in the case of MS DLC coated, quenched, and tempered and plasma nitrided AISI H11 hot work tool steel [43]. Extended (up to 100%) tool life and improved surface quality of the plastic product were ensured by the DLC coating deposited on a plasma nitrided Vanadis 4 P/M austenitic stainless steel forming tool due to the outstanding anti-sticking properties and wear resistance of the DLC layer [44].

The major drawback suffered by these coatings on steel substrate is the occasionally poor adhesion that causes cracking under an external load. Doping with non-metal or metal is a suitable method to improve the thermal stability and enhance the tribological properties of the coating due to better adhesion [8,45]. The efficiency of a novel HiPIMS technique with positive pulses used for depositing DLC coating on K360, Vanadis 4, and Vancron steels was demonstrated in regard to higher wear resistance due to improved adherence of the coating [46]. The synergetic effect of low-pressure arc plasma-assisted nitriding (PAN) of M2 type steel and the PECVD technique on the adhesion of the DLC coatings was analysed by wear and scratch tests demonstrating the importance of the process parameters of both treatments on the adhesion and, consequently, the tribological behaviour of the coating [47].

The adhesion mechanism at the coating/substrate interface can be realised by mechanical interlocking and physical or chemical bonding or a combination [47]. While the effect of plasma nitriding on the improved adhesion properties of the DLC coatings is still being investigated, explanations by some researchers refer to the larger thickness of the intermixed layer at the interface and lower level of graphitisation—due to an increased bias voltage—during plasma nitriding [48,49]. These favourable effects may be further enhanced by an increased polarisation tension in the PECVD deposition process leading to better adhesion of the DLC coating to the M2 steel substrate [47].

When the coating is exposed to high loadings, plastic deformation begins in the vicinity of the coating/substrate interface. The applied coating can become damaged at this interface due to poor adhesion to the substrate or if the cohesive strength of the coating is poor. It was shown that a large plastic zone first develops in the substrate, followed by the initiation of plastic deformation in the coating. Thus, the substrate properties play a vital role in improving the load-bearing capacity of the coating-substrate system [50], which is evaluated quantitatively by the force required to remove the coating from the substrate, known as the critical force.

The instrumented scratch test is a simple, cost-effective, and widely used conventional method of determining the adhesive strength of the coatings to the interface. The result depends on several intrinsic factors, such as substrate mechanical properties, coating thickness, and interfacial bond strength, as well as extrinsic parameters such as scratching speed, applied normal load, type of indenter, tip radius, etc. The morphology of the scratch track and cracks initiating from the scratch groove with different orientations provide information on the controlling damage mechanisms such as cohesive or adhesive failure of coating; interfacial, tensile, or conformal cracking; brittle chipping, etc. The progressive load scratch tests provide the highest amount of information on the cooperative behaviour of the coated systems; therefore, such test results are extensively reported in the literature [51–54] for various combinations of the substrate/coating pairs.

Wear is defined as the irreversible material loss of interacting surfaces in relative motion. Physical and chemical elementary processes within the area of contact of a sliding pair leading, subsequently, to the change in material structure and geometry of the friction partners are known as wear mechanisms [55,56]. The most common ones are abrasion, adhesion erosion, fretting, and cavitation. Wear tests provide relevant qualitative and quantitative information on the wear resistance; the mechanism, severity, and mode of the damage; and materials performance among friction and wear type loadings that are useful during the design of the components to reduce material loss due to wearing [57]. A wear test configuration applied in engineering practice can be of great variety. It may realise, e.g., a point contact (ball on plate, ball on disk), a linear contact (block on plate), or a plane contact (block on ring, pin on disk) [58] to model the actual loading condition at the most adequate.

The substrate material in this work is X42Cr13 steel, which has a very similar chemical composition to that of the AISI 420-type steels. The literature research on plasma nitriding of X42Cr13 steel is found to a limited extent, while research on the AISI 420 martensitic stainless steel (MSS) is extensively reported [59,60]. The requirements of high surface hardness and simultaneous wear- and corrosion resistance is required for a wide range of applications. For low alloy steel, plasma nitriding is usually advantageous in improving corrosion resistance. In contrast, this treatment can have either positive or negative effects in the case of Fe-based stainless steels. The low-temperature plasma nitriding was suggested as a possible solution to enhance wear resistance without altering the excellent anti-corrosion properties that require a homogeneous distribution of the high Cr content. It can be assured by a low-temperature or short-duration nitriding operation, allowing the CrN precipitation in the nitrided layer to be avoided [61].

Another solution for enhancing wear resistance with simultaneous reservation of the corrosion resistance of high Cr steels is represented by the deposition of anti-wear ceramic coatings (such as DLC) on the previously nitrided steel substrate. In this regard, we report the first wear test results obtained on duplex-treated multilayer DLC/X42Cr13 tribosystems compared to those obtained on the control set of the un-nitrided and simple DLC-coated samples.

The primary objective of the current research work consists of comparing the tribological performance—in terms of the adhesion and wear behaviour—of two coated material systems, which have the same multilayer DLC top layer deposited onto the X42Cr13 plastic mould tool steel with two different surface conditions. We address the key issue of CrN precipitation in the tool steel at high-temperature nitriding, limiting the applicability of the duplex treatment in the case of this grade of tool steel. The reciprocating wear test configuration provides a good basis to elaborate on the harmful effects of CrN precipitation in the nitrided layer on the wear resistance of the coated system.

2. Materials and Methods

The tribological tests were carried out on the samples made of high-alloy plastic mould tool steel X42Cr13 (DIN) Nr. 1.2083. The chemical composition (in wt%) is C = 0.38–0.45, Si $\leq$ 1.0, Mn $\leq$ 1.0, P $\leq$ 0.030, S $\leq$ 0.030, Cr = 12.00–13.50, V~0.3, and the remainder is Fe. Disc-shaped samples with dimensions of Φ30 mm $\times$ 10 mm were prepared.

The two types of the investigated coated systems are denoted by Sample (S) and Sample (D), where the identical DLC top layer is deposited on two different substrates. In the case of Sample (S), the substrate is bulk heat treated, that is, secondary-hardened steel, while in the case of Sample (D), this bulk heat treatment is followed by a subsequent plasma nitriding. Thus, Sample (S) is a simple DLC-coated system, while Sample (D) is a duplex-treated one. The technological parameters of the bulk heat treatment were as follows. Austenitisation was carried out at T_{aust} = 1020 °C, for t = 20 min, followed by cooling in a vacuum. The temperature and holding time of tempering was T_{temp} = 580 °C and t = 2 h in air.

The reason for this treatment is, on the one hand, to provide the best dimensional stability of the product, on the other hand, to increase the toughness, playing a vital role, for example, in low temperature and cryogenic applications, or dynamic loadings characteristic for tool materials. These treatment parameters represent a non-conventional treatment resulting in a lower substrate hardness than in most cases when this steel is used as tool steel.

The temperature of the plasma nitriding was 520 °C, with a holding time of 8 h, a voltage of 600 V, and a pressure of 2 mbar. The source of nitrogen was decomposed ammonia (N_2:H_2 = 1:3).

The DLC top layer deposited on Samples (S) and (D) by plasma enhanced chemical vapour deposition (PECVD) is a multilayer coating consisting of CrN + WC + aC:HW + aC:H layers. The coating is deposited using the equipment TSD 550 (from HEF Durferrit M&S SAS, Andrézieux-Bouthéon, France).

The process in the furnace starts with cleaning the samples by ion-etching using ionised Ar atoms to bombard the substrate for 60 min with 65 sccm argon (purity ~99.95%) under vacuum (pressure = 1×10^{-3} Pa). This operation was executed using a microwave generator as an auxiliary plasma booster with 1000 W and 2.45 GHz frequency. The first step of the coating process is the deposition of the CrN and WC-Co underlayers.

The first sublayer was the CrN deposited at 150 °C for 30 min. The Cr target was sputtered using a power of 6.5 kW, and the gas flow of N_2 was controlled by PLC via light measurement of plasma. The second sublayer was the WC-Co interlayer created by magnetron sputtering. The target was a typical cemented carbide constituting of WC and 6% Co. For the deposition of the a-C:H top layer, acetylene gas with a flow rate of 200 sccm was utilised. The microwave generator served as an auxiliary plasma booster with 500 W and 2.45 GHz frequency.

The instrumented scratch tests were accomplished on a scratch tester (model SP15, Institute of Materials Science and Technology, the University of Miskolc, manufacturer: Sunplant, Miskolc, Hungary) [62]), conforming with the requirements of the ASTM C1624-22 standard [63]. During the applied progressive loading scratch test, a standard Rockwell C stylus (120° vertex angle and 0.2 mm radius) was drawn over the sample surface with normal force, increasing from an initial normal load of 2 N to a maximum normal load of 150 N until the coating was removed from the substrate. The gradient of the normal force was 10 N/mm, and the velocity of the table supporting the sample was 5 mm/min. The total scratch length was 15 mm. Tests were made on two pieces of the simple- and duplex-coated batches producing three scratches per sample. The diamond stylus was cleaned by manual polishing after each scratching using a polish cloth (grade Struers MD-NAP) soaked with acetone. Then, the tool was sonicated in acetone for 3 min to remove the scratch-test-originated residuals. The integrity of the stylus was checked by OM after each test. The critical loads were obtained from the scratch diagrams registered by the test machine, while subcritical loads and damage characteristics were evaluated by optical microscopy based on the ASTM C1624-22 standard.

The wear tests were carried out in the laboratory of Rtec Instruments, Switzerland using a multi-functional tribometer (type MFT-5000, from Rtec Instruments, Yverdon-les-Bains, Switzerland) equipped with a load cell (200 N) and an imaging system using a confocal microscope. Two samples of simple- and duplex-coated types were tested, carrying out two measurements on one piece. To model the fretting-type loading, generally occurring during the operation of forming tools, a reciprocating wear test configuration was chosen to characterise the wear behaviour of the samples. This type of test regime is widely used for tool coatings. Wear tests were accomplished with a normal load of 25 N, a reciprocating frequency of 5 Hz, a sliding distance of 10 mm, and a cycle number of 7900 at room temperature, in dry sliding conditions, at a relative humidity of RH = 50%. The material of the frictional counterpart was an Al_2O_3 ball of 6 mm in diameter, using a new ball for each measurement.

The coating hardness was characterised by the composite hardness of the coated systems using a standard micro-Vickers method with $F = 0.5$ N loading force. This test was performed with $F = 0.25$ N normal load to define the hardness profile in the nitrided layer to obtain the nitrided layer depth (Mitutoyo HVK-H1 hardness tester from Mitutoyo Corp., Kanagawa, Japan). Two samples from each group (simple and duplex-treated) were tested by performing ten valid measurements—having no cracks from the corners—by sample. Three profiles on each of the two duplex-treated samples were determined.

To evaluate the thickness of coatings, the ball cratering method was used (Compact CAT^2c Calotester from Anton Paar GmbH, Graz, Austria) by creating three craters on two samples of both the simple- and the duplex-coated groups [64,65].

The structure of the coatings and the substrate were examined by scanning electron microscopy (Zeiss Evo MA10, form Zeiss, Jena, Germany) equipped with an energy-dispersive X-ray spectroscopy. Optical microscopy (Zeiss, Axio Observer D1m from Zeiss,

Jena, Germany) using the Axio Vision imaging software (Zeiss, Jena, Germany) was applied for the morphological analysis of the scratch grooves.

3. Results and Discussion

3.1. Coating Thickness and Structure

The architecture of the multilayer coatings is shown in Figure 1. They are built up with a CrN bottom layer (thickness of 0.7 µm for both samples), an a-C:HW intermediate layer (0.8 and 0.9 µm for the simple and duplex coating, respectively), and an a-CH top layer (thickness of 1.9 and 2 µm, respectively). The thickness of coatings, defined by the ball-cratering method on the simple and duplex coatings, were 3.4 ± 0.07 and 3.6 ± 0.06 µm, respectively.

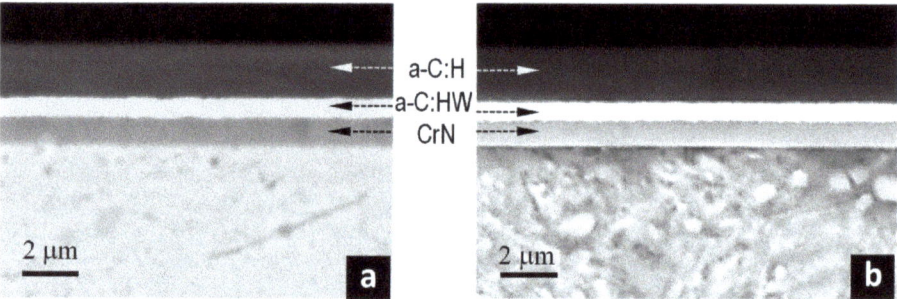

Figure 1. The architecture of the DLC coating on the simple-coated (**a**) and duplex-treated (**b**) samples taken by optical microscopy.

3.2. Microhardness

The average depth of the nitrided layer is 0.14 mm for the duplex-treated sample, which was defined based on the hardness profile below the surface (Figure 2).

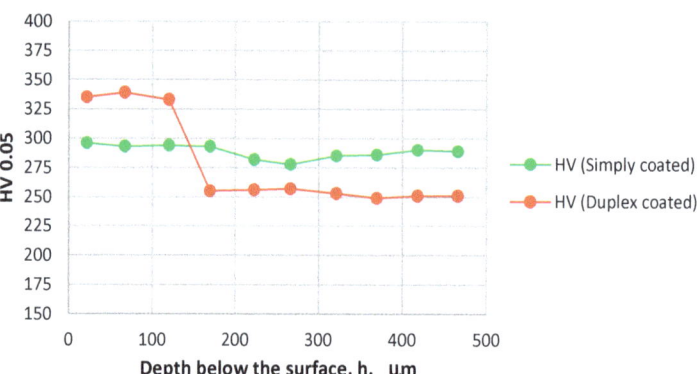

Figure 2. Microhardness profile of the simple-coated and duplex-treated sample.

The coating characteristics are summarised in Table 1. It is seen that DLC coating caused no changes in the substrate hardness (see HV of the substrate surface and substrate core for Sample (S). At the same time, plasma nitriding increased the surface hardness while decreasing the core hardness of the secondary hardened substrate in the case of Sample (D).

Table 1. Sample nomination and characteristics of the multilayered DLC coating.

Sample		Substrate Condition	Coating				
			Thickness, μm		Hardness, $HV_{0.05}$		
Nomi-Nation	Type		DLC Coating	Nitrided Layer	DLC Coating *	Substrate Surface	Substrate Core
S	simple-coated	un-nitrided	3.4	–	1386	295	290
D	duplex-treated	nitrided	3.6	140	2077	335	250

* The coating hardness is characterised by the composite hardness of the coated system.

3.3. Microstructure Characterisation

A magnified SEM cross-sectional image of the nitrided layer produced at 520 °C for 8 h holding time is shown in Figure 3. The image was captured using the backscattered electron (BSE) mode to visualise the microstructural features and compositional variations in the nitrided surface. Elemental analysis by EDX in three characteristic points was made. Chemical composition at Spot 1, close to the spherical precipitate, and at Spot 3, which was taken in the middle of the grain, at a somewhat higher depth (~3 μm) below the surface, was almost identical. The increased Cr and N content suggest the presence of nitrides, which appear as dark grey areas around the spherical precipitation and at spot 3 in the BSD imaging mode. Spot 2, taken along the grain boundary (GB) region, shows higher Cr, N content compared to Spots 1 and 3, which give strong evidence of the path of CrN precipitation along the GB.

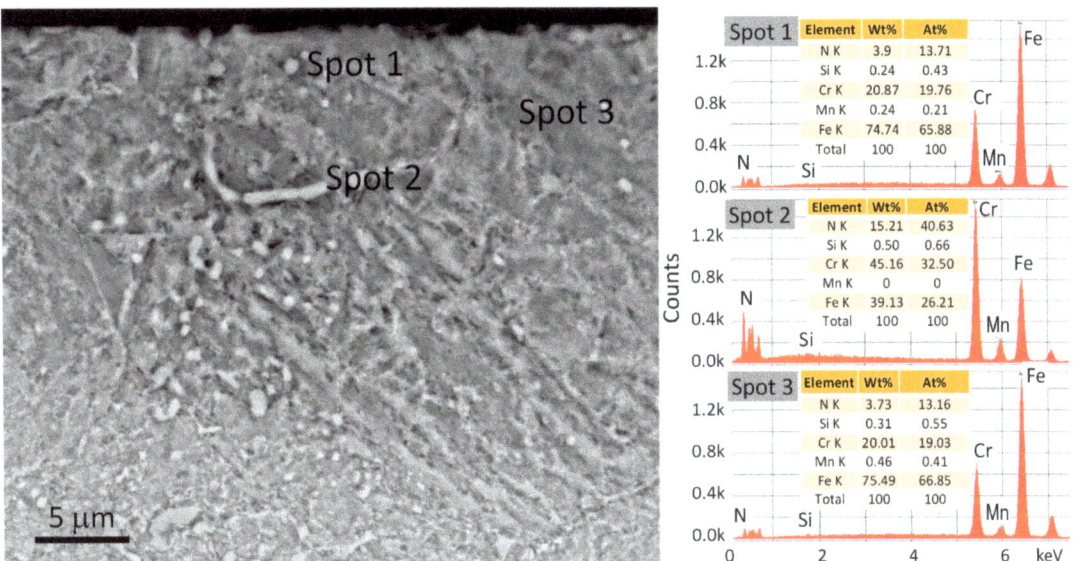

Figure 3. The SEM image of the nitrided zone of the (S) sample with the EDX spectra taken in three characteristic locations of the layer: Spot 1—close to a precipitate in the near-surface region inside the grain, Spot 2—a grain boundary precipitation, Spot 3—the grain volume at a slightly deeper region below the surface.

The GBs are clearly seen due to the heterogeneous deformation among the grains, resulting from the compressive residual stresses generated by the diffusion of nitrogen from the surface to the core, accompanied by simultaneous nitride precipitation [66].

3.4. Scratch Resistance

The scratch diagrams for the simple-coated (S) and duplex-treated (D) samples recorded during the progressive loading scratch tests are shown in Figure 4. which provides a comparative analysis of the friction coefficient (μ) vs. scratch length for the two coated systems with the differently treated substrate. In the first approximation, it is seen that the critical (L_{c3}) and subcritical (L_{c1} and L_{c2}) loading forces are higher for the duplex-treated sample.

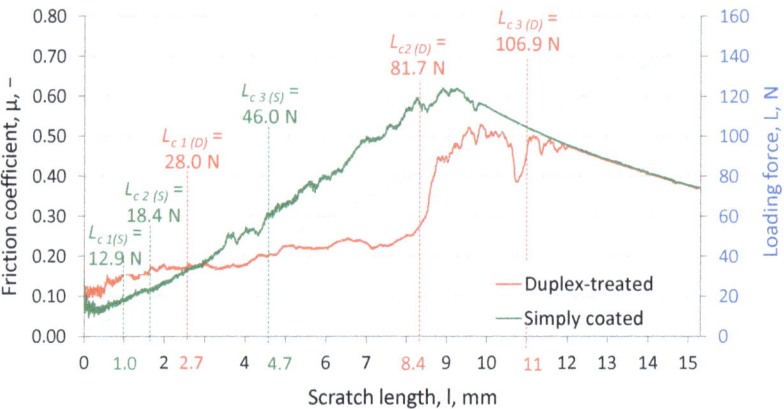

Figure 4. Scratch test diagrams for the simple-coated and duplex-treated samples.

These load levels indicate the initiation of specific characteristic damage processes identified using the optical micrographs shown in Figure 5.

The friction coefficient curve for the simple-coated sample (Figure 4) illustrates that the initial micro-cracking (Figure 5a) takes place at 1.0 mm at a loading force of $L_{c1\,(S)}$ = 12.9 N, followed by buckling spallation (Figure 5b) observed at the scratch length of 1.7 mm and a normal load of $L_{c2\,(S)}$ = 18.4 N, where μ = 0.110. At the critical force of $L_{c3\,(S)}$ = 46.0 N, when the friction coefficient is μ = 0.217, the coating becomes detached from the substrate, making the highly deformed, periodically torn substrate visible (Figure 5c).

Once the coating is removed, there are traces of detached coating material beyond the groove. It is indicative of gross spallation (Figure 5c), a characteristic failure mechanism of coatings with low adhesion strength [63]. The reason behind the increasing value of μ is the presence of the small coating's particles removed already by the subcritical loads, which cling in the vicinity of the stylus, creating additional resistance to the motion of the stylus and thereby increasing the μ value.

Regarding the friction coefficient curve obtained for the duplex-treated sample, the critical force at which the coating delamination begins is significantly higher, that is, $L_{c3\,(D)}$ = 106.9 N. The $L_{c1\,(D)}$ and $L_{c2\,(D)}$ subcritical forces are 28.0 and 81.7 N, respectively, which initiate microcracking (Figure 5d) and arc tensile cracking (Figure 5e) of the coating. Beyond the $L_{c3\,(D)}$ loading, there are traces of rounded regions of DLC coating that are removed laterally from the edges of the scratch groove. It is known as chipping [63], which was observed in the case of the duplex sample. It suggests a better scratch resistance of the coating and strong adherence [63] to the substrate. Thus, it can be concluded that the duplex sample possesses a higher critical load, higher scratch resistance accompanied by different damage mechanisms, and stronger adherence to the substrate compared to the simple-coated sample. The improvement is 117, 344, and 132% in terms of the L_{c1}, L_{c2}, and L_{c3} subcritical and critical loads, respectively. The authors of the current work experienced a similar improvement in a recent round robin test [67], as well as found by others [40,53] for hydrogenated DLC coating if duplex treatment was applied.

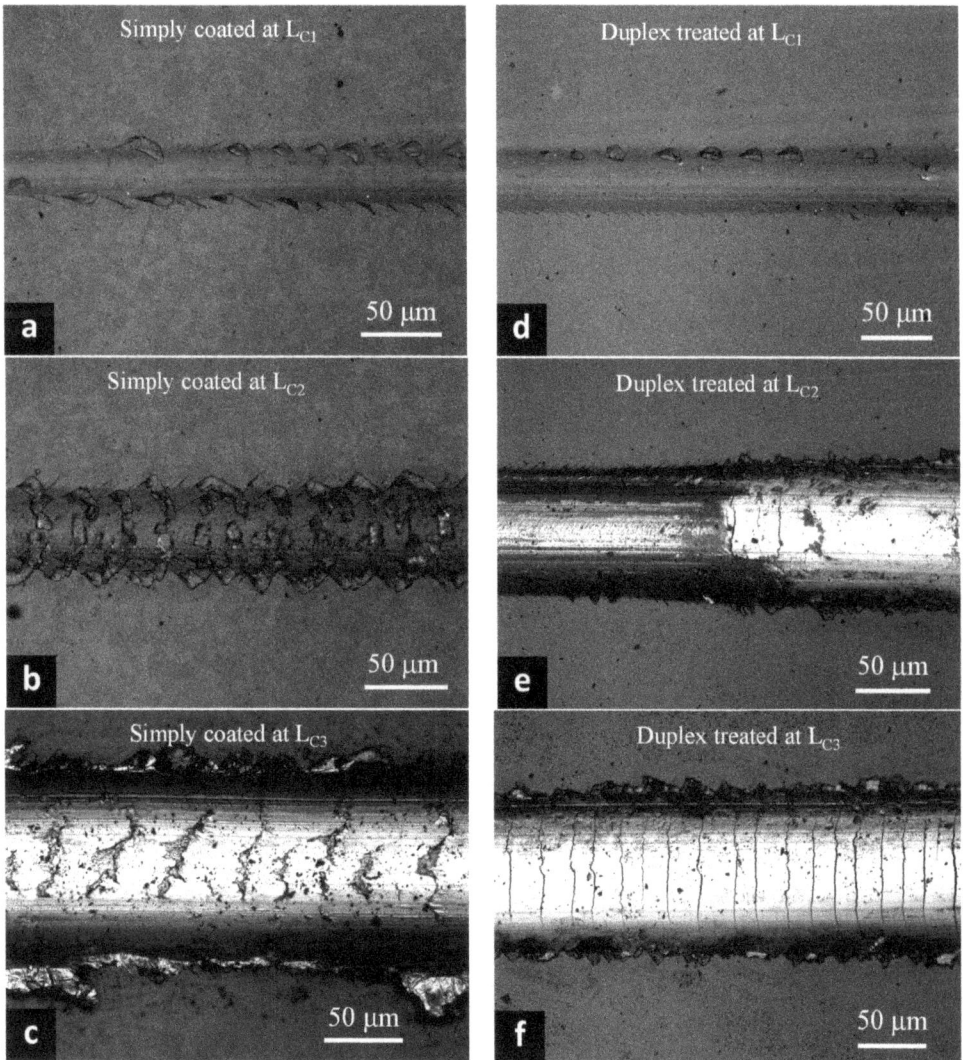

Figure 5. Optical micrographs illustrating the characteristic damage mechanisms that occurred at the subcritical and critical loadings in the simple DLC coated system (**a–c**) and the duplex-coated DLC system (**d–f**).

3.5. Wear Resistance of the Simple and Duplex-Treated Coatings

The friction coefficient curves for Sample (S) and (D), obtained during the reciprocating wear test, are shown in Figure 6. Surprisingly, the steady state friction coefficient is twice as high for the duplex-treated sample ($\mu_{(S)}$ = 0.049) than the simple-coated one ($\mu_{(D)}$ = 0.020). At the beginning of the test, that is, from 162 to 300 s of running, an abrupt increase in the friction coefficient—from 0.06 to 0.22—is also seen in the case of the duplex coating.

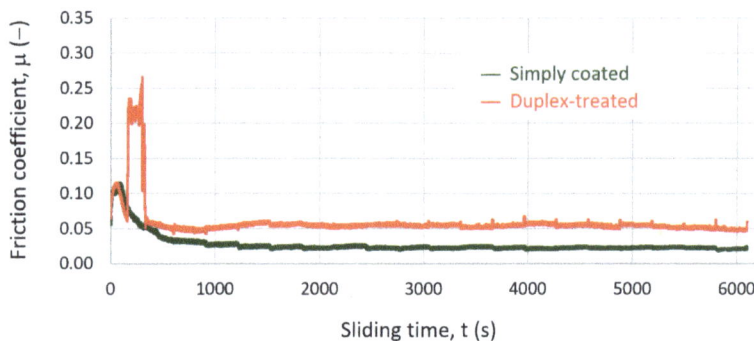

Figure 6. Friction coefficient vs. time curves obtained during the wear testing of the simple-coated and duplex-treated samples.

The 2D and 3D profilometry images for the disks (Figure 7) and the balls (Figure 8) provide qualitative and quantitative analysis of the wear damage produced on the two different coating systems. The worn track is significantly thinner for the simple coating showing negligible debris outside the worn path. In contrast, a large amount of debris decorates the two sides of the wear track in the case of the duplex coating.

Regarding the quantitative characteristics, the worn width values are $w_S = 165$ µm and $w_D = 245$ µm, and the worn depth values are $d_S = 0.44$ µm and $d_D = 0.86$ µm on the simple-coated and duplex-coated samples, respectively. Similar differences in the worn scar diameters, that is, $x_S = 155$ µm and $x_D = 247$ µm for the simple and duplex coating, were measured on the worn alumina balls (Figure 8). These results represent a 48%, 93%, and 59% increase in wear width, wear depth, and ball scar diameter, respectively. Thus, we can conclude that during reciprocating wearing, a significantly higher amount of material loss occurred in the case of the duplex coating, that is, the wear resistance decreased considerably. Nevertheless, the integrity of the DLC coatings was kept in both coating systems in terms of the wear depth remaining below the thickness of the a-:CH top layer.

In analysing the reason for this unexpected tribological response of the duplex-treated sample, the following considerations can be made. The $F = 25$ N normal load caused a reversal motion of the sample and the ball in every test cycle of the reciprocating test. How can this cause more significant damage to the duplex coating? The most probable answer is given by the more enhanced Bauschinger effect (BE) in the secondary hardened and nitrided substrate material. Ellermann and Scholtz studied the BE in different heat treatment conditions of 42CrMo4 steel and found that the larger amount of carbide precipitates causes larger BE [68]. Queyreau and Devincre made dislocation dynamics simulations to analyse the Bauschinger effect in precipitation-strengthened materials. They established that the most effective contribution to BE is given by the Orowan-looping around the II. phase particles, interacting with the mobile dislocations. Thus, an increasing volume fraction of unshearable particles contributes extensively to the larger BE. In addition to the volume fraction, the density, size, and shear strength of these precipitates also have a definite role in the BE [69]. Kostryzhev et al. demonstrated that a higher amount of larger precipitates in differently treated C-Nb and C-Nb-V steel plates are responsible for a higher BE stress parameter [70]. The BE is strongly related to the residual stresses [68,71], that is, it may be enhanced by the large tensile stresses usually forming around the precipitates [72].

These observations support the hypothesis that the underlying duplex-treated substrate material below the hard DLC coating was softened to a greater extent during the reversed shear plastic loading caused by the friction force. It can be explained by the higher amount of precipitates present in the nitrided layer due to the high-temperature nitriding applied.

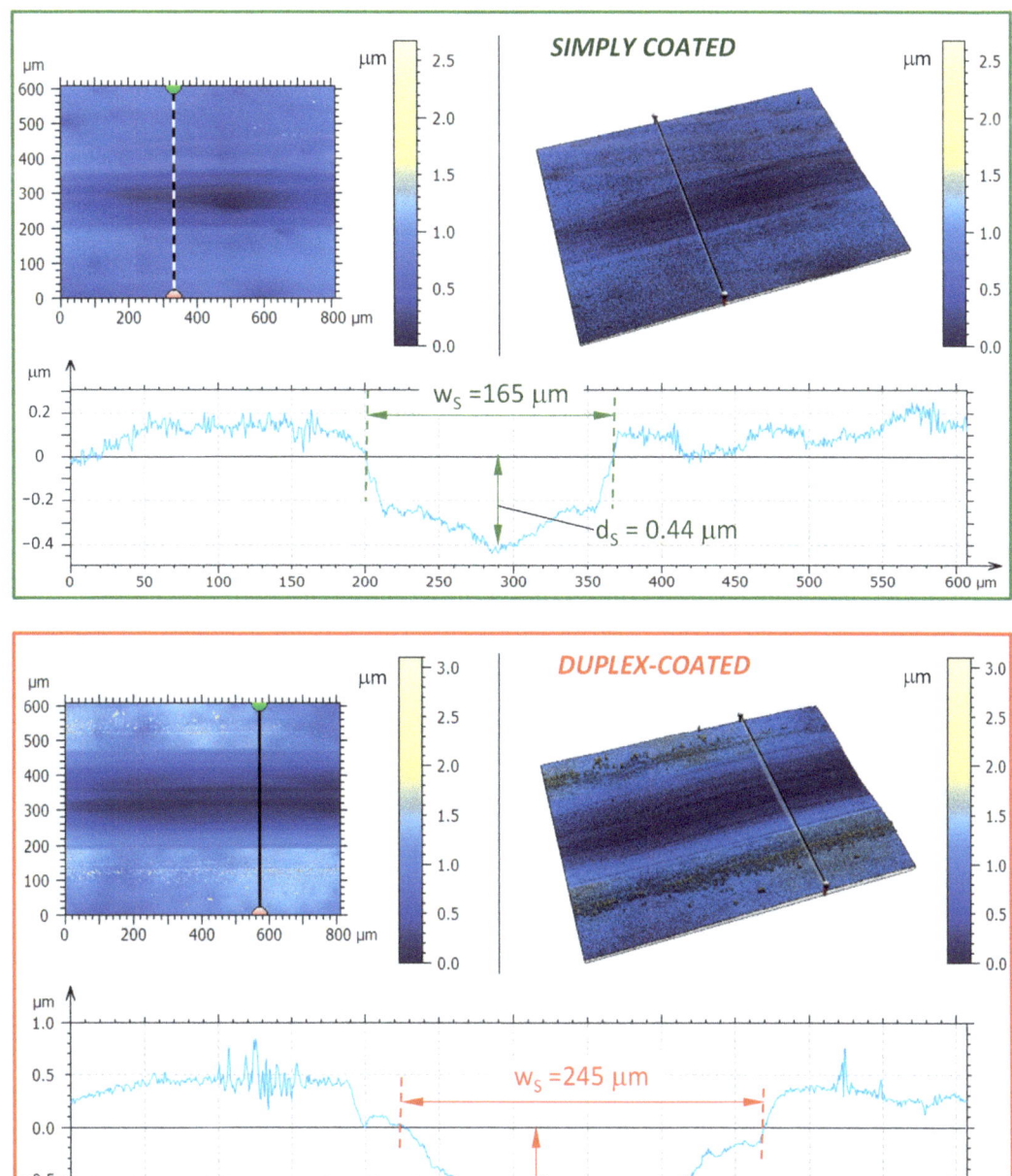

Figure 7. 2D and 3D profile of the wear track on the simple coated (**up**) and the duplex-coated (**down**) discs.

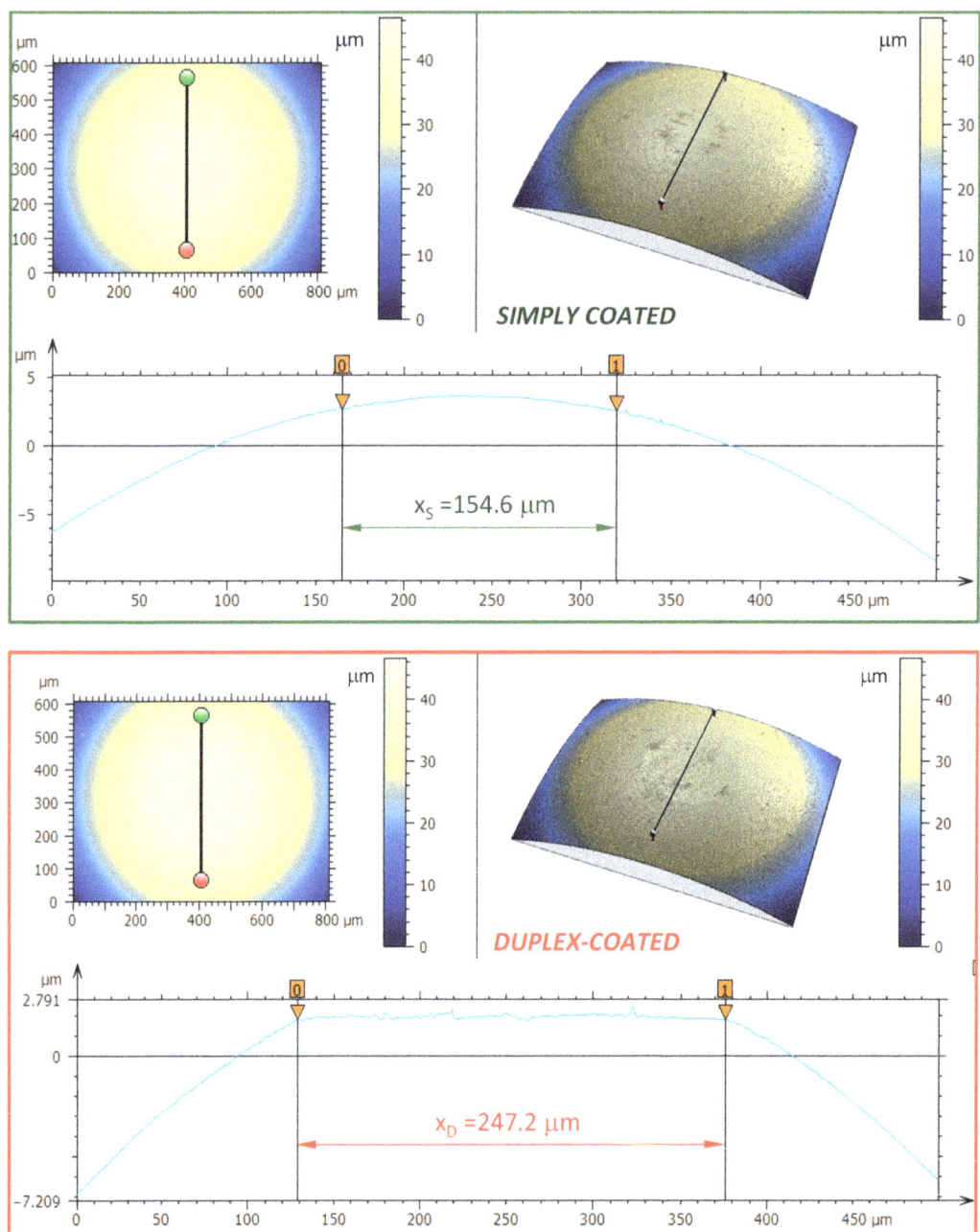

Figure 8. 2D and 3D profile of the wear track of the ball counterpart of the simple-coated (**up**) and the duplex-coated (**down**) discs. Consider the different scales along the vertical axes of the profile curves.

Similarly, an unexpected decrease in the hardness below the worn surface was reported by Mussa et al. during the reciprocating wear test of the case-hardened martensitic

22NiCrMo12–F steel. At the same time, they observed strain hardening for the same material in unidirectional loading of the ball-on-disc test [73].

It should also be noted that the nitrided layer, providing generally better support for hard coatings, was in this case, relatively thin due to the high Cr content of the steel preventing efficient N diffusion into the surface (see Figure 2). Thus, a compositionally and microstructurally heterogeneous subsurface layer structure was developed, unfavourable regarding the cyclic fatigue during the wear test. As a result, the hard coating was easily broken on the softened substrate producing hard-wear debris particles, enhancing and accelerating the damage process and leading to severe wear due to a three-body abrasive wear mechanism.

The main reason for the failure of the DLC coating altogether is the softening of the substrate material, resulting in an intensive plastic deformation in the metallic substrate during both the fro- and the reversal motion of the ball. The intensive shear deformation of the soft substrate towards the two sides of the groove may contribute to the high tensile stresses in the DLC coating. The accumulation of the plastic deformation during the subsequent passages of the loading reveals itself in the ploughing mechanism, traces of which are visible on the worn surface, as shown by the 3D OM image of the wear track of the duplex-coated specimen in Figure 7.

The load condition is further complicated by residual stresses around the precipitates in the high-Cr nitrided steel, which can initiate intensive cracking in the base material. Their stress-concentration effect can also contribute to the increase in tensile stresses arising in the coating.

A considerable contribution to the amount of abrasive particles may be given by the broken particles from the alumina ball, the amount of which was significantly higher while wearing the duplex-coated system.

The magnified image of the initial region of the friction coefficient vs. time diagram reveals that the two curves of the simple-coated and duplex system moved close to each other for approximately 160 s (Figure 9). The friction coefficient of the duplex-treated sample decreased slightly faster than that of the single-treated sample after the local maximum associated with the initial stick-slip behaviour. At 160 s, an abrupt increase in the friction coefficient indicated the premature failure of the duplex-treated coating. The very hard ceramic coating particles caused an increase in the friction coefficient to about four times greater than the friction coefficient, exceeding the 0.2 value. The detached particles of the extremely hard coating caused a three-body abrasive wear until they became chopped due to the repeated rubbing of the mating surfaces. After approximately 300 s, the coefficient of friction abruptly fell and showed a similar character over time as the curve of the simple-coated sample; however, its value always moved above it.

If neglecting the deviations in the initial stage, the friction coefficient curves typically show a decreasing trend with time for both coating systems, which indicates the tribochemical nature of the wear of the DLC coatings.

The explanation for the fact that the friction coefficient is higher, approximately double for the duplex-coated specimen compared to the simple-coated one, is the substantial softening of the substrate due to the more considerable plastic deformation in the reversed direction. It contributes to a much more intensive crack formation and propagation, consequently, a more significant amount of debris formation in the DLC top layer, which accompanies the entire wear process, as seen in Figure 6.

The presented tribological investigations show that the applied duplex treatment—namely, the high-temperature plasma nitriding of the steel substrate before the coating deposition—is inefficient in offering higher wear resistance during reciprocating sliding compared to that of the simple-coated samples.

It should be noted that ball-on-disk wear tests—realising unidirectional circumferential loading conditions—on the same batch of simple and duplex-coated samples prevailed better wear resistance of the duplex-coated tribosystems [74]. The similar wear behaviour of cemented 22NiCrMo12–F substrate in ball-on-disc, as well as reciprocating wear test configuration, was observed by the authors of the work [73].

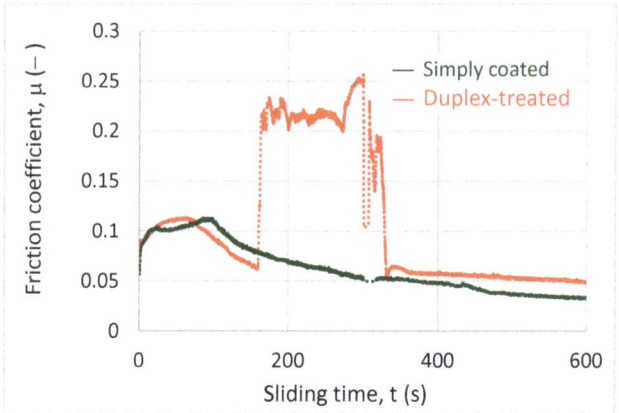

Figure 9. The initial part of the friction coefficient vs. time diagram illustrating the different wear behaviour of the DLC coating on the simple- (S) and duplex-treated substrate.

At the same time, we could demonstrate that during the scratch tests, where only unidirectional loading occurs, plastic softening of the substrate in the duplex-treated specimens did not happen, and the favourable effect of the combined bulk and surface treatments prevailed.

These experiences draw attention to the fact that the problem of high-temperature nitriding of steels with a high Cr content is not yet solved. The optimal technological parameters must be developed for the given steel grade based on further experimental work. In addition, our hypothesis on the modified BE behaviour of the duplex-coated system, in reciprocal and unidirectional wear test conditions, has to be supported by further, purposefully designed comparative wear tests supplemented with detailed microstructural and phase analysis of the worn surfaces.

4. Summary

The current study evaluates the adhesion and wear resistance of two hard coating systems represented by the identical multilayer DLC top coatings deposited on nitrided and un-nitrided steel substrates. The substrate material was a secondary hardened X42Cr13 tool steel involving tempering to high toughness and dimensional stability. Based on the results of the tribological tests performed, the following conclusions can be drawn.

The applied duplex treatment benefits the tested DLC coating system in terms of scratch resistance. The L_{c3} critical load was improved more than 130% and the L_{c1} and L_{c3} subcritical loads were doubled and tripled, respectively.

In contrast, the applied surface treatment cannot be efficiently used for the tested combination of substrate material/heat treatment/multilayer coating in respect of the reciprocating wear test accomplished with the given test circumstances.

Despite the better adhesion strength of the coating for the duplex sample, shown by the progressive loading scratch test, the duplex-coated system offers poor wear resistance in the reciprocating wear test configuration compared to the simple-coated ones. The reason behind the unfavourable behaviour is the intense CrN precipitation during the nitriding process and particle coarsening, which on the one hand, induces an intensive plastic softening of the substrate due to the enhanced Bauschinger effect, a characteristic of the reversed motion during reciprocating sliding. On the other hand, precipitates in the nitrided steel subsurface region may induce high residual tensile stresses, contributing to the tensile stresses developing in the DLC coating due to the shear deformation of the softened steel substrate. As a result, a more enhanced cracking and failure of the DLC coating occur, accompanied by a more intensive debris formation in the reciprocating wearing of the duplex-treated coating.

These results raise the need for a more profound overview of these novel findings, during which both purposefully designed new comparative wear tests, deeper microstructural and phase analyses of the worn substrate, compositional and morphological analysis of the debris particles, as well as estimation of the developed stress state by simulation software (e.g., ABAQUS, 3D-FEM) have to be taken into account as tools of the solution.

The direction of future work is to analyse the effect of the different loading conditions embodied by the other test methods on the wear behaviour of the duplex-treated coating with particular attention to the mentioned features. In addition, the influence of the core hardness of the substrate material—plastic mould tool steel—controlled by the tempering temperature during bulk heat treatment should be optimised for better wear performance of the coated systems.

Author Contributions: Conceptualisation, methodology, M.B.M.; test execution, data processing, S.A.S.; writing—original draft preparation, S.A.S.; writing—review and editing, M.B.M.; investigation, data evaluation and analysis, S.A.S. and M.B.M.; visualisation: S.A.S. and M.B.M.; supervision: M.B.M. All authors have read and agreed to the published version of the manuscript.

Funding: This research received no external funding.

Institutional Review Board Statement: Not applicable.

Informed Consent Statement: Not applicable.

Data Availability Statement: The data presented in this study are available on request from the corresponding author.

Acknowledgments: The authors thank the support of the Stipendium Hungaricum scholarship by the Tempus foundation, Hungary. The authors express their gratitude to Attila Széll and Gyula Juhász at T.S. Hungary; Andrea Szilágyiné Biró at Oerlikon Balzers GmbH Kft., Székesfehérvár, Hungary; and Norbert Luczai at Voestalpine High-Performance Metals Hungary Kft. for their assistance in heat treatment, coating, and valuable professional consultations. We also thank the colleagues at the University of Miskolc, namely, Csaba Felhő for the profilometry analyses, Árpád Kovács for the SEM and EDX measurements, and Gregory Favaro and Vishal K. Khosla at Rtec Instruments, Switzerland for the reciprocating wear tests. The authors thank the Institute of Materials Science and Technology of UM for the financial support covering the costs of the mechanical, tribological, and microstructural testing.

Conflicts of Interest: The authors declare no conflict of interest.

References

1. Bell, T.; Dong, H.; Sun, Y. Realising the potential of duplex surface engineering. *Tribol. Int.* **1998**, *31*, 127–137. [CrossRef]
2. Hakami, F.; Pramanik, A.; Basak, A. Duplex surface treatment of steels by nitriding and chromizing. *Aust. J. Mech. Eng.* **2015**, *15*, 55–72. [CrossRef]
3. Lin, N.; Liu, Q.; Zou, J.; Guo, J.; Li, D.; Yuan, S.; Ma, Y.; Wang, Z.; Wang, Z.; Tang, B. Surface Texturing-Plasma Nitriding Duplex Treatment for Improving Tribological Performance of AISI 316 Stainless Steel. *Materials* **2016**, *9*, 875. [CrossRef]
4. Haftlang, F.; Habibolahzadeh, A. Influence of Treatment Sequence on Tribological Performance of Duplex Surface-Treated AISI 1045 Steel. *Acta Met. Sin. (Engl. Lett.)* **2019**, *32*, 1227–1236. [CrossRef]
5. Das, K.; Alphonsa, J.; Ghosh, M.; Ghanshyam, J.; Rane, R.; Mukherjee, S. Influence of pretreatment on surface behavior of duplex plasma treated AISI H13 tool steel. *Surf. Interfaces* **2017**, *8*, 206–213. [CrossRef]
6. Sun, Y.; Bell, T. Combined Plasma Nitriding and PVD Treatments. *Trans. IMF* **1992**, *70*, 38–44. [CrossRef]
7. Dalibon, E.L.; Trava-Airoldi, V.; Pereira, L.A.; Cabo, A.; Brühl, S.P. Wear resistance of nitrided and DLC coated PH stainless steel. *Surf. Coat. Technol.* **2014**, *255*, 22–27. [CrossRef]
8. Dalibon, E.L.; Escalada, L.; Simison, S.; Forsich, C.; Heim, D.; Brühl, S.P. Mechanical and corrosion behavior of thick and soft DLC coatings. *Surf. Coat. Technol.* **2017**, *312*, 101–109. [CrossRef]
9. Holmberg, K.; Ronkainen, H.; Matthews, A. Tribology of thin coatings. *Ceram. Int.* **2000**, *26*, 787–795. [CrossRef]
10. Wood, R.J.; Neville, A. Second International Conference on Erosive and Abrasive Wear (ICEAW II). *Wear* **2005**, *258*, 1. [CrossRef]

11. Chung, K.H.; Lee, Y.H.; Kim, Y.T.; Kim, D.E.; Yoo, J.; Hong, S. Nano-tribological characteristics of PZT thin film investigated by atomic force microscopy. *Surf. Coat. Technol.* **2007**, *201*, 7983–7991. [CrossRef]
12. Robertson, J. Diamond-like amorphous carbon. *Mater. Sci. Eng. R Rep.* **2002**, *37*, 129–281. [CrossRef]
13. Chung, K.H.; Kim, D.E. Wear characteristics of diamond-coated atomic force microscope probe. *Ultramicroscopy* **2007**, *108*, 1–10. [CrossRef]
14. Wilson, G.M.; Sullivan, J.L. An investigation into the effect of film thickness on nanowear with amorphous carbon-based coatings. *Wear* **2009**, *266*, 1039–1043. [CrossRef]
15. Hainsworth, S.V.; Uhure, N.J. Diamond like carbon coatings for tribology: Production techniques, characterisation methods and applications. *Int. Mater. Rev.* **2007**, *52*, 153–174. [CrossRef]
16. Casiraghi, C.; Piazza, F.; Ferrari, A.C.; Grambole, D.; Robertson, J. Bonding in hydrogenated diamond-like carbon by Raman spectroscopy. *Diam. Relat. Mater.* **2005**, *14*, 1098–1102. [CrossRef]
17. Hauert, R. An overview on the tribological behavior of diamond-like carbon in technical and medical applications. *Tribol. Int.* **2004**, *37*, 991–1003. [CrossRef]
18. Wang, C.T.; Escudeiro, A.; Polcar, T.; Cavaleiro, A.; Wood, R.J.; Gao, N.; Langdon, T.G. Indentation and scratch testing of DLC-Zr coatings on ultrafine-grained titanium processed by high-pressure torsion. *Wear* **2013**, *306*, 304–310. [CrossRef]
19. Tiainen, V.M. Amorphous carbon as a bio-mechanical coating—Mechanical properties and biological applications. In Proceedings of the 3rd Specialist Meeting on Amorphous Carbon, Mondovi, Italy, 31 August–1 September 2000; Volume 10, pp. 153–160.
20. Treutler, C.P.O. Industrial use of plasma-deposited coatings for components of automotive fuel injection systems. In Proceedings of the 32nd International Conference on Metallurgical Coatings and Thin Film, San Diego, CA, USA, 2–6 May 2005; Volume 200, pp. 1969–1975.
21. Joyce, T.J. Examination of failed ex vivo metal-on-metal metatarsophalangeal prosthesis and comparison with theoretically determined lubrication regimes. In Proceedings of the 16th International Conference on Wear of Materials, Montreal, QC, Canada, 15–17 April 2007; Volume 263, pp. 1050–1054.
22. Falub, C.V.; Müller, U.; Thorwarth, G.; Parlinska-Wojtan, M.; Voisard, C.; Hauert, R. In vitro studies of the adhesion of diamond-like carbon thin films on CoCrMo biomedical implant alloy. *Acta Mater.* **2011**, *59*, 4678–4689. [CrossRef]
23. Hauert, R.; Thorwarth, G.; Müller, U.; Stiefel, M.; Falub, C.V.; Thorwarth, K.; Joyce, T.J. Analysis of the in-vivo failure of the adhesive interlayer for a DLC coated articulating metatarsophalangeal joint. *Diam. Relat. Mater.* **2012**, *25*, 34–39. [CrossRef]
24. Navabpour, P.; Teer, D.; Hitt, D.; Gilbert, M. Evaluation of non-stick properties of magnetron-sputtered coatings for moulds used for the processing of polymers. *Surf. Coat. Technol.* **2006**, *201*, 3802–3809. [CrossRef]
25. Delaney, K.D.; Bissacco, G.; Kennedy, D. A structured review and classification of demolding issues and proven solutions. *Int. Polym. Process.* **2012**, *27*, 77–90. [CrossRef]
26. Sasaki, T.; Koga, N.; Shirai, K.; Kobayashi, Y.; Toyoshima, A. An experimental study on ejection forces of injection molding. *Precis. Eng.* **2000**, *24*, 270–273. [CrossRef]
27. Martins, L.C.; Ferreira, S.C.; Martins, C.I.; Pontes, A.J. Study of ejection forces in injection moulding of thin-walled tubular mouldings. In Proceedings of the PMI 2014 International Conference on Polymers and Moulds Innovations, Guimarães, Portugal, 10–12 September 2014; pp. 281–286.
28. Sorgato, M.; Masato, D.; Lucchetta, G. Tribological effects of mold surface coatings during ejection in micro injection molding. *J. Manuf. Process.* **2018**, *36*, 51–59. [CrossRef]
29. Lucchetta, G.; Masato, D.; Sorgato, M.; Crema, L.; Savio, E. Effects of different mould coatings on polymer filling flow in thin-wall injection moulding. *CIRP Ann.* **2016**, *65*, 537–540. [CrossRef]
30. Liu, J.; Li, L.; Wei, B.; Wen, F.; Cao, H.; Pei, Y. Effect of sputtering pressure on the surface topography, structure, wettability and tribological performance of DLC films coated on rubber by magnetron sputtering. *Surf. Coat. Technol.* **2019**, *365*, 33–40. [CrossRef]
31. Ye, Y.; Wang, Y.; Ma, X.; Zhang, D.; Wang, L.; Li, X. Tribocorrosion behaviors of multilayer PVD DLC coated 304L stainless steel in seawater. *Diam. Relat. Mater.* **2017**, *79*, 70–78. [CrossRef]
32. Santiago, J.; Fernández-Martínez, I.; Sánchez-López, J.; Rojas, T.; Wennberg, A.; Bellido-González, V.; Molina-Aldareguia, J.; Monclús, M.; González-Arrabal, R. Tribomechanical properties of hard Cr-doped DLC coatings deposited by low-frequency HiPIMS. *Surf. Coat. Technol.* **2019**, *382*, 124899. [CrossRef]
33. Wang, H.; Wang, L.; Wang, X. Structure characterization and antibacterial properties of Ag-DLC films fabricated by dual-targets HiPIMS. *Surf. Coat. Technol.* **2021**, *410*, 126967. [CrossRef]
34. Bouabibsa, I.; Lamri, S.; Alhussein, A.; Minea, T.; Sanchette, F. Plasma investigations and deposition of Me-DLC (Me = Al, Ti or Nb) obtained by a magnetron sputtering-RFPECVD hybrid process. *Surf. Coat. Technol.* **2018**, *354*, 351–359. [CrossRef]
35. Bouabibsa, I.; Lamri, S.; Sanchette, F. Structure, mechanical and tribological properties of Me-doped diamond-like carbon (DLC)(Me= Al, Ti, or Nb) hydrogenated amorphous carbon coatings. *Coatings* **2018**, *8*, 370. [CrossRef]
36. Lugo, D.; Silva, P.; Ramirez, M.; Pillaca, E.; Rodrigues, C.; Fukumasu, N.; Corat, E.; Tabacniks, M.; Trava-Airoldi, V. Characterization and tribologic study in high vacuum of hydrogenated DLC films deposited using pulsed DC PECVD system for space applications. *Surf. Coat. Technol.* **2017**, *332*, 135–141. [CrossRef]
37. Capote, G.; Ramírez, M.; da Silva, P.; Lugo, D.; Trava-Airoldi, V. Improvement of the properties and the adherence of DLC coatings deposited using a modified pulsed-DC PECVD technique and an additional cathode. *Surf. Coat. Technol.* **2016**, *308*, 70–79. [CrossRef]

38. Chang, S.H.; Tang, T.C.; Huang, K.T.; Liu, C.M. Investigation of the characteristics of DLC films on oxynitriding-treated ASP23 high speed steel by DC-pulsed PECVD process. *Surf. Coat. Technol.* **2015**, *261*, 331–336. [CrossRef]
39. Sanchette, F.; El Garah, M.; Achache, S.; Schuster, F.; Chouquet, C.; Ducros, C.; Billard, A. DLC-Based Coatings Obtained by Low-Frequency Plasma-Enhanced Chemical Vapor Deposition (LFPECVD) in Cyclohexane, Principle and Examples. *Coatings* **2021**, *11*, 1225. [CrossRef]
40. Kovacı, H.; Baran, Ö.; Yetim, A.F.; Bozkurt, Y.B.; Kara, L.; Çelik, A. The friction and wear performance of DLC coatings deposited on plasma nitrided AISI 4140 steel by magnetron sputtering under air and vacuum conditions. *Surf. Coat. Technol.* **2018**, *349*, 969–979. [CrossRef]
41. Ghiotti, A.; Bruschi, S. Tribological behaviour of DLC coatings for sheet metal forming tools. *Wear* **2011**, *271*, 2454–2458. [CrossRef]
42. Toboła, D.; Kania, B. Phase composition and stress state in the surface layers of burnished and gas nitrided Sverker 21 and Vanadis 6 tool steels. *Surf. Coat. Technol.* **2018**, *353*, 105–115. [CrossRef]
43. Tillmann, W.; Dias, N.F.L.; Stangier, D. Influence of plasma nitriding pretreatments on the tribo-mechanical properties of DLC coatings sputtered on AISI H11. *Surf. Coat. Technol.* **2018**, *357*, 1027–1036. [CrossRef]
44. Podgornik, B.; Hogmark, S.; Sandberg, O.; Leskovsek, V. Wear resistance and anti-sticking properties of duplex treated forming tool steel. *Wear* **2003**, *254*, 1113–1121. [CrossRef]
45. Czyzniewski, A. Optimising deposition parameters of W-DLC coatings for tool materials of high speed steel and cemented carbide. *Vacuum* **2012**, *86*, 2140–2147. [CrossRef]
46. Claver, A.; Jiménez-Piqué, E.; Palacio, J.F.; Almandoz, E.; Fernández de Ara, J.F.; Fernández, I.; Santiago, J.A.; Barba, E.; García, J.A. Comparative Study of Tribomechanical Properties of HiPIMS with Positive Pulses DLC Coatings on Different Tools Steels. *Coatings* **2021**, *11*, 28. [CrossRef]
47. Moreno-Bárcenas, A.; Alvarado-Orozco, J.M.; Carmona, J.M.G.; Mondragón-Rodríguez, G.C.; González-Hernández, J.; García-García, A. Synergistic effect of plasma nitriding and bias voltage on the adhesion of diamond-like carbon coatings on M2 steel by PECVD. *Surf. Coat. Technol.* **2019**, *374*, 327–337. [CrossRef]
48. An, X.; Wu, Z.; Liu, L.; Shao, T.; Xiao, S.; Cui, S.; Lin, H.; Fu, R.K.; Tian, X.; Chu, P.K.; et al. High-ion-energy and low-temperature deposition of diamond-like carbon (DLC) coatings with pulsed kV bias. *Surf. Coat. Technol.* **2018**, *365*, 152–157. [CrossRef]
49. Sheeja, D.; Tay, B.K.; Lau, S.P.; Shi, X. Tribological properties and adhesive strength of DLC coatings prepared under different substrate bias voltages. *Wear* **2001**, *249*, 433–439. [CrossRef]
50. Sun, Y.; Bloyce, A.; Bell, T. Finite element analysis of plastic deformation of various TiN coating/substrate systems under normal contact with a rigid sphere. *Thin Solid Film.* **1995**, *271*, 122–131. [CrossRef]
51. Łępicka, M.; Grądzka-Dahlke, M.; Pieniak, D.; Pasierbiewicz, K.; Niewczas, A. Effect of mechanical properties of substrate and coating on wear performance of TiN- or DLC-coated 316LVM stainless steel. *Wear* **2017**, *382–383*, 62–70. [CrossRef]
52. Li, Z.; Guan, X.; Wang, Y.; Li, J.; Cheng, X.; Lu, X.; Wang, L.; Xue, Q. Comparative study on the load carrying capacities of DLC, GLC and CrN coatings under sliding-friction condition in different environments. *Surf. Coat. Technol.* **2017**, *321*, 350–357. [CrossRef]
53. Kovacı, H.; Yetim, A.F.; Baran, Ö.; Çelik, A. Tribological behavior of DLC films and duplex ceramic coatings under different sliding conditions. *Ceram. Int.* **2018**, *44*, 7151–7158. [CrossRef]
54. Zawischa, M.; Supian, M.M.A.B.M.; Makowski, S.; Schaller, F.; Weihnacht, V. Generalized approach of scratch adhesion testing and failure classification for hard coatings using the concept of relative area of delamination and properly scaled indenters. *Surf. Coat. Technol.* **2021**, *415*, 127118. [CrossRef]
55. Arnell, D. Mechanisms and laws of friction and wear. In *Tribology Dynamic Engine Powertrain Fundamentals, Applications and Future Trends*; Elsevier Ltd.: Amsterdam, The Netherlands, 2010; pp. 41–72, ISBN 9781845693619.
56. Williams, J. *Engineering Tribology*; Cambridge University Press: Cambridge, UK, 2005; ISBN 9780511805905.
57. Hutchings, I.M. *Tribology: Friction and Wear of Engineering Materials*; Edward Arnold A Division of Hodder & Stoughton: London, UK, 1992; pp. 77–78, ISBN 0-340-56184-x.
58. Balla, V.K.; Das, M. Advances in Wear and Tribocorrosion Testing of Artificial Implants and Materials: A Review. *Trends Biomater. Artif. Organs* **2017**, *31*, 150–163.
59. Pinedo, C.E.; Monteiro, W.A. Surface hardening by plasma nitriding on high chromium alloy steel. *J. Mater. Sci. Lett.* **2001**, *20*, 147–150. [CrossRef]
60. Brühl, S.P.; Charadia, R.; Sanchez, C.; Staia, M.H. Wear behavior of plasma nitrided AISI 420 stainless steel. *Int. J. Mater. Res.* **2008**, *99*, 779–786. [CrossRef]
61. Dalibon, E.; Charadia, R.; Cabo, A.; Brühl, S.P. Short Time Ion Nitriding of AISI 420 Martensitic Stainless Steel to Improve Wear and Corrosion Resistance. *Mater. Res.* **2019**, *22*. [CrossRef]
62. Czél, G.; Baán, M.; Makk, P.; Raffay, C.; Fancsali, J.; Janovszky, D. Research Scene to Examine Mechanical Strength of Coatings on Metallic Substrate. In Proceedings of the Fourth International Symposium on Measurement Technology and Intelligent Instruments, ISMT II '98, ISMTII 1998, Miskolc-Lillafured, Hungary, 2–4 September 1998; Dudas, I., Ed.; Springer: Budapest, Hungary, 1998; pp. 335–338, ISBN 963-8455-578.
63. ASTM C1624-22; Standard Test Method for Adhesion Strength and Mechanical Failure Modes of Ceramic Coatings by Quantitative Single Point Scratch Testing. ASTM International: West Conshohocken, PA, USA, 2022.

64. *EN ISO 26423:2016*; Fine Ceramics (Advanced Ceramics, Advanced Technical Ceramics)—Determination of Coating Thickness by Crater-Grinding Method. International Organisation for Standardization: Geneva, Switzerland, 2016.
65. Kocsisné Baán, M.; Marosné, B.M.; Szilágyiné, B.A. *(Szerk): Nitridálás-Korszerű Eljárások és Vizsgálati Módszerek*; Miskolci Egyetem: Miskolc, Hungary, 2015; p. 296, ISBN 978-963-3580-806.
66. Riazi, H.; Ashrafizadeh, F.; Hosseini, S.R.; Ghomashchi, R.; Liu, R. Characterization of simultaneous aged and plasma nitrided 17-4 PH stainless steel. *Mater. Charact.* **2017**, *133*, 33–43. [CrossRef]
67. Siddiqui, S.A.; Favaro, G.; Berkes Maros, M. Investigation of the Damage Mechanism of CrN and Diamond-Like Carbon Coatings on Precipitation-Hardened and Duplex-Treated X42Cr13/W Tool Steel by 3D Scratch Testing. *J. Mater. Eng. Perform.* **2022**, *31*, 7830–7842. [CrossRef]
68. Ellermann, A.; Scholtes, B. The Bauschinger Effect in Different Heat Treatment Conditions of 42CrMo4. *Int. J. Struct. Changes Solids-Mech. Appl.* **2011**, *3*, 1–13.
69. Queyreau, S.; Devincre, B. Bauschinger effect in precipitation-strengthened materials: A dislocation dynamics investigation. *Philos. Mag. Lett.* **2009**, *89*, 419–430. [CrossRef]
70. Kostryzhev, A.G.; Strangwood, M.; Davis, C.L. Bauschinger effect in Nb and V alloyed line-pipe steels. *Ironmak. Steelmak.* **2009**, *36*, 186–192. [CrossRef]
71. Roostaei, A.A.; Jahed, H. (Eds.) *Cyclic Plasticity of Metals*; Elsevier Inc.: Amsterdam, The Netherlands, 2022; ISBN 978-0-12-819293-1. [CrossRef]
72. Bell, T.; Sun, Y. Load bearing capacity of plasma nitrided steel under rolling–sliding contact. *Surf. Eng.* **1990**, *6*, 133–139. [CrossRef]
73. Mussa, A.; Krakhmalev, P.; Bergström, J. Sliding wear and fatigue cracking damage mechanisms in reciprocal and unidirectional sliding of high-strength steels in dry contact. *Wear* **2019**, *444–445*, 203119. [CrossRef]
74. Siddiqui, S.A. Enhancing the Tribological Performance of X42cr13 Steel by Simple and Duplex Treated CrN and DLC Coating. Ph.D. Thesis, University of Miskolc, Miskolc, Hungary, 2022; pp. 1–117. Available online: http://193.6.1.94:9080/JaDoX_Portlets/documents/document_40884_section_38533.pdf (accessed on 3 December 2022).

Article

Optimization of Plasma Electrolytic Oxidation Technological Parameters of Deformed Aluminum Alloy D16T in Flowing Electrolyte

Liubomyr Ropyak [1,*], Thaer Shihab [2,3], Andrii Velychkovych [4], Vitalii Bilinskyi [1], Volodymyr Malinin [5] and Mykola Romaniv [1]

1. Department of Computerized Engineering, Ivano-Frankivsk National Technical University of Oil and Gas, 76019 Ivano-Frankivsk, Ukraine
2. Medical Instruments Techniques Engineering Department, Technical College of Engineering, Al-Bayan University, Baghdad 10070, Iraq
3. Department of Welding, Ivano-Frankivsk National Technical University of Oil and Gas, 076019 Ivano-Frankivsk, Ukraine
4. Department of Construction and Civil Engineering, Ivano-Frankivsk National Technical University of Oil and Gas, 15 Karpatska Str., 076019 Ivano-Frankivsk, Ukraine
5. Department of Physical Principles for Surface Engineering, G.V. Kurdyumov Institute for Metal Physics, National Academy of Sciences of Ukraine, 36 Academician Vernadsky Boulevard, 003142 Kyiv, Ukraine
* Correspondence: l_ropjak@ukr.net

Citation: Ropyak, L.; Shihab, T.; Velychkovych, A.; Bilinskyi, V.; Malinin, V.; Romaniv, M. Optimization of Plasma Electrolytic Oxidation Technological Parameters of Deformed Aluminum Alloy D16T in Flowing Electrolyte. *Ceramics* 2023, 6, 146–167. https://doi.org/10.3390/ceramics6010010

Academic Editors: Amirhossein Pakseresht and Kamalan Kirubaharan Amirtharaj Mosas

Received: 6 November 2022
Revised: 29 December 2022
Accepted: 5 January 2023
Published: 10 January 2023

Copyright: © 2023 by the authors. Licensee MDPI, Basel, Switzerland. This article is an open access article distributed under the terms and conditions of the Creative Commons Attribution (CC BY) license (https://creativecommons.org/licenses/by/4.0/).

Abstract: The prospects of plasma electrolytic oxidation (PEO) technology applied for surface hardening of aluminum alloys are substantiated. The work aims to optimize the technological process of PEO for aluminum in flowing electrolyte. The design of the equipment and the technological process of the PEO for aluminum deformed alloy D16T in flowing silicate–alkaline electrolyte have been developed. Oxide coatings were formed according to various technological parameters of the PEO process. The properties of the oxide coatings were evaluated, respectively, by measurements of coating thickness, geometric dimensions of the samples, microhardness, wear tests, and optical and scanning electron microscopy. To study the influence of the technological parameters of the PEO process of forming oxide coatings on geometrical, physical, and mechanical properties, planning of the experiment was used. According to the results of the conducted experiments, a regression equation of the second order was obtained and the response surfaces were constructed. We determined the optimal values of the technological parameters of the PEO process: component concentration ratio (Na_2SiO_3/KOH), current density, flow rate, and electrolyte temperature, which provide the oxide coating with minimal wear and sufficiently high physical and mechanical properties and indicators of the accuracy of the shape of the parts. The research results showed that the properties of oxide coatings mainly depend on almost all constituent modes of the PEO process. Samples with Al_2O_3 oxide coating were tested during dry friction according to the "ring–ring" scheme. It was established that the temperature in the friction zone of aluminum samples with an oxide coating is lower compared to steel samples without a coating, and this indicates high frictional heat resistance of the oxide coating.

Keywords: plasma electrolytic oxidation; aluminum; coating; technological process; experiment planning; microhardness; wear and tear; cone-likeness; friction heat resistance of materials

1. Introduction

Today, aluminum alloys are widely used in the automotive, aerospace, and radio electronics industries, nuclear engineering, mining, oil and gas production, as well as in construction and other areas of modern engineering [1–3]. Relatively high ratios between strength characteristics and specific weight, high thermal conductivity, good machinability by cutting and plastic deformation, and high corrosion resistance are the most important properties that make aluminum alloys technologically attractive and cost-effective structural materials [4–6].

However, in many cases, insufficient wear resistance, heat resistance, and vulnerability to thermal shock remain the restraining factors that significantly limit the scope of application of aluminum alloys [7–9]. Aluminum parts are usually operated at temperatures up to 230 °C because, at higher temperatures (T > 0.4 T_m, where T_m is the melting temperature of the alloy), diffusion processes occur, especially grain boundary diffusion, which begins to play an important role, which leads to extensive growth grains during deformation [10,11]. Attention should also be paid to the problem of hydrogen embrittlement of aluminum alloys, which occurs in modern tanks with high-pressure hydrogen [12], the danger of sulfide corrosion cracking of critical parts of drilling, oil and gas industrial and pumping equipment [13], as well as problems regarding the interaction of biological environments with metal implants [14–16].

In order to take advantage of the significant advantages and eliminate certain disadvantages of aluminum alloys for specific practical applications, engineers use constructive, operational, and technological methods.

Design methods include selection of rational forms of parts from aluminum alloys [17], rational choice of the alloy grade [18,19], as well as conducting stress state studies [20], temperature calculations of layered compositions [21,22], and use of monitoring of damage to coatings [23–25].

Some customers are skeptical about protective coatings of working surfaces of products because, in practice, application of coatings sometimes faces the problem of their premature destruction [26,27]. Abnormal conditions during operation can cause accelerated reduction in the service life of machine components with protective coatings [28,29]. Therefore, "base material–coating" compositions should always be required to combine special properties (for example, heat resistance and wear resistance) with a sufficient margin of strength [30–32]. Here, first, it is necessary to evaluate the strength of the "base material–coating" composition as a two-layer deformable body under the action of operational loads [33,34].

Thus far, significant progress has been made in the field of mechanics of thin films, coatings, and overlays in the presence of singular stress fields caused by sharp defects or localized loads. The influence of the curvature and shape of the damaged surface on the strengthening effect of the applied coating was studied in works [35,36] using the theory of thin shells. Stress concentration near crack-like defects in coating itself was the subject of studies described in publications [37,38]. 1D [39–41] and 2D [42,43] models are used to develop analytical methods for assessing the stress state of layered coatings under local loading. An example of such an analysis of a ceramic–aluminum coating under an arbitrarily oriented load concentrated along a line is [44].

Among the technological methods, the choice of methods of mechanical processing of parts to reduce manufacturing errors, including considering technological heredity [45] to ensure long-term operation throughout the life cycle of products [46] with coatings, deserves attention. To protect aluminum alloys from wear corrosion at elevated temperatures, various methods of surface strengthening are used [47,48], namely titanium modification during ultrasonic shock treatment [49], silicon carbide during laser surface treatment [50], formed coating by electric spark alloying method [51], high-speed oxygen-fuel HVOF coating [52], electrochemical chromium plating [53], hard anodizing [54], and plasma electrolytic oxidation (PEO), which is also known as micro-arc oxidation (MAO) [55–58].

It is known that, among metal coatings, the most widespread are chrome, and, among non-metallic coatings, oxide. PEO can be a promising alternative [59] to replace use of environmentally harmful chrome plating processes for both aluminum [53] and other metals [60–62].

Among several strengthening methods, PEO should be singled out, which is intensively developed and is currently one of the most popular, environmentally friendly, and fairly cost-effective technologies for forming oxide layers on aluminum and its alloys [55–58], as well as other metals, for example, titanium [63,64], magnesium [65], and steels [66]. In addition, PEO is one of the fundamental ways to improve the operational properties of products by modifying the working surfaces of aluminum parts in order to transform the surface layer into a hard wear-resistant and heat-resistant oxide ceramic [55–58]. PEO is a technological process of forming ceramic layers on the surface of aluminum alloys in an electrolyte under high voltages in the mode of spark and micro-arc discharges, which enables obtaining higher-quality oxide coatings that are firmly attached to the base of the part, comparable to hard anodizing and plasma spray ceramic [67]. The PEO process is accompanied by discharges that develop under the influence of a strong electric field in a system consisting of a part (substrate), an oxide layer, a vapor–gas envelope, an electrolyte, and an electrode (usually a galvanic bath made of stainless steel) [58]. Electrical breakdown in this system leads to emergence of a plasma state in the discharge channel, in which, during plasma–chemical reactions, the substrate material is transformed into chemical compounds consisting of the substrate material itself (including alloying elements), oxygen, and electrolyte components [56,57]. When applying rational technological parameters of the PEO process, we obtain a multifunctional wear-resistant and heat-resistant oxide ceramic coating that has a reliable chemical and metallurgical connection with the main material of the part [55–58,63–65,67].

Aluminum deformed alloy D16T has fairly high mechanical properties compared to other aluminum alloys, and its products are widely used in industry and everyday life. This alloy belongs to the aluminum alloys of the Al–Cu–Mg system, and it is subject to hardening and natural aging. Intermetallics formed in the microstructure are the main factor in the strengthening mechanism of this alloy. However, these intermetallics lead to microelectrochemical inhomogeneity of the alloy, which leads to pitting corrosion, intergranular corrosion, or delamination [68–70]. Therefore, deformed alloys of such a system are used clad with a layer of aluminum and/or subjected to PEO [71]. Predicting the phase composition of multicomponent Al-based alloys during PEO within certain thermodynamic approaches [72–74] or first-principles calculations [75,76] is complicated due to formation of metastable gradient structures, whose redistribution requires detailed statistical analysis.

Microstructural properties, part surface morphology, wear resistance, heat resistance, frictional heat resistance, and other operational properties of oxide ceramic coatings formed by the PEO method depend on the component composition of the electrolyte [56–58,77] and technological process parameters [56–58,78–80]. In particular, in [81], the influence of technological parameters of PEO on corrosion properties of oxide coatings was investigated. Researchers [82–84] studied the tribological characteristics of coatings formed by PEO under various types of lubrication and established high wear resistance during friction in a pair with carbon steel and revealed the effect of hydrogen on the wear resistance of steels in contact with PEO layers synthesized on aluminum alloys during tests in mineral lubricant [85]. A number of studies are devoted to study of thermal conductivity, thermal protective properties [56,86–89], and thermal shock resistance [90] of oxide ceramic coatings on various materials, the results of which demonstrate low thermal conductivity and high protective properties of these coatings, including taking into account the thermal conductivity of the substrate [91]. Papers [92,93] report on use of aluminum oxide to reinforce composite coatings to improve their protective properties.

Thus, depending on the selected material of the substrate, electrical parameters of the PEO process (current density, ratio of anode and cathode currents, voltage, frequency, time), chemical composition, concentration, and temperature of the electrolyte, it is possible to obtain oxide ceramic coatings with a complex of physical and mechanical properties that most fully satisfy the specific operational requirements of consumers [55–58]. The complexity of choosing rational technological parameters of PEO is due to the diversity of the chemical composition of aluminum alloys, as well as the component composition of electrolytes, temperature, and electrical regimes of coating formation. However, there is practically no information in the literature about the tribological and thermal characteristics of parts with oxide coatings and the processes of formation of PEO coatings in the flowing electrolyte. Therefore, development of PEO strengthening processes for formation of oxide ceramic coatings on the surface of parts made of heterogeneous alloys is an urgent problem. Use of the experiment planning methodology during development of various technologies enables significantly reducing the number of experiments and establishing optimal process modes [94–100]. The technological process of PEO should be optimized specifically for the selected brand of deformed aluminum alloy and depending on the field of application and the desired operational properties of the parts covered with enveloping ceramic coatings.

The study aims to determine the optimal technological parameters of the PEO process and the component composition of the electrolyte to ensure formation of a wear-resistant oxide ceramic coating on aluminum deformed alloy D16T in the flowing electrolyte.

To achieve the goal, the following tasks should be addressed:

- Develop a mathematical model of oxide coating forming and study the influence of the PEO technological parameters on the physical and mechanical properties of oxide coatings and the geometric dimensions of parts;
- Determine the optimal technological parameters to provide the maximum microhardness, minimal wear with low cone-likeness of the cylindrical surface of the coated part, as well as justify the placement of parts taking into account their shape in the electrochemical cell during PEO;
- Conduct microscopic studies of the oxide coating;
- Investigate the frictional heat resistance of the oxide coating.

2. Materials and Methods

2.1. Research Materials and Equipment

The authors studied the strengthening process of PEO samples made of aluminum deformed alloy D16T (State Standart GOST 4784–2019 Aluminum and aluminum alloys are deformable. Marks (ISO 209:2007, NEQ)) of the Al–Cu–Mg system, hardened and naturally aged). Chemical composition and properties are presented in Tables 1 and 2, respectively.

Table 1. Chemical composition of the aluminum deformed alloy D16T, mass.%.

Si	Fe	Cu	Mn	Mg	Cr	Zn	Ti	Ti+Zr	Others Elements	Al
0.50	0.50	3.8–4.9	0.3–0.9	1.2–1.8	0.10	0.25	0.15	0.20	0.15	The rest

Table 2. Physical and mechanical properties of the aluminum deformed alloy D16T.

σ_{UTS}, MPa	σ_Y, MPa	δ_5	E, GPa	HB	α, degree^{-1}	λ, W/(m·degree)	ρ, kg/m^3	C, J/(kg·degree)
390–420	255–275	10–12	72	105	0.0000229	130	2770	0.922

Manufacturing of parts with operational surfaces strengthened with PEO coatings was carried out by the developed technological process (Figure 1).

Based on a review of scientific and technical literature, patents, and the results of our own preliminary studies of the PEO process, we chose a water-based silicate–alkaline electrolyte for experiments. For preparation of electrolytes, chemically pure components were used: potassium hydroxide (KOH), State Standart GOST 9285−78 and sodium silicate (Na_2SiO_3), and State Standart GOST 13078−81. The electrolyte was prepared by simply mixing the components in distilled water, which was obtained using water distiller DE-4-2.

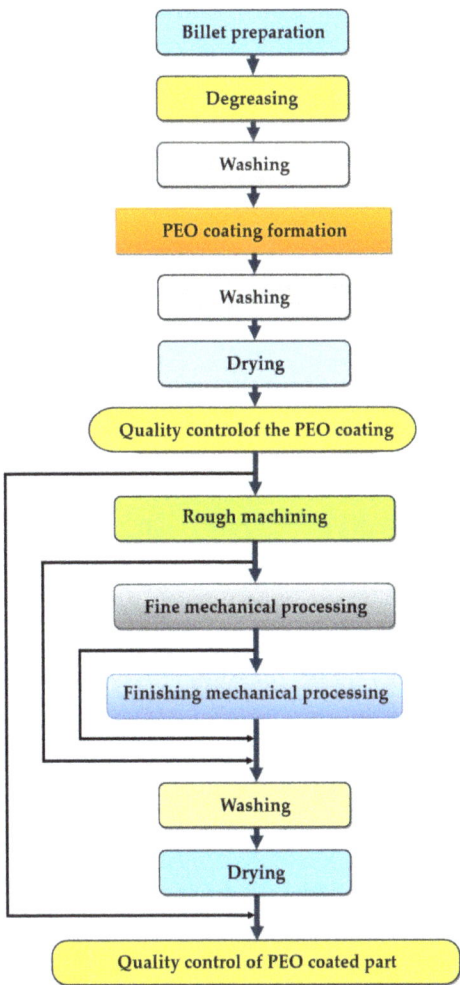

Figure 1. Structural diagram of manufacturing of parts with operational surfaces strengthened with PEO coatings.

PEO of aluminum deformed alloy D16T was carried out on a device developed by us, which enables formation of oxide coatings on parts in a flowing electrolyte. The installation includes an alternating current power source (voltage up to 1000 V), a system for measuring voltage and current values in the cathodic and anodic periods, an electrolyte circulation system, variable electrochemical cells with stainless steel electrodes, an electrolyte cooling system, an exhaust ventilation system, and a damage blocking system electric current. Before forming the PEO coating, the aluminum alloy samples were subjected to mechanical processing to obtain a surface roughness of $Ra = 0.8–1.1$ μm. In the studies, the average value of $Ra = 0.9$ μm was taken. During formation of the oxide coating, the cylindrical sample was placed vertically, coaxially with the axis of the electrochemical cell. Technological modes of coating formation are shown in Table 3. The thickness of the formed working layer of the oxide coating was approximately 300 μm.

Table 3. Coding of factors and levels of their variation during PEO experiments.

Levels of Factors	Coded Values				Natural Values			
	x_1	x_2	x_3	x_4	Concentration Ratio (Na_2SiO_3/KOH), C	Current Density, i, A/dm^2	Flow Rate, v, cm/s	Electrolyte Temperature, T, °C
The main level	0	0	0	0	4.4	5	75	50
Interval of variation	1	1	1	1	1.3	2	40	15
The upper level	+1	+1	+1	+1	5.7	7	115	65
The lower level	−1	−1	−1	−1	3.1	3	45	35
Star points (+)	+1.4826	+1.4826	+1.4826	+1.4826	6.3	8	141.7	72.2
Star points (−)	−1.4826	−1.4826	−1.4826	−1.4826	2.6	2	8.3	27.8

2.2. Microstructure Observation and Mechanical Properties Measurement

The microhardness of coatings and the base was measured on transverse microsands using a PMT-3 hardness tester. The load on the diamond Vickers pyramid was 0.5 N and 1.0 N with the holding time of 12 s to 15 s.

The cone-likeness of the coated part is the deviation of the profile of the longitudinal section for which the components are rectilinear, but not parallel, determined from the dependence:

$$Y_C = (d_{max} - d_{min})/2, \quad (1)$$

where d_{max} and d_{min} are the maximum and minimum diameter of the coated part, respectively. Measurements were carried out using a clock-type indicator with a division price of 1 μm.

The total thickness of oxide coatings was determined on cross-sectional microsections with an optical microscope MIM-7 and PMT-3 hardness tester using an eyepiece micrometer with a resolution of 1 μm.

We conducted the microscopic studies of oxide coatings, PEO-formed on aluminum alloy D16T, at the Center for collective use of scientific instruments "Center for Electron Microscopy and X-ray Microanalysis" of the Karpenko Physico-mechanical Institute of the National Academy of Sciences of Ukraine, Lviv. We used a ZEISS Stemi 2000-C (Germany) optical microscope, and ZEISS EVO 40XVP (ZEISS Group, Jena, Germany) scanning electron microscope with a micro X-ray spectral analysis system and an INCA ENERGY 350 (Oxford Instruments, Abingdon, UK) energy dispersive X-ray spectrometer. Before conducting electron microscopic studies, we applied a monolayer of gold to the surface of the samples to increase their conductivity.

2.3. Wear Test

Wear tests of samples with PEO coatings were carried out on a bench for researching friction pairs during reciprocating motion, which simulates the real operating conditions of elements of metal–rubber friction pairs of piston pumps: cylinder sleeve–piston seal; piston rod–seal; plunger–seal [101]. The amount of weight loss due to wear samples was determined by the gravimetric method: weighing the test specimens on the analytical axis of VLA-200g-M (weight accuracy of 0.1 mg) before and after the test. Before each weighing, the samples were thoroughly wiped with ethyl alcohol and dried. The arithmetic mean value from three tests for wear of the PEO ceramic coatings was chosen as the resulting value for determining the wear resistance.

Evaluation of the frictional heat resistance of oxide coatings formed by PEO was carried out according to recommendations (State Standart GOST 23.210–80 Ensuring the wear resistance of products. The method of evaluating the frictional heat resistance of materials). Experiments were carried out on the machine for testing materials for friction UMT-1 (PA "Tochprilad", Ivanovo, USSR) during dry friction of samples according to the "ring–ring" face scheme. Dimensions of the samples: outer diameter D = 28 mm, inner diameter d = 20 mm, h = 15 mm, area of the nominal friction surface of the stationary sample A = 2.9 cm^2. The rotation frequency of the moving sample is 1000 min^{-1}, with an accuracy of ±5%. The load on the sample is 200 N, with an accuracy of ±4%. The temperature in the friction contact zone was measured using an automatic electronic potentiometer of accuracy class 0.5 using a chromel-copel (E) thermocouple of the ChK type, the measurement range of which is from −50 to +600 °C. The essence of the method is that the rotating and stationary ring samples of the studied pair of materials are installed coaxially, pressed against each other by end working surfaces with a given axial force, and the temperature of frictional heating is controlled. A stationary ring sample is mounted in a hinge assembly to ensure self-alignment. For comparative tests, structural steel alloyed 40ChN (State Standart GOST 4543−2016), λ = 44 W/(m·degree) was used. The frictional heat resistance of materials (coatings) is judged by the dependence of the temperature change on the test time. The duration of the tests, after the samples were run-in, was monitored with a stopwatch.

2.4. Methods of Planning Experimental Research

Experiment planning was used to optimize the technological parameters of the PEO process in the flowing electrolyte.

To conduct experimental studies, the results of which allow to calculate the coefficients of the regression equations, that is, the dependences for microhardness Y_H, wear Y_W, and the cone-likeness Y_C:

$$Y_H = f(C, i, v, T); Y_W = f(C, i, v, T); Y_C = f(C, i, v, T) \qquad (2)$$

of PEO coatings on D16T aluminum alloy from the technological parameters of the process: the mass ratio of the concentration of the components of the electrolyte Na_2SiO_3/KOH—sodium silicate and potassium hydroxide (C), current density (i, A/dm^2), electrolyte flow rate (v, cm/s), and electrolyte temperature (T, °C) chose an orthogonal central composite plan of the second order.

In our case, for four experimental factors k = 4 and two experiments in the center of plan n_0 = 2, the total number of experiments was N = 2^4 + 2 × 4 + 2 = 26.

Orthogonal central composite planning of PEO experiments was carried out on five coded levels with a step equal to α = 1.4826: −α; −1; 0; +1; +α.

The selected factors meet all the requirements for them. The accuracy of maintaining the technological parameters of the PEO process was 3–5%. The intervals of variation are given in Table 3.

The order of experiments was chosen from a series of random numbers to exclude the influence of unregulated and uncontrollable factors.

To simplify the calculations, when determining the coefficients of the regression equations and their evaluation, a transition was made from the natural values of the factors to coded values (Table 3). Each experiment was repeated three times at the same level of factors and the arithmetic mean value of the initial parameter in each experiment was calculated (Table 4).

A second-order polynomial was used to describe the response function (optimization parameter) of the investigated technological process of forming PEO coatings in the flowing electrolyte:

$$Y = b_0 + \sum_{i=1}^{k} b_i x_i + \sum_{i \neq j}^{k} b_{ij} x_i x_j + \sum_{i=1}^{k} b_{ii} x_i^2, \qquad (3)$$

where b_0, b_i, b_{ij}, b_{ii} regression coefficients, x_i factors of the experiment, k = 4.

The coefficients of the regression equation were calculated in matrix form according to known formulas.

Table 4. Plan-matrix of experimental results of the PEO process.

Experiment Number	Levels of Factors					The Average Value of the Optimization Parameters		
	x_0	x_1	x_2	x_3	x_4	Microhardness Y_H, GPa	Wear Y_W, g	Cone-likeness Y_C, μm
1	+1	−1	−1	−1	−1	13.93	0.277	15.8
2	+1	−1	−1	−1	+1	14.84	0.287	15.3
3	+1	−1	−1	+1	−1	15.11	0.260	10.4
4	+1	−1	−1	+1	+1	14.99	0.272	10.9
5	+1	−1	+1	−1	−1	13.89	0.238	13.1
6	+1	−1	+1	−1	+1	17.73	0.215	11.4
7	+1	−1	+1	+1	−1	16.32	0.234	10.4
8	+1	−1	+1	+1	+1	19.00	0.207	8.2
9	+1	+1	−1	−1	−1	15.37	0.241	13.1
10	+1	+1	−1	−1	+1	13.53	0.258	15.3
11	+1	+1	−1	+1	−1	17.44	0.230	9.3
12	+1	+1	−1	+1	+1	15.47	0.245	8.7
13	+1	+1	+1	−1	−1	19.40	0.201	10.4
14	+1	+1	+1	−1	+1	19.86	0.198	11.4
15	+1	+1	+1	+1	−1	21.95	0.182	7.1
16	+1	+1	+1	+1	+1	20.29	0.196	6.5
17	+1	−1.4826	0	0	0	13.91	0.268	12.1
18	+1	+1.4826	0	0	0	19.65	0.207	7.4
19	+1	0	−1.4826	0	0	14.08	0.287	11.1
20	+1	0	+1.4826	0	0	21.51	0.192	8.8
21	+1	0	0	−1.4826	0	18.91	0.213	14.2
22	+1	0	0	+1.4826	0	19.86	0.205	7.6
23	+1	0	0	0	−1.4826	18.68	0.215	7.7
24	+1	0	0	0	+1.4826	19.63	0.207	8.8
25	+1	0	0	0	0	21.51	0.192	8.3
26	+1	0	0	0	0	21.98	0.188	8.4

The significance of the obtained coefficients of the regression equations was checked by Student's t-test. The coefficient was considered statistically significant under the condition that $|b_s| > t \cdot S(b_i)$, where the t-test, selected from the reference table for the significance level of 0.05 and degrees of freedom f_E; $S(b_i)$—coefficient error (here, b_i is considered b_0, b_i, b_{ij}, b_{ii}). Statistically insignificant coefficients were discarded and regression equations were refined. The resulting regression equations were tested for adequacy using Fisher's F-test.

The calculated value F_c was compared to that selected from the reference table (F_{tabl}) for the f_{ad} and f_E degrees of freedom using 0.05 significance level.

If the condition of regression equation $F_c < F_{tabl}$ is fulfilled, it can be considered adequate at the accepted level of significance 0.05.

On the basis of the obtained regression equations, response surfaces were constructed and the levels of the same output of parameter Y on the horizontal plane were calculated. To determine the optimal values of the technological parameters of the PEO process in the flowing electrolyte, we took the partial derivatives from the regression equations, equated them to zero, and solved the system of equations. Note that more general systems of equations with partial derivatives are considered in the article [102]. Based on the obtained regression equations, the response surfaces of the optimization parameters (microhardness Y_H, wear Y_W, and cone-likeness Y_C) were constructed from two variable technological parameters, with the other two fixed at the basic level (C, i, v, and T).

3. Results and Discussion

3.1. Optimization of the PEO Process Technological Parameters

Ensuring the quality of the working surface of cylindrical parts and improving the operational properties of products is achieved by using PEO in the flow electrolyte. In order to substantiate the technological parameters of the PEO process in the flowing electrolyte and to establish the laws of their influence on the physical and mechanical properties of the coating, parameters of the shape of the part and laboratory experimental studies of samples with coatings formed according to the developed PEO technology were conducted. We studied the change in microhardness, wear, and cone-likeness of the cylindrical surface of the part with PEO coating depending on the mass ratio of the concentrations of the electrolyte components C, the current density i, the electrolyte flow rate v, and the temperature of the electrolyte T. In order to evaluate the quality indicators of the PEO coatings, the microhardness of the coating was measured, tested on wear during reciprocating motion, and the cone-likeness of cylindrical parts was determined.

The unknown coefficients of the regression equation were determined by the matrix method in coded form. The significance of the coefficients of the regression equation was checked at the 0.05 level using Student's t-test. Statistically insignificant coefficients were discarded. After that, they were recoded according to a known method into natural values.

Response functions (optimization parameter) reflecting the dependence of the microhardness of the oxide coating Y_H, wear Y_W, and cone-likeness Y_C of the cylindrical surface depending on the mass ratio of the concentrations of the electrolyte components C, the current density i, and the electrolyte flow rate v, the temperature of the electrolyte T, based on the results of the orthogonal of the central composite planning of the experiment in the natural values of the variable factors, the regression equation of the second order takes the following form for:

- microhardness Y_H:

$$Y_H = -33.1213 + 12.2770C + 2.2724i + 0.5159T - 1.2249C^2 \\ -0.3538i^2 - 0.0037T^2 + 0.2905Ci - 0.0411CT + 0.0174iT; \quad (4)$$

- wear Y_W:

$$Y_W = 46.4174 - 7.1806C - 3.0236i - 0.1534v \\ +0.6687C^2 + 0.2752i^2 + 0.0007v^2; \quad (5)$$

- cone-likeness Y_C:

$$Y_C = 0.7054 - 0.1151C - 0.0470i - 0.0002iT + 0.00115C^2 + 0.0041i^2. \quad (6)$$

After checking the adequacy of the obtained second-order models for this PEO technological process, the Fisher test was applied. Therefore, as $F_c < F_{tabl}$, it satisfies the requirements for the hypothesis about the adequacy of the regression equations, which was accepted.

The obtained second-order regression equations in natural values can be directly used to calculate microhardness Y_H, wear Y_W, and cone-likeness Y_C of parts with an oxide coating depending on the change in the mass ratio of concentrations of electrolyte components C, current density i, A/dm^2, electrolyte flow rate v, cm/s, and electrolyte temperature T, °C, which are, respectively, within the following limits:

$$2.6 \leq C \leq 6.3; 2 \leq i \leq 8; 8.3 \leq v \leq 141.7; 8 \leq T \leq 72.2$$

To analyze the influence of the technological parameters of the PEO process in the flow electrolyte (variable factors) on the optimization parameters, response surfaces and their two-dimensional sections were constructed depending on two variable factors (the other two factors were at a constant basic level) (Figures 2–4).

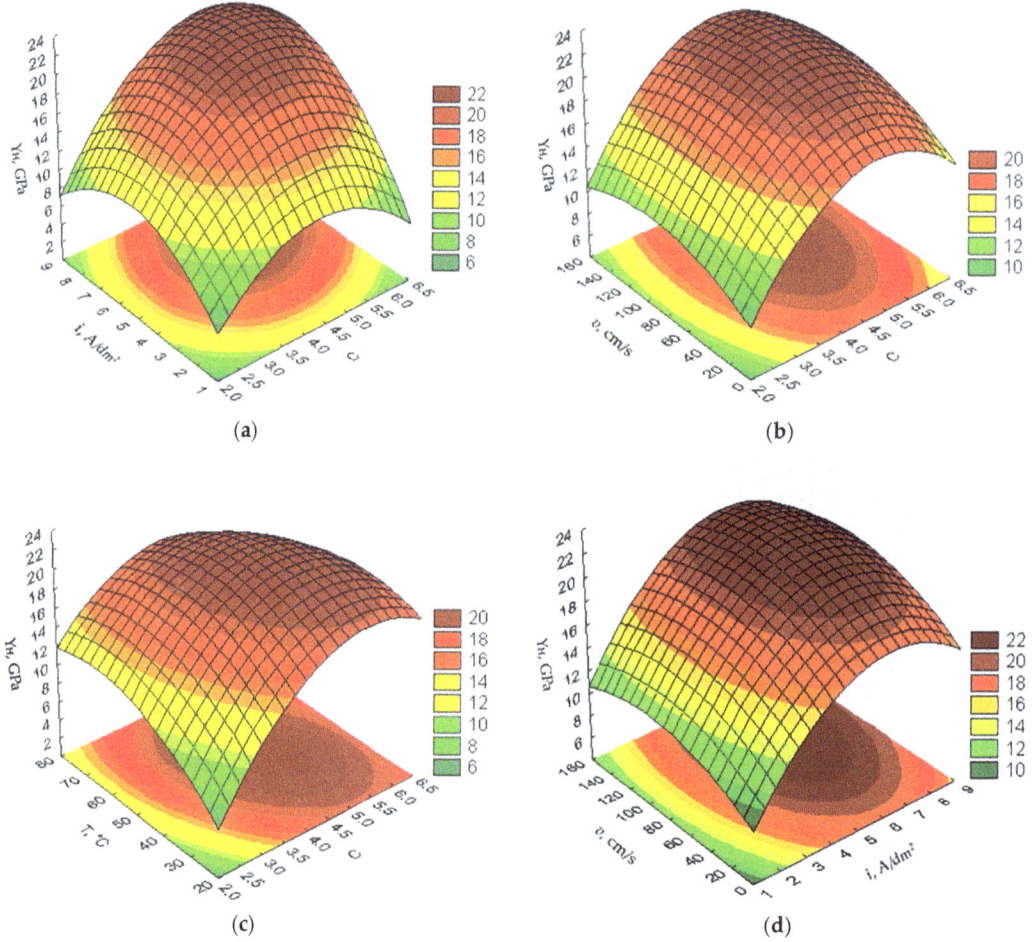

Figure 2. *Cont.*

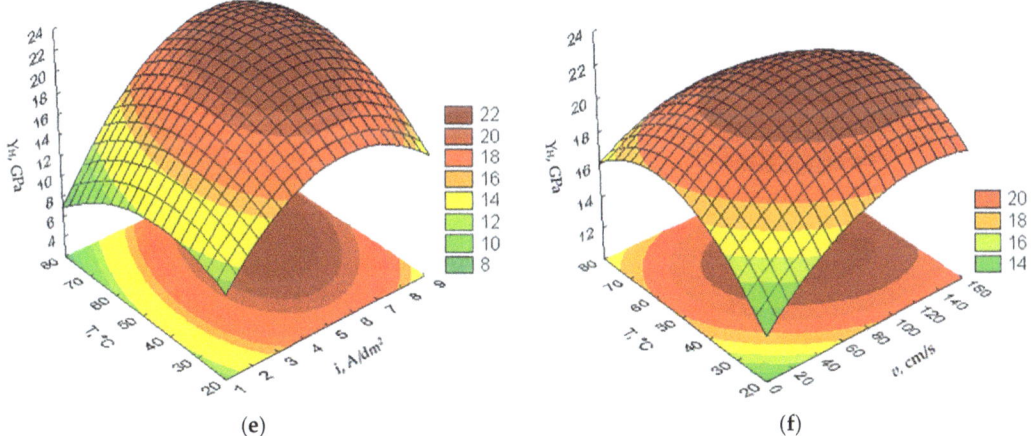

Figure 2. Response surfaces of dependence of the microhardness Y_H coating on the technological parameters of the PEO process in the flowing electrolyte: (**a**)—C, i, $v = 75$ cm/s, $T = 50$ °C; (**b**)—C, v, $i = 5$ A/dm^2, $T = 50$ °C; (**c**)—C, T, $i = 5$ A/dm^2, $v = 75$ cm/s; (**d**)—i, v, $C = 4.4$, $T = 50$ °C; (**e**)—i, T, $C = 4.4$, $v = 75$ cm/s; (**f**)—v, T, $C = 4.4$, $i = 5$ A/dm^2.

Analysis of the obtained second-order regression Equations (4)–(6) and constructed response surfaces (Figures 2–4) shows that the values of the optimization parameters microhardness, wear, and cone-likeness for the oxide coating depend on almost all the technological parameters of the PEO process C, i, v, T.

In addition, the optimal values of the technological parameters of the PEO process in the flow electrolyte for the D16T alloy were determined to ensure the maximum microhardness, minimal wear of the coating, and minimal cone-likeness of the cylindrical surface of the PEO-coated part, which are presented in Table 5.

Table 5. Optimal values of technological parameters of the PEO process.

Optimization Parameters		Technological Parameters			
Name	Optimal Value	Concentration Ratio (Na$_2$SiO$_3$/KOH), C	Current Density, i, A/dm^2	Flow Rate, v, cm/s	Electrolyte Temperature, T, °C
		Minimum Values			
		3.1	3	45	35
		Optimal Values			
Microhardness Y_H, GPa	18.56	5.04	6.63	106.77	49.92
Wear Y_W, g	0.185	4.98	6.61	103.96	53.59
Cone – likeness Y_C, µm	10.6	5.07	5.82	119.52	55.16
		Maximum Values			
	–	5.7	7	115	65

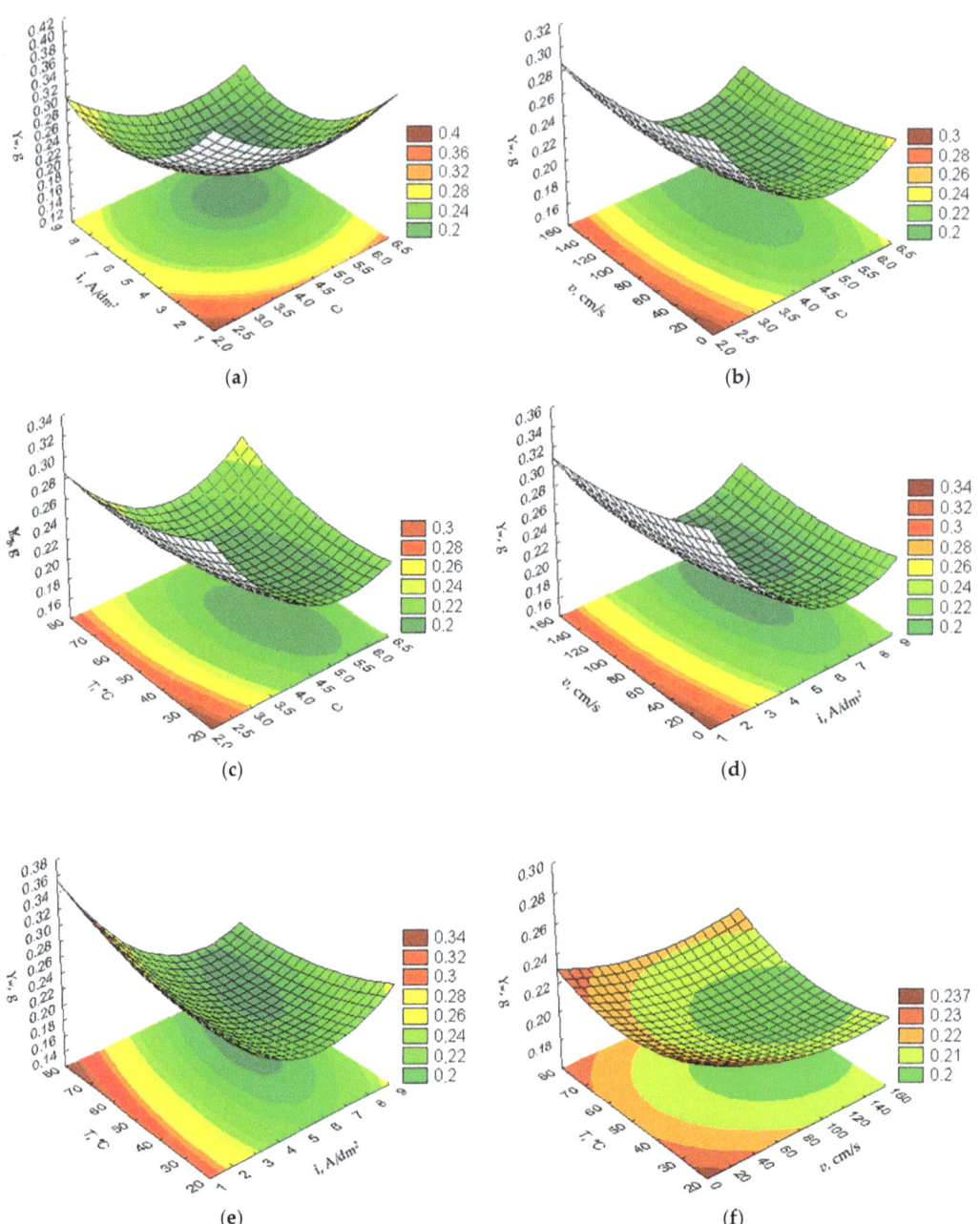

Figure 3. Response surfaces of dependence of the wear Y_W coating on the technological parameters of the PEO process in the flowing electrolyte: (**a**)—C, i, $v = 75$ cm/s, $T = 50$ °C; (**b**)—C, v, $i = 5$ A/dm^2, $T = 50$ °C; (**c**)—C, T, $i = 5$ A/dm^2, $v = 75$ cm/s; (**d**)—i, v, $C = 4.4$, $T = 50$ °C; (**e**)—i, T, $C = 4.4$, $v = 75$ cm/s; (**f**)—v, T, $C = 4.4$, $i = 5$ A/dm^2.

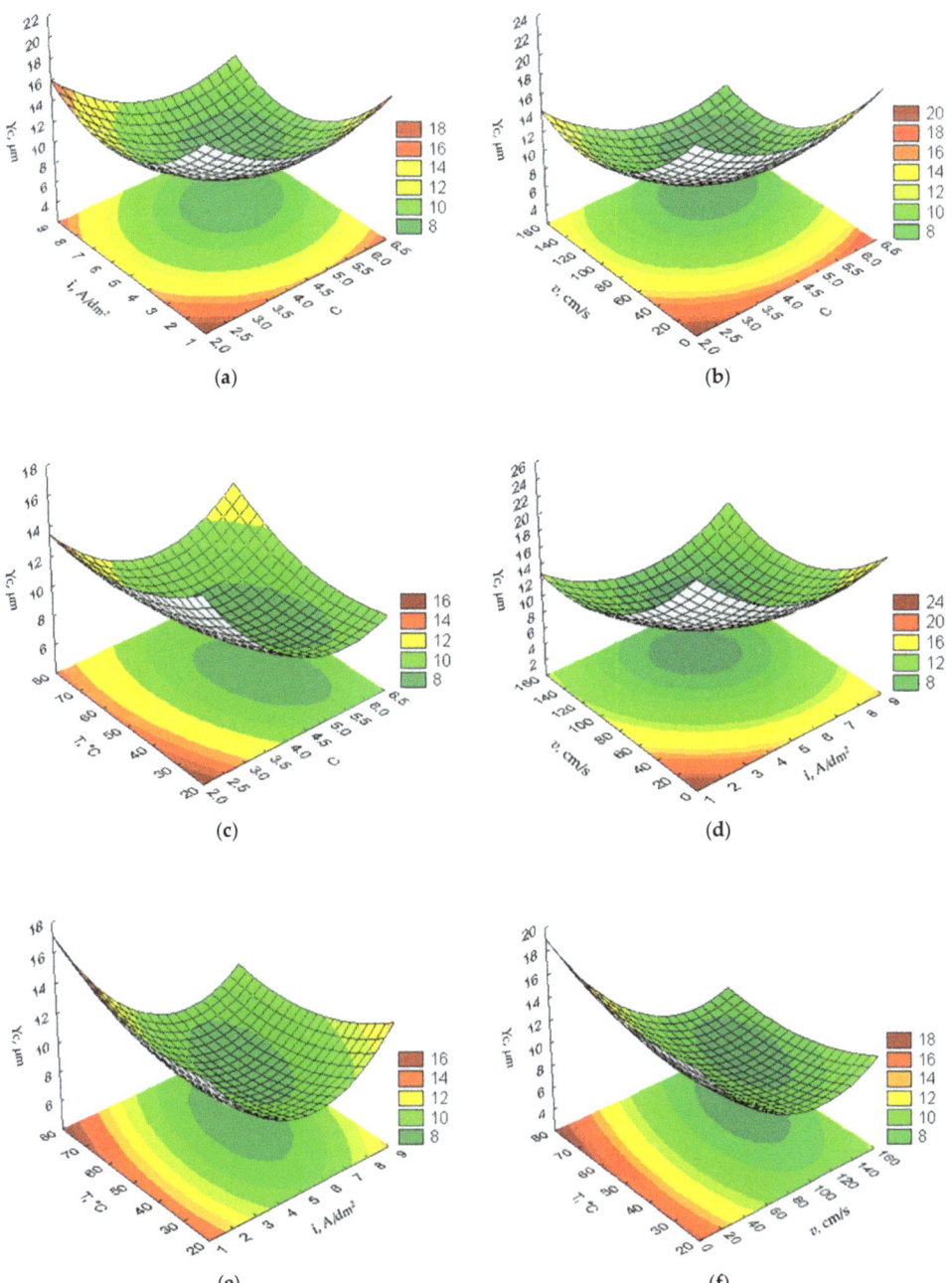

Figure 4. Response surfaces of dependence of the cone-likeness Y_C surface coating on the technological parameters of the PEO process in the flowing electrolyte: (**a**)—C, i, v = 75 cm/s, T = 50 °C; (**b**)—C, v, i = 5 A/dm², T = 50 °C; (**c**)—C, T, i = 5 A/dm², v = 75 cm/s; (**d**)—i, v, C = 4.4, T = 50 °C; (**e**)—i, T, C = 4.4, v = 75 cm/s; (**f**)—v, T, C = 4.4, i = 5 A/dm².

It should be noted that, at the maximum value of microhardness of the formed oxide coating, its minimal wear is ensured, and the optimal values of the technological parameters of the PEO process are within the factor space (Table 5). For the optimization parameters of cone-likeness, the optimal values of technological parameters of the process are also within the factor space.

Analysis of the results of planning the experiment provided in Table 5 shows that the optimal values of the technological parameters of PEO in the flow electrolyte for the optimization parameters microhardness of the oxide coating Y_H, wear Y_W, and cone-likeness Y_C differ from each other. For the first two parameters of optimization–microhardness and wear, the optimal values of the technological parameters of the process practically coincide and are within the range of variation of the factors. That is, with maximum microhardness, minimal wear of the oxide ceramic coating is ensured. The optimal values of the technological parameters to ensure the minimum cone-likeness are slightly different from the optimal technological parameters to ensure the maximum microhardness and minimal wear. At the same time, in order to obtain the minimum value of Y_C, the optimal value of the electrolyte flow rate is beyond the range of variation of the factors, even taking into account the value of the star arm.

Premchand, C. et al. [103] studied the influence of the ratio of electrolyte components on the corrosion resistance of PEO coatings and found that excellent corrosion resistance is achieved when the coating is formed in an aqueous electrolyte with the same concentration ratio of the components $(Na_2SiO_3/KOH) = 1$.

Since wear resistance is an important operational characteristic for products that are operated in abrasive environments, the optimal values of the technological parameters of the PEO process in the flowing electrolyte were taken to be those that ensure the minimum amount of wear of parts $Y_W = 0.185$ g: $C = 4.98$; and $i = 6.61$ A/dm^2; $v = 103.96$ cm/s; $T = 53.59$ °C. By substituting these values of the optimal technological parameters of PEO into the Formulas (4)–(6), we obtained a calculated value of microhardness $Y_H = 17.96$ GPa; i.e., its value is slightly smaller than that obtained during the optimization $Y_H = 18.56$ GPa and cone-likeness $Y_C = 10.9$ µm; i.e., its value is slightly greater than that obtained during the optimization $Y_C = 10.6$ µm (Tables 5 and 6). The necessary cone-likeness of the cylindrical surface in accordance with the technical requirements for production of parts, depending on their functional purpose, is expediently obtained during further machining operations–diamond grinding.

Table 6. Calculated values of the optimization parameters: Y_H, Y_C according to the optimal values of the technological parameters of the PEO process, which ensure minimal wear Y_W.

Optimization Parameters		Technological Parameters				
Name	Optimal Value	Concentration Ratio (Na_2SiO_3/KOH), C	Current Density, i, A/dm^2	Flow Rate, v, cm/s	Electrolyte Temperature, T, °C	Optimization Parameters Deviation, %
		Optimal Values				
		4.98	6.61	103.96	53.59	
		The Values of the Optimization Parameters are Calculated				
Microhardness Y_H, GPa	18.56		17.96			3.35
Wear Y_W, g	0.185		0.185			0
Cone – likeness Y_C, µm	10.6		10.9			2.91

It should be noted that, during PEO in the flowing electrolyte of cylindrical parts in a vertical position, the maximum increase in diameter for the outer surfaces of the shafts and a decrease for the inner surfaces of the bushings was observed in the lower part of the parts. To eliminate cone-likeness and increase the accuracy of cylindrical surfaces, considering technological heredity, during strengthening with oxide ceramic coatings, the parts must be measured and mounted in a vertical position in the electrochemical cell before starting PEO in such a way that the minimum diameter of the shaft is from below, and, for the sleeve during application coating on the inner surface, its maximum diameter was from below.

Analysis of the results presented in Table 6 shows that the calculated values of the optimization parameters for microhardness and cone-likeness, obtained for the optimal values that ensure the minimum amount of wear, differ from their optimal values by 3.35% and 2.91%, respectively.

Thus, the obtained results of the study show the possibility of targeted control of the technological parameters of the PEO process in the flowing electrolyte by changing the technological parameters considering the technological heredity of manufacturing blanks of aluminum deformed alloy parts.

The obtained research results prove the possibility of practical application of the optimal values of the technological parameters of the PEO process in the flow electrolyte for serial production of parts, such as bodies of rotation, for example, cylinder sleeves and pump plungers at machine-building enterprises.

3.2. Microscopic Studies

We formed PEO coating (Figure 5) according to the obtained optimal values of technological parameters. According to Figure 5, PEO coatings have a uniform structure and do not have visible defects, such as pores, microcracks, and others.

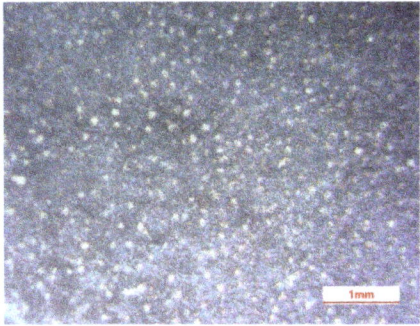

Figure 5. Optical micrograph of the PEO coating after finishing mechanical processing.

Figure 6 shows the results of studying the morphology and composition of PEO coatings by SEM/EDS methods.

The interface between the base metal (D16T) and the PEO coating is shown in Figure 6a. As can be seen from the figure, the interface zone is characterized by integrity and the absence of significant delamination, indicating a high level of chemical bonding. It is caused by direct synthesis of the oxide phase from the base metal due to electrolyte elemental composition during the PEO process.

As can be seen from Figure 6b, the coating has the layered pancake-shaped structure typical of PEO coatings (Sikdar, S. et al.) [58]. The chemical composition analysis indicates the presence of key components for oxide formation (Al and O) and the components of the electrolyte (K, Na, Si) (Figure 6c). At the grain boundaries, there are many evenly distributed small pores formed as a result of gas migration during plasma discharge. The thickness of the PEO coatings was 180–210 μm, which determines the heat-protective properties of oxide ceramic coatings. The mechanism of formation of PEO coatings on aluminum alloys is described in publications [55–58].

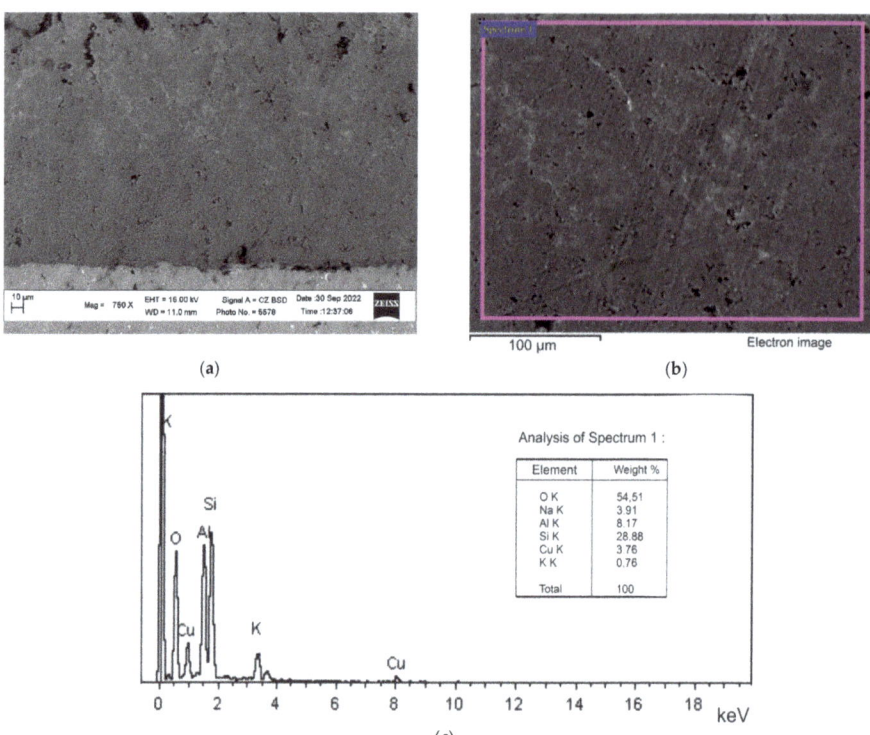

Figure 6. SEM micrograph of the interface between the PEO coating and the base alloy D16T (**a**) and the EDS analysis of the PEO coating: (**b**)—area of EDS analysis; (**c**)—resulting spectrum and chemical composition at the analyzed area.

3.3. Frictional Heat Resistance

The results of evaluation of the frictional heat resistance of PEO coatings during the tests of the samples using the "ring–ring" end friction scheme are presented in Figure 7.

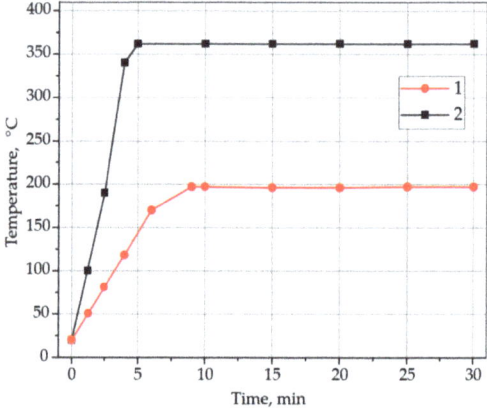

Figure 7. Dependence of the temperature in the friction zone according to the "ring–ring" scheme on the duration of the tests: 1—D16T alloy with PEO coating; 2—steel 40ChN.

From the graphic dependences (Figure 7) of temperature changes in the friction zone on the duration of the tests, the temperature in the contact zone of the samples with the PEO coating stabilizes approximately 9–10 min after the start of the tests, reaches values from 192 to 197 °C, and practically does not change during further wear oxide coating (30 min). During testing of samples made of 40ChN steel, a higher temperature was established in the contact zone, which reached rather high values from 345 to 362 °C, in a less short period of time from 4 min to 6 min after the start of the tests. This attests to the high heat-shielding properties of the oxide ceramic coating and is consistent with the results of research by Curran J. et al. [88], where a low coefficient of thermal conductivity of PEO coatings is indicated, the value of which is from 0.5 to 1.5 $W \cdot m^{-1} \cdot K^{-1}$. Oxides in the composition of the ceramic coating formed by PEO provide high heat-shielding properties of the composition and protection of the base made of aluminum deformed alloy D16T from temperature effects and wear.

Further research is planned to study the effect of heating on the occurrence of thermal stresses in PEO ceramic coatings.

4. Conclusions

As a result of studies of the PEO technological process PEO in the flowing electrolyte of aluminum deformed alloy D16T:

- A mathematical model of the PEO process was built, based on which it was established that the optimal values of the technological parameters to ensure obtaining the maximum microhardness and minimal wear practically coincide, and, for cone-likeness, they do not coincide either with the optimal technological parameters for microhardness and wear or with each other;
- It was established that, in order to ensure high operational performance of parts with an oxide coating, optimization should be carried out according to minimum wear, which is achieved with the following values of the optimal technological parameters of the PEO process: mass ratio of concentrations of electrolyte components $C = 4.98$; current density $i = 6.61$ A/dm^2; electrolyte flow rate in the electrochemical cell $v = 103.96$ cm/s; temperature of the electrolyte $T = 53.59$ °C; and necessary cone-likeness of the surface of the aluminum part should be obtained during further machining operations;
- Vertical mounting of cylindrical parts in the electrochemical cell is justified in order to eliminate the cone-likeness in the PEO process that arose during the previous operations of mechanical processing of the workpiece; for shafts, the minimum diameter of the surface on which the oxide coating is formed is from below, and, for bushings, respectively, from below, the maximum internal diameter to increase the accuracy of parts with oxide coatings since during formation of the oxide coating there is a maximum increase in diameter for external surfaces and a decrease for internal surfaces from bottom to top;
- The results of the SEM analysis of the microstructure of the oxide ceramic coating indicate the presence in the composition of oxides formed from the chemical elements of the alloy material and electrolyte components, the presence of pores, as well as the absence of through cracks;
- Research results showed high friction heat resistance of the oxide ceramic coating, which provides protection of aluminum alloy parts during dry friction, while the temperature stabilization time in the friction zone for a pair of samples with an oxide ceramic coating on the D16T alloy is twice as long as for a pair of samples with steel 40ChN.

Author Contributions: Conceptualization, L.R., A.V. and V.B.; methodology, L.R., T.S., A.V. and V.B.; software, L.R., A.V. and V.B.; validation, L.R., T.S., A.V., V.B. and M.R.; formal analysis, L.R., A.V., V.M. and M.R.; investigation, L.R., A.V., V.M., V.B. and M.R.; resources, L.R., T.S. and V.B.; data curation, L.R. and A.V.; writing—original draft preparation, L.R., A.V. and V.B.; writing—review and editing, L.R. and A.V.; visualization, L.R., T.S., A.V. and V.B.; supervision, L.R. and A.V.; project administration, L.R. and A.V.; funding acquisition, L.R. and A.V. All authors have read and agreed to the published version of the manuscript.

Funding: This research was funded by the Ministry of Science and Education of Ukraine for the grant to implement projects 0121U109591 and 0122U002082.

Institutional Review Board Statement: Not applicable.

Informed Consent Statement: Not applicable.

Data Availability Statement: Data are contained within the article.

Acknowledgments: The authors are grateful for help in statistical data processing hab. inż. Michał Bembenek, AGH, Department of Manufacturing Systems, Faculty of Mechanical Engineering and Robotics, AGH University of Science and Technology, Krakow, Poland. The authors are also grateful for thorough consultations with Vasyl Vytvytskyi and Tetiana Pryhorovska, Department of Engineering and Computer Graphics, Ivano-Frankivsk National Technical University of Oil and Gas. The authors thank the team of the Center for collective use of scientific instruments "Center for Electron Microscopy and X-ray Microanalysis" of the Karpenko Physico-mechanical Institute, National Academy of Sciences of Ukraine, Lviv for promptly conducting microscopic studies of PEO ceramic coatings on aluminum. The team of authors express their gratitude to the reviewers for valuable recommendations that have been taken into account to significantly improve the quality of this paper.

Conflicts of Interest: The authors declare no conflict of interest.

References

1. Cooke, K.O. Introductory Chapter: Structural Aluminum Alloys and Composites. In *Aluminium Alloys and Composites*; Cooke, K.O., Ed.; IntechOpen: London, UK, 2020; pp. 1–14. [CrossRef]
2. Bhatta, L.; Pesin, A.; Zhilyaev, A.P.; Tandon, P.; Kong, C.; Yu, H. Recent Development of Superplasticity in Aluminum Alloys: A Review. *Metals* **2020**, *10*, 77. [CrossRef]
3. Georgantzia, E.; Gkantou, M.; Kamaris, G. Aluminium alloys as structural material: A review of research. *Eng. Struct.* **2021**, *227*, 111372. [CrossRef]
4. Vlasiy, O.; Mazurenko, V.; Ropyak, L.; Rogal, O. Improving the aluminum drill pipes stability by optimizing the shape of protector thickening. *East. Eur. J. Enterp. Technol.* **2017**, *1*, 25–31. [CrossRef]
5. Varshney, D.; Kumar, K. Application and use of different aluminium alloys with respect to workability, strength and welding parameter optimization. *Ain Shams Eng. J.* **2021**, *12*, 1143–1152. [CrossRef]
6. Bazaluk, O.; Velychkovych, A.; Ropyak, L.; Pashechko, M.; Pryhorovska, T.; Lozynskyi, V. Influence of Heavy Weight Drill Pipe Material and Drill Bit Manufacturing Errors on Stress State of Steel Blades. *Energies* **2021**, *14*, 4198. [CrossRef]
7. Czerwinski, F. Thermal Stability of Aluminum Alloys. *Materials* **2020**, *13*, 3441. [CrossRef]
8. Timelli, G.; Fabrizi, A.; Vezzù, S.; De Mori, A. Design of Wear-Resistant Diecast AlSi$_9$Cu$_3$(Fe) Alloys for High-Temperature Components. *Metals* **2020**, *10*, 55. [CrossRef]
9. Udoye, N.E.; Fayomi, O.S.I.; Inegbenebor, A.O. Assessment of Wear Resistance of Aluminium Alloy in Manufacturing Industry-A Review. *Procedia Manuf.* **2019**, *35*, 1383–1386. [CrossRef]
10. Chokshi, A.H.; Meyers, M. The prospects for superplasticity at high strain rates: Preliminary considerations and an example. *Scr. Metall. Mater.* **1990**, *24*, 605–610. [CrossRef]
11. Sergueeva, A.V.; Mara, N.A.; Mukherjee, A.K. Grain Size Distribution Effect on Mechanical Behavior of Nanocrystalline Materials. *MRS Online Proc. Libr.* **2004**, *821*, 336–342. [CrossRef]
12. Chen, Y.; Zhao, S.; Ma, H.; Wang, H.; Hua, L.; Fu, S. Analysis of Hydrogen Embrittlement on Aluminum Alloys for Vehicle-Mounted Hydrogen Storage Tanks: A Review. *Metals* **2021**, *11*, 1303. [CrossRef]
13. Shatskyi, I.; Vytvytskyi, I.; Senyushkovych, M.; Velychkovych, A. Modelling and improvement of the design of hinged centralizer for casing. *IOP Conf. Ser. Mater. Sci. Eng.* **2019**, *564*, 12073. [CrossRef]
14. Barrak, F.N.; Li, S.; Muntane, A.M. Particle release from implantoplasty of dental implants and impact on cells. *Int. J. Implant. Dent.* **2020**, *6*, 50. [CrossRef] [PubMed]
15. Bazaluk, O.; Chuzhak, A.; Sulyma, V.; Velychkovych, A.; Ropyak, L.; Vytvytskyi, V.; Mykhailiuk, V.; Lozynskyi, V. Determining the Tightrope Tightening Force for Effective Fixation of the Tibiofibular Syndesmosis during Osteomeatal Synthesis of Fibula Injuries. *Appl. Sci.* **2022**, *12*, 4903. [CrossRef]

16. Pelekhan, B.; Dutkiewicz, M.; Shatskyi, I.; Velychkovych, A.; Rozhko, M.; Pelekhan, L. Analytical Modeling of the Interaction of a·Four Implant-Supported Overdenture with Bone Tissue. *Materials* **2022**, *15*, 2398. [CrossRef] [PubMed]
17. Zhang, Z.; Sun, C.; Xu, X.; Liu, L. Surface quality and forming characteristics of thin-wall aluminium alloy parts manufactured by laser assisted MIG arc additive manufacturing. *Int. J. Lightweight Mater. Manuf.* **2018**, *1*, 89–95. [CrossRef]
18. Zinchenko, A.; Baiul, K.; Krot, P.; Khudyakov, A.; Vashchenko, S.; Banasiewicz, A.; Wróblewski, A. Materials Selection and Design Options Analysis for a Centrifugal Fan Impeller in a Horizontal Conveyor Dryer. *Materials* **2021**, *14*, 6696. [CrossRef]
19. Levchuk, K.G.; Moisyshyn, V.M.; Tsidylo, I.V. Influence of mechanical properties of a material on dynamics of the stuck drilling pipes. *Metallofiz. Noveishie Tekhnologii* **2016**, *38*, 1655–1668. [CrossRef]
20. Dean, J.; Gu, T.; Clyne, T.W. Evaluation of residual stress levels in plasma electrolytic oxidation coatings using a curvature method. *Surf. Coat. Technol.* **2015**, *269*, 47–53. [CrossRef]
21. Tatsiy, R.M.; Pazen, O.Y.; Vovk, S.Y.; Ropyak, L.Y.; Pryhorovska, T.O. Numerical study on heat transfer in multilayered structures of main geometric forms made of different materials. *J. Serb. Soc. Comput. Mech.* **2019**, *13*, 36–55. [CrossRef]
22. Tatsiy, R.; Stasiuk, M.; Pazen, O.; Vovk, S. Modeling of Boundary-Value Problems of Heat Conduction for Multilayered Hollow Cylinder. In Proceedings of the 2018 International Scientific-Practical Conference on Problems of Infocommunications Science and Technology, PIC S and T 2018, Kharkiv, Ukraine, 9–12 October 2018; pp. 21–25.
23. Shaloo, M.; Schnall, M.; Klein, T.; Huber, N.; Reitinger, B. A Review of Non-Destructive Testing (NDT) Techniques for Defect Detection: Application to Fusion Welding and Future Wire Arc Additive Manufacturing Processes. *Materials* **2022**, *15*, 3697. [CrossRef] [PubMed]
24. Bembenek, M.; Mandziy, T.; Ivasenko, I.; Berehulyak, O.; Vorobel, R.; Slobodyan, Z.; Ropyak, L. Multiclass Level-Set Segmentation of Rust and Coating Damages in Images of Metal Structures. *Sensors* **2022**, *22*, 7600. [CrossRef] [PubMed]
25. Student, M.M.; Ivasenko, I.B.; Posuvailo, V.M.; Veselivs'ka, H.H.; Pokhmurs'kyi, A.Y.; Sirak, Y.Y.; Yus'kiv, V.M. Influence of the Porosity of a Plasma-Electrolytic Coating on the Corrosion Resistance of D16 Alloy. *Mater. Sci.* **2019**, *54*, 899–906. [CrossRef]
26. Chen, Z.; Zhou, K.; Lu, X.; Lam, Y.C. A review on the mechanical methods for evaluating coating adhesion. *Acta Mech.* **2014**, *225*, 431–452. [CrossRef]
27. Abadias, G.; Chason, E.; Keckes, J.; Sebastiani, M.; Thompson, G.B.; Barthel, E.; Doll, G.L.; Murray, C.E.; Stoessel, C.H.; Martinu, L. Review Article: Stress in thin films and coatings: Current status, challenges, and prospects. *J. Vac. Sci. Technol. A* **2018**, *36*, 020801. [CrossRef]
28. Ropyak, L.Y.; Velychkovych, A.S.; Vytvytskyi, V.S.; Shovkoplias, M.V. Analytical study of "crosshead-slide rail" wear effect on pump rod stress state. *J. Phys. Conf. Ser.* **2021**, *1741*, 12039. [CrossRef]
29. Dubei, O.Y.; Tutko, T.F.; Ropyak, L.Y.; Shovkoplias, M.V. Development of Analytical Model of Threaded Connection of Tubular Parts of Chrome-Plated Metal Structures. *Metallofiz. Noveishie Tekhnologii* **2022**, *44*, 251–272. [CrossRef]
30. Szczepankowski, A.; Przysowa, R.; Perczyński, J.; Kułaszka, A. Health and Durability of Protective and Thermal Barrier Coatings Monitored in Service by Visual Inspection. *Coatings* **2022**, *12*, 624. [CrossRef]
31. Duriagina, Z.A.; Kulyk, V.V.; Filimonov, O.S.; Trostianchyn, A.M.; Sokulska, N.B. The role of stress–strain state of gas turbine engine metal parts in predicting their safe life. *Prog. Phys. Met.* **2021**, *22*, 643–677. [CrossRef]
32. Dolgov, N.A. Analytical Methods to Determine the Stress State in the Substrate–Coating System Under Mechanical Loads. *Strength Mater.* **2016**, *48*, 658–667. [CrossRef]
33. Wheeler, J.M.; Curran, J.A.; Shrestha, S. Microstructure and multi-scale mechanical behavior of hard anodized and plasma electrolytic oxidation (PEO) coatings on aluminum alloy 5052. *Surf. Coat. Technol.* **2012**, *207*, 480–488. [CrossRef]
34. Shi, Y.; Wang, Y.; Cheng, C. Properties and Structure of PEO Treated Aluminum Alloy. *J. Wuhan Univ. Technol. Mat. Sci. Ed.* **2021**, *36*, 424–432. [CrossRef]
35. Shatskyi, I.P.; Makoviichuk, M.V.; Shcherbii, A.B. Equilibrium of cracked shell with flexible coating. In *Shell Structures: Theory and Applications*; CRC Press: Leiden, The Netherlands, 2018; Volume 4, pp. 165–168. [CrossRef]
36. Shatskyi, I.P.; Makoviichuk, M.V.; Shcherbii, A.B. Influence of flexible coating on the limit equilibrium of a spherical shell with meridional crack. *Mater. Sci.* **2020**, *55*, 484–491. [CrossRef]
37. Shats'kyi, I.P.; Makoviichuk, M.V. Contact interaction of the crack edges in the case of bending of a plate with elastic support. *Mater. Sci.* **2003**, *39*, 371–376. [CrossRef]
38. Nobili, A.; Radi, E.; Lanzoni, L. A cracked infinite Kirchhoff plate supported by a two-parameter elastic foundation. *J. Eur. Ceram. Soc.* **2014**, *34*, 2737–2744. [CrossRef]
39. Shatskyi, I.P.; Ropyak, L.Y.; Makoviichuk, M.V. Strength optimization of a two-layer coating for the particular local loading conditions. *Strength Mater.* **2016**, *48*, 726–730. [CrossRef]
40. Ropyak, L.Y.; Makoviichuk, M.V.; Shatskyi, I.P.; Pritula, I.M.; Gryn, L.O.; Belyakovskyi, V.O. Stressed state of laminated interference-absorption filter under local loading. *Funct. Mater.* **2020**, *27*, 638–642. [CrossRef]
41. Bembenek, M.; Makoviichuk, M.; Shatskyi, I.; Ropyak, L.; Pritula, I.; Gryn, L.; Belyakovskyi, V. Optical and Mechanical Properties of Layered Infrared Interference Filters. *Sensors* **2022**, *22*, 8105. [CrossRef]
42. Shyrokov, V.V.; Maksymuk, O.V. Analytic Methods of Calculation of the Contact Interaction of Thin-Walled Structural Elements (Review). *Mater. Sci.* **2002**, *38*, 62–73. [CrossRef]
43. Kul'chyts'kyi-Zhyhailo, R.; Bajkowski, A. Elastic Coating with Inhomogeneous Interlayer Under the Action of Normal and Tangential Forces. *Mater. Sci.* **2014**, *49*, 650–659. [CrossRef]

44. Ropyak, L.Y.; Shatskyi, I.P.; Makoviichuk, M.V. Influence of the Oxide-Layer Thickness on the Ceramic–Aluminium Coating Resistance to Indentation. *Metallofiz. Noveishie Tekhnol.* **2017**, *39*, 517–524. [CrossRef]
45. Kusyi, Y.M.; Kuk, A.M. Investigation of the technological damageability of castings at the stage of design and technological preparation of the machine Life Cycle. *J. Phys. Conf. Ser.* **2020**, *1426*, 12034. [CrossRef]
46. Kusyi, Y.; Onysko, O.; Kuk, A.; Solohub, B.; Kopei, V. Development of the Technique for Designing Rational Routes of the Functional Surfaces Processing of Products. In *New Technologies, Development and Application V*; Lecture Notes in Networks and Systems; Springer: Berlin/Heidelberg, Germany, 2022; Volume 472, pp. 135–143. [CrossRef]
47. Sathish, M.; Radhika, N.; Saleh, B. A critical review on functionally graded coatings: Methods, properties, and challenges. *Compos. Part B Eng.* **2021**, *225*, 109278. [CrossRef]
48. Kusyj, J.; Kuk, A. Acentrifugal vibration strengthening method devised to improve technological reliability of machine parts. *East. Eur. J. Enterp. Technol.* **2015**, *1*, 41–51. [CrossRef]
49. Mordyuk, B.N.; Silberschmidt, V.V.; Prokopenko, G.I.; Nesterenko, Y.V.; Iefimov, M.O. Ti particle-reinforced surface layers in Al: Effect of particle size on microstructure, hardness and wear. *Mater. Charact.* **2010**, *61*, 1126–1134. [CrossRef]
50. Pokhmurs'ka, H.V.; Student, M.M.; Chervins'ka, N.R.; Smetana, K.R.; Wank, A.; Hoenig, T.; Podlesak, H. Structure and properties of aluminum alloys modified with silicon carbide by laser surface treatment. *Mater. Sci.* **2005**, *41*, 316–323. [CrossRef]
51. Gaponova, O.; Kundera, C.; Kirik, G.; Tarelnyk, V.; Martsynkovskyy, V.; Konoplianchenko, I.; Dovzhyk, M.; Belous, A.; Vasilenko, O. Estimating qualitative parameters of aluminized coating obtained by electric spark alloying method. In *Advances in Thin Films, Nanostructured Materials, and Coatings*; Lecture Notes in Mechanical Engineering; Springer: Berlin/Heidelberg, Germany, 2019; pp. 249–266. [CrossRef]
52. Hutsaylyuk, V.; Student, M.; Zadorozhna, K.; Student, O.; Veselivska, H.; Gvosdetskii, V.; Maruschak, P.; Pokhmurska, H. Improvement of wear resistance of aluminum alloy by HVOF method. *J. Mater. Res. Technol.* **2020**, *9*, 16367–16377. [CrossRef]
53. Peltier, F.; Thierry, D. Review of Cr-Free Coatings for the Corrosion Protection of Aluminum Aerospace Alloys. *Coatings* **2022**, *12*, 518. [CrossRef]
54. Student, M.M.; Pohrelyuk, I.M.; Hvozdetskyi, V.M.; Veselivska, H.H.; Zadorozhna, K.R.; Mardarevych, R.S.; Dzioba, Y.V. Influence of the Composition of Electrolyte for Hard Anodizing of Aluminum on the Characteristics of Oxide Layer. *Mater. Sci.* **2021**, *57*, 240–247. [CrossRef]
55. Mohedano, M.; Lu, X.; Matykina, E.; Blawert, C.; Arrabal, R.; Zheludkevich, M.L. Plasma electrolytic oxidation (PEO) of metals and alloys. In *Encyclopedia of Interfacial Chemistry: Surface Science and Electrochemistry*; Wandelt, K., Ed.; Elsevier: Amsterdam, The Netherlands, 2018; pp. 423–438. [CrossRef]
56. Clyne, T.W.; Troughton, S.C. A review of recent work on discharge characteristics during plasma electrolytic oxidation of various metals. *Int. Mater. Rev.* **2019**, *64*, 127–162. [CrossRef]
57. Simchen, F.; Sieber, M.; Kopp, A.; Lampke, T. Introduction to Plasma Electrolytic Oxidation—An Overview of the Process and Applications. *Coatings* **2020**, *10*, 628. [CrossRef]
58. Sikdar, S.; Menezes, P.V.; Maccione, R.; Jacob, T.; Menezes, P.L. Plasma Electrolytic Oxidation (PEO) Process—Processing, Properties, and Applications. *Nanomaterials* **2021**, *11*, 1375. [CrossRef]
59. Qin, J.; Shi, X.; Li, H.; Zhao, R.; Li, G.; Zhang, S.; Ding, L.; Cui, X.; Zhao, Y.; Zhang, R. Performance and failure process of green recycling solutions for preparing high degradation resistance coating on biomedical magnesium alloys. *Green Chem.* **2022**, *24*, 8113–8130. [CrossRef]
60. Protsenko, V.S.; Bobrova, L.S.; Baskevich, A.S.; Korniy, S.A.; Danilov, F.I. Electrodeposition of chromium coatings from a choline chloride based ionic liquid with the addition of water. *J. Chem. Technol. Metall.* **2018**, *53*, 906–915.
61. Protsenko, V.S.; Bobrova, L.S.; Golubtsov, D.E.; Korniy, S.A.; Danilov, F.I. Electrolytic Deposition of Hard Chromium Coatings from Electrolyte Based on Deep Eutectic Solvent. *Russ. J. Appl. Chem.* **2018**, *91*, 1106–1111. [CrossRef]
62. Bazaluk, O.; Dubei, O.; Ropyak, L.; Shovkoplias, M.; Pryhorovska, T.; Lozynskyi, V. Strategy of Compatible Use of Jet and Plunger Pump with Chrome Parts in Oil Well. *Energies* **2022**, *15*, 83. [CrossRef]
63. Tang, Q.; Qiu, T.; Ni, P.; Zhai, D.; Shen, J. Soft Sparking Discharge Mechanism of Micro-Arc Oxidation Occurring on Titanium Alloys in Different Electrolytes. *Coatings* **2022**, *12*, 1191. [CrossRef]
64. Rokosz, K.; Hryniewicz, T.; Dudek, Ł. Phosphate Porous Coatings Enriched with Selected Elements via PEO Treatment on Titanium and Its Alloys: A Review. *Materials* **2020**, *13*, 2468. [CrossRef]
65. Shi, X.; Wang, Y.; Li, H.; Zhang, S.; Zhao, R.; Li, G.; Zhang, R.; Sheng, Y.; Cao, S.; Zhao, Y.; et al. Corrosion resistance and biocompatibility of calcium-containing coatings developed in near-neutral solutions containing phytic acid and phosphoric acid on AZ31B alloy. *J. Alloys Compd.* **2020**, *823*, 153721. [CrossRef]
66. Wang, Y.L.; Jiang, Z.H.; Yao, Z.P. Microstructure, bonding strength and thermal shock resistance of ceramic coatings on steels prepared by plasma electrolytic oxidation. *Appl. Surf. Sci.* **2009**, *256*, 650–656. [CrossRef]
67. Mahmoud, E.R.I.; Algahtani, A.; Tirth, V. Study on Microstructure Characterisation of Three Different Surface Coating Techniques on 6082-T6 Aluminum Alloy. *Met. Mater. Int.* **2021**, *27*, 4002–4013. [CrossRef]
68. Berladir, K.; Hovorun, T.; Gusak, O.; Reshetniak, Y.; Khudaybergenov, D. Influence of Modifiers-Ligatures on the Properties of Cast Aluminum Alloy AK5M2 for the Automotive Industry. In *Advances in Design, Simulation and Manufacturing III, DSMIE 2020*; Ivanov, V., Trojanowska, J., Pavlenko, I., Zajac, J., Peraković, D., Eds.; Lecture Notes in Mechanical Engineering; Springer: Cham, Switzerland, 2020; pp. 473–482. [CrossRef]

69. Pokhmurskii, V.I.; Zin, I.M.; Vynar, V.A.; Khlopyk, O.P.; Bily, L.M. Corrosive wear of aluminium alloy in presence of phosphate. *Corros. Eng. Sci. Technol.* **2012**, *47*, 182–187. [CrossRef]
70. Vynar, V.A.; Pokhmurs'kyi, V.I.; Zin', I.M.; Vasyliv, K.B.; Khlopyk, O.P. Determination of the Mechanism of Tribocorrosion of D16T Alloy According to the Electrode Potential. *Mater. Sci.* **2018**, *53*, 717–723. [CrossRef]
71. Wang, Z.J.; Nie, X.Y.; Hu, H.; Hussein, R.O. In situ fabrication of blue ceramic coatings on wrought Al alloy 2024 by plasma electrolytic oxidation. *J. Vac. Sci. Technol. A* **2012**, *30*, 021302. [CrossRef]
72. Melnick, A.B.; Soolshenko, V.K.; Levchuk, K.H. Thermodynamic Prediction of Phase Composition of Transition Metals High-Entropy Alloys. *Metallofiz. Noveishie Tekhnologii* **2020**, *42*, 1387–1400. [CrossRef]
73. Radchenko, T.M.; Tatarenko, V.A.; Zapolsky, H.; Blavette, D. Statistical-thermodynamic description of the order-disorder transformation of D0(19)-type phase in Ti-Al alloy. *J. Alloy. Compd.* **2008**, *452*, 122–126. [CrossRef]
74. Shihab, T.; Prysyazhnyuk, P.; Semyanyk, I.; Anrusyshyn, R.; Ivanov, O.; Troshchuk, L. Thermodynamic Approach to the Development and Selection of Hardfacing Materials in Energy Industry. *Manag. Syst. Prod. Eng.* **2020**, *28*, 84–89. [CrossRef]
75. Prysyazhnyuk, P.; Shlapak, L.; Semyanyk, I.; Kotsyubynsky, V.; Troshchuk, L.; Korniy, S.; Artym, V. Analysis of the effects of alloying with Si and Cr on the properties of manganese austenite based on AB INITIO modelling. *East. Eur. J. Enterp. Technol.* **2020**, *6*, 28–36. [CrossRef]
76. Karnaukhov, I.M.; Levchuk, K.H. Electronic band structure of Dirac materials with Hubbard interaction. *Metallofiz. Noveishie Tekhnologii* **2022**, *5*, 565–585. [CrossRef]
77. Wang, S.; Liu, X.; Yin, X.; Du, N. Influence of electrolyte components on the microstructure and growth mechanism of plasma electrolytic oxidation coatings on 1060 aluminum alloy. *Surf. Coat. Technol.* **2020**, *381*, 125214. [CrossRef]
78. Lee, J.H.; Jung, K.H.; Kim, S.J. Characterization of ceramic oxide coatings prepared by plasma electrolytic oxidation using pulsed direct current with different duty ratio and frequency. *Appl. Surf. Sci.* **2020**, *516*, 146049. [CrossRef]
79. Li, X.-J.; Zhang, M.; Wen, S.; Mao, X.; Huo, W.-G.; Guo, Y.-Y.; Wang, Y.-X. Microstructure and wear resistance of micro-arc oxidation ceramic coatings prepared on 2A50 aluminum alloys. *Surf. Coat. Technol.* **2020**, *394*, 125853. [CrossRef]
80. Rogov, A.B.; Huang, Y.; Shore, D.; Matthews, A.; Yerokhin, A. Toward rational design of ceramic coatings generated on valve metals by plasma electrolytic oxidation: The role of cathodic polarisation. *Ceram. Int.* **2021**, *47*, 34137–34158. [CrossRef]
81. Mengesha, G.A.; Chu, J.P.; Lou, B.-S.; Lee, J.-W. Effects of Processing Parameters on the Corrosion Performance of Plasma Electrolytic Oxidation Grown Oxide on Commercially Pure Aluminum. *Metals* **2020**, *10*, 394. [CrossRef]
82. Pokhmurs'kyi, V.I.; Dovhunyk, V.M.; Student, M.M.; Klapkiv, M.D.; Posuvailo, V.M.; Kytsya, A.R. Influence of silver nanoparticles added to lubricating oil on the tribological behavior of combined metal-oxide ceramic layers. *Mater. Sci.* **2013**, *48*, 636–641. [CrossRef]
83. Student, M.M.; Dovhunyk, V.M.; Posuvailo, V.M.; Koval'chuk, I.V.; Hvozdets'kyi, V.M. Friction Behavior of Iron-Carbon Alloys in Couples with Plasma-Electrolytic Oxide-Ceramic Layers Synthesized on D16T Alloy. *Mater. Sci.* **2017**, *53*, 359–367. [CrossRef]
84. Kotsyubynsky, V.; Shyyko, L.; Shihab, T.; Prysyazhnyuk, P.; Aulin, V.; Boichuk, V. Multilayered MoS2/C nanospheres as high performance additives to lubricating oils. *Mater. Today Proc.* **2019**, *35*, 538–541. [CrossRef]
85. Hutsaylyuk, V.; Student, M.; Dovhunyk, V.; Posuvailo, V.; Student, O.; Maruschak, P.; Koval'chuck, I. Effect of Hydrogen on the Wear Resistance of Steels upon Contact with Plasma Electrolytic Oxidation Layers Synthesized on Aluminum Alloys. *Metals* **2019**, *9*, 280. [CrossRef]
86. Shen, X.; Nie, X.; Hu, H.; Tjong, J. Effects of coating thickness on thermal conductivities of alumina coatings and alumina/aluminum hybrid materials prepared using plasma electrolytic oxidation. *Surf. Coat. Technol.* **2012**, *207*, 96–101. [CrossRef]
87. Akatsu, T.; Kato, T.; Shinoda, Y.; Wakai, F. Thermal barrier coating made of porous zirconium oxide on a nickel-based single crystal superalloy formed by plasma electrolytic oxidation. *Surf. Coat. Technol.* **2013**, *223*, 47–51. [CrossRef]
88. Curran, J.A.; Kalkanci, H.; Magurova, Y.; Clyne, T.W. Mullite-rich plasma electrolytic oxide coatings for thermal barrier applications. *Surf. Coat. Technol.* **2006**, *201*, 8683–8687. [CrossRef]
89. Shirani, A.; Joy, T.; Rogov, A.; Lin, M.; Yerokhin, A.; Mogonye, J.-E.; Korenyi-Both, A.; Aouadi, S.M.; Voevodin, A.A.; Berman, D. PEO-Chameleon as a potential protective coating on cast aluminum alloys for high-temperature applications. *Surf. Coat. Technol.* **2020**, *397*, 126016. [CrossRef]
90. Shen, D.J.; Wang, Y.L.; Nash, P.; Xing, G.Z. Microstructure, temperature estimation and thermal shock resistance of PEO ceramic coatings on aluminum. *J. Mater. Process. Technol.* **2008**, *205*, 477–481. [CrossRef]
91. Dai, W.; Zhang, X.; Li, C.; Yao, G. Effect of thermal conductivity on micro-arc oxidation coatings. *Surf. Eng.* **2022**, *38*, 44–53. [CrossRef]
92. Pakseresht, A.H.; Saremi, M.; Omidvar, H.; Alizadeh, M. Micro-structural study and wear resistance of thermal barrier coating reinforced by alumina whisker. *Surf. Coat. Technol.* **2019**, *366*, 338–348. [CrossRef]
93. Kamalan Kirubaharan, A.M.; Kuppusami, P. Corrosion behavior of ceramic nanocomposite coatings at nanoscale. In *Corrosion Protection at the Nanoscale*; Micro and Nano Technologies; Rajendran, S., Nguyen, T.A., Kakooei, S., Yeganeh, M., Li, Y., Eds.; Elsevier: Amsterdam, The Netherlands, 2020; pp. 295–314. [CrossRef]
94. Matykina, E.; Arrabal, R.; Skeldon, P.; Thompson, G.E. Optimisation of the plasma electrolytic oxidation process efficiency on aluminium. *Surf. Interface Anal.* **2010**, *42*, 221–226. [CrossRef]
95. Vakili-Azghandi, M.; Fattah-alhosseini, A.; Keshavarz, M.K. Optimizing the electrolyte chemistry parameters of PEO coating on 6061 Al alloy by corrosion rate measurement: Response surface methodology. *Measurement* **2018**, *124*, 252259. [CrossRef]

96. Bembenek, M.; Popadyuk, O.; Shihab, T.; Ropyak, L.; Uhryński, A.; Vytvytskyi, V.; Bulbuk, O. Optimization of Technological Parameters of the Process of Forming Therapeutic Biopolymer Nanofilled Films. *Nanomaterials* **2022**, *12*, 2413. [CrossRef] [PubMed]
97. Pan, S.; Tu, X.; Yu, J.; Zhang, Y.; Miao, C.; Xu, Y.; Fu, R.; Li, J. Optimization of AZ31B Magnesium Alloy Anodizing Process in NaOH-Na_2SiO_3-$Na_2B_4O_7$ Environmental-Friendly Electrolyte. *Coatings* **2022**, *12*, 578. [CrossRef]
98. An, L.; Ma, Y.; Yan, X.; Wang, S.; Wang, Z. Effects of electrical parameters and their interactions on plasma electrolytic oxidation coatings on aluminum substrates. *Trans. Nonferrous Met. Soc. China* **2020**, *30*, 883–895. [CrossRef]
99. Kaseem, M.; Dikici, B. Optimization of Surface Properties of Plasma Electrolytic Oxidation Coating by Organic Additives: A Review. *Coatings* **2021**, *11*, 374. [CrossRef]
100. Songur, F.; Arslan, E.; Dikici, B. Taguchi optimization of PEO process parameters for corrosion protection of AA7075 alloy. *Surf. Coat. Technol.* **2022**, *434*, 128202. [CrossRef]
101. Hladkyi, S.I.; Palazhchenko, S.P. Development of equipment for the study of friction nodes operating in the mode of reciprocating motion. *Sci. Bull. Ivano-Frankivsk Natl. Tech. Univ. Oil Gas* **2015**, *2*, 89–100.
102. Bandura, A.; Skaskiv, O. Analog of Hayman's theorem and its application to some system of linear partial differential equations. *J. Math. Phys. Anal. Geom.* **2019**, *15*, 170–191. [CrossRef]
103. Premchand, C.; Hariprasad, S.; Saikiran, A.; Lokeshkumar, E.; Manojkumar, P.; Ravisankar, B.; Venkataraman, B.; Rameshbabu, N. Assessment of Corrosion and Scratch Resistance of Plasma Electrolytic Oxidation and Hard Anodized Coatings Fabricated on AA7075-T6. *Trans. Indian Inst. Met.* **2021**, *74*, 1991–2002. [CrossRef]

Disclaimer/Publisher's Note: The statements, opinions and data contained in all publications are solely those of the individual author(s) and contributor(s) and not of MDPI and/or the editor(s). MDPI and/or the editor(s) disclaim responsibility for any injury to people or property resulting from any ideas, methods, instructions or products referred to in the content.

Increasing the Flow Stress during High-Temperature Deformation of Aluminum Matrix Composites Reinforced with TiC-Coated CNTs

Artemiy V. Aborkin *, Dmitriy V. Bokaryov, Sergey A. Pankratov and Alexey I. Elkin

Department of Mechanical Engineering Technology, Vladimir State University Named after Alexander and Nikolay Stoletovs, 600000 Vladimir, Russia
* Correspondence: aborkin@vlsu.ru

Abstract: In this work, composites based on AA5049 aluminium alloy reinforced with multiwalled carbon nanotubes (CNTs) and multiwalled TiC-coated CNTs were prepared by powder metallurgy. For the first time, the effect of TiC coating on the CNT surface on the flow stress of aluminum matrix composites under compressive conditions at 300–500 °C has been investigated. It was found that composites reinforced with TiC-coated CNTs have a higher flow stress during high-temperature deformation compared to composites reinforced with uncoated CNTs. Moreover, with an increasing temperature in the 300–500 °C range, the strengthening effect increases from 14% to 37%. Compared to the reference sample of the matrix material without reinforcing particles, obtained by the same technological route, the composites reinforced with CNTs and CNT-hybrid structures had a 1.8–2.9 times higher flow stress during high-temperature deformation. The presented results show that the modification of the CNTs surface with ceramic nanoparticles is a promising structure design strategy that improves the heat resistance of aluminum matrix composites. This extends the potential range of application of aluminum matrix composites as a structural material for operation at elevated temperatures.

Keywords: carbon nanotubes; TiC coating; hot deformation; aluminum matrix composite; compression; properties; flow stress

1. Introduction

The modern industry needs lightweight yet tenacious materials. Aluminum alloys have met these requirements for many years. However, despite the economic availability of these materials, the limiting factor for their use is often their low strength at elevated temperatures. Typically, the operating temperature limit for aluminum alloys does not exceed 200–300 °C [1]. It is possible to increase the operating temperature range of aluminum alloys by creating particulate-reinforced composites on their basis by controlling the properties by selecting the type, size and shape of reinforcing particles. Liquid-phase or solid-phase methods [2–4] for creating bulk composites, as well as various methods of depositing coatings and surface modification [5–10], can be used for this purpose. Great progress in this direction has been achieved in the production of so-called sintered aluminum powders (SAP's) [1,11–14], which are, in fact aluminum matrix composites reinforced with Al_2O_3 particles. These materials exhibit relatively high strength properties even at ~0.85 Tm of the matrix alloy. For example, in the study [1], the in situ formation of an Al_2O_3 shell on an aluminum surface provided the preservation of the strength of a material in the range of 140 to 120 MPa at 0.66–0.88 Tm (350–520 °C). In [14], it was shown that the addition of 1.5 wt.% Al_2O_3 facilitates the measurement of the compression strength of the Al7075 alloy at the level of 260–125 MPa at 0.61–0.82 Tm (300–500 °C).

One of the promising types of reinforcing particles, due to their unique physical and mechanical properties, are carbon nanotubes (CNTs). The Young moduli of CNTs, depending on their chirality, number of walls, diameter and defectiveness of their structure, can reach from ~1 to 4 TPa [15–19], and the axial tensile strength is 63–110 GPa [18,19]. Despite this, the effective utilization of the load-bearing potential of CNTs in aluminum matrix composites at high temperature is hindered by their tendency to react chemically with the matrix aluminum alloy. This leads to the in situ formation of the Al_4C_3 phase, compromising the structural integrity of CNTs. For example, in [20], it was shown for the 2009 Al alloy composite reinforced with 1.5 vol.% CNTs that, after increasing temperature from 20 to 300 °C, the yield strength increment decreases from ~100 to ~30 MPa, and in the case of 4.5 vol.% CNTs, a decrease in the yield strength of the composite material at 300 °C was noted even in comparison with the matrix material without reinforcement. It should be noted that this is also characteristic of aluminum matrix composites reinforced with carbon microfibers. For instance, paper [21] reports that AA7075 alloy reinforced with 15 vol.% carbon fibers with an average diameter of 10 μm shows a 5-fold decrease in compressive flow stress from 750 MPa to 150 MPa at an increasing temperature from 25 °C to 300 °C. A more recent work [22] shows an increase in the strengthening effect when the CNTs volume fraction increases from 2.5 to 5 vol.%, which can be explained by a rather homogeneous distribution of nanotubes in the aluminum matrix. At the same time, the strengthening effect at 200 °C was greater than at 400 °C. The authors attributed this to the higher recrystallization resistance of the composite compared to the matrix material. The reduction of the strengthening effect with increasing test temperature from 200 °C to 400 °C is attributed to the fact that, at this temperature, the CNTs could no longer so effectively prevent grain boundary movement.

Considering the above, it seems promising to use CNT-hybrid nanostructures with ex situ modified CNTs as reinforcing additives. Such a modified surface of CNTs will contribute to the formation of an interfacial layer at the Al–CNT interface [23–28]. The presence of an ex situ interfacial layer, for example, TiC carbide ceramics, on the surface of CNTs should improve the interfacial interaction by increasing adhesion [29,30] and also prevent unwanted chemical reactions with the matrix leading to the structural degradation of CNTs [31].

The purpose of this work is to investigate the effect of TiC coating on the surface of CNTs on increasing the flow stress at high temperature of aluminum matrix composites.

2. Materials and Methods

The initial multiwall CNTs were synthesized by the MOCVD method with ferrocene and toluene used as precursors under argon flow in a tubular reactor equipped with a tubular furnace at 825 °C, as described in detail elsewhere [32]. The deposition of titanium carbide nanoparticles on the surface of CNTs was performed by the thermal decomposition of titanocene dichloride vapors according to the procedure described in detail earlier [33].

The initial matrix material was in the state of granules of aluminum alloy AA5049 1–2 mm in size. The elemental composition of the granules measured using an ARL ADVANT'X (Thermo Scientific) sequential X-ray fluorescence spectrometer was as follows (wt.%): Al 96.11; Mg 2.42; Mn 0.55; Fe 0.26; Si 0.37; Zn up to 0.1; Ti up to 0.1; Cu to 0.1.

Granules of aluminum alloy with reinforcing additives (named as CNT or TiC/CNT for simplicity hereafter) were mixed by high-energy ball milling (HEBM) using a PULVERISETTE 6 planetary mill (Fritsch, Idar-Oberstein, Germany). For this purpose, aluminum alloy granules, reinforcing particles (CNT or TiC/CNT, 1 wt.%), and a process control agent (stearic acid, 0.8 wt.%) were placed in a 250 mL stainless steel vessel in which grinding balls of hardened steel 100 × 6 with a diameter of 8 mm were also added. The ball-to-powder weight ratio was 15:1. The processing speed and time were 600 rpm and 6 h, respectively. Thus, composite powders named A_CNT or A_TiC/CNT for simplicity hereafter and a reference sample of matrix alloy powder without reinforcing particles named A_m were obtained. The milled composite powders were hot-pressed in a special heated steel die at 450 °C and 350 MPa to obtain compacts with a diameter of 17 mm and a height of 12 mm. The consolidation modes were adopted, considering the recommendations of [34].

The surface morphology of the initial CNTs, TiC-coated CNTs and synthesized composite powders were studied using a Zeiss 55 Ultra scanning electron microscope (Carl Zeiss AG, Jena, Germany). The particle size analysis of the powders was carried out with a Microsizer 201C laser analyzer (Moscow, Russia).

The phase composition of CNTs, CNT-hybrid structures, composite powders and bulk samples was studied with the XRD method using a D8 ADVANCE (BRUKER, Bremen, Germany) with Cu Kα radiation (λ = 1.5148 Å, 40 kV and 40 mA) and linear detection at angles of 25–85°. The crystallite size was calculated from XRD data using the Scherer relation for the main diffraction peak of aluminum (111). The instrumental broadening was determined using a standard α-Al_2O_3 sample.

The bulk samples were investigated by Raman spectroscopy Integra Spectra (NT-MDT, Moscow, Russia) at 473 nm with no more than 50 mW. The Raman spectra were measured at more than 10 different points on each sample.

The surface porosity was determined on metallographic sections using an Altami Met 1-C optical microscope (Moscow, Russia). The quantitative analysis of surface porosity was performed using ImageJ software. At least 10 images were studied (the surface area was ~3.5 mm^2).

Compression tests at elevated temperatures were performed on a universal testing machine WDW-100E (Time Group Inc., Beijing, China) in a special thermal cell at temperatures of 300, 400 and 500 °C and a compression rate of 0.1 s^{-1}. Cylindrical samples for compression testing at elevated temperatures with a diameter of 6 mm and a height of 12 mm were obtained by electroerosive cutting on a Mitsubishi BA8 machine. Graphite was used to reduce friction on the end surfaces of the specimens. The choice of the lower boundary of the test temperature interval corresponds to the limiting operating temperature of aluminum alloys. The upper boundary of the range is the limit in most of the published investigations. The compression rate was taken for the possibility of comparing the data of this work with those published in other sources as one of the most frequently used.

The results of calculating the volume fraction of reinforcing nanostructures using the model given in [27] show that the volume fraction of the reinforcement in CNT-reinforced composite was 1.2 vol.%, and in TiC/CNT-reinforced composites, it was 0.65 vol.%.

3. Results and Discussion

The results of the characterization of the structural-phase composition of reinforcing nanoparticles of both types used in this study using scanning electron microscopy and X-ray diffractometry are shown in Figure 1.

The SEM images show that the initial CNTs have a diameter of ~60–100 nm and a length of a few micrometers (see Figure 1a). After TiC coating, the CNTs diameter increases to ~160–180 nm (see Figure 1b). The carbide coating on the CNTs' surface has a relatively smooth and continuous morphology. Figure 1b,d shows the XRD results of initial CNT and TiC/CNT. The initial CNTs are characterized by the presence of (002), (100), (101), and (004) carbon lines. The TiC/CNT structures contain, similarly to the original CNT, a carbon phase and fcc-TiC, for which (111), (200), (220), and (311) peaks are clearly recorded. Thus, the data presented show that, during MOCVD treatment, a TiC coating is formed on the surface of the initial CNT.

Typical SEM images of the morphology of powders obtained by the HEBM method are shown in Figure 2. The synthesized powders are characterized by a rounded shape. The average particle size was 50–60 μm. The results of the particle size analysis of the synthesized powders are shown in Figure 2d. Combined analysis of the SEM images and particle size data shows that the reinforcing particles contribute to a slightly more intense refinement of the matrix alloy particles. For example, the average particle size of the matrix powder without reinforcing particles was 60.5 μm. The particle size of composite powder reinforced with 1 wt.% CNT and TiC/CNT was 51.3 and 54.8 μm, respectively. Therefore, it can be concluded that the average particle size of the composite powder decreases with the increasing volume fraction of reinforcing nanostructures.

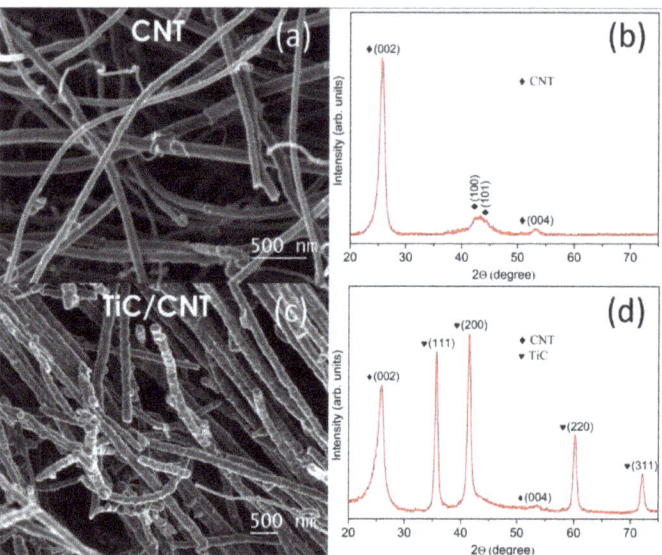

Figure 1. SEM image of pristine CNTs (**a**), TiC/CNT (**b**), XRD of pristine CNTs (**c**), TiC/CNT (**d**).

Figure 2. SEM images of matrix alloy powders after HEBM (**a**), CNT-reinforced composite (**b**) and TiC/CNT-reinforced composite (**c**), and particle size analysis of these powders (**d**).

The XRD results of the powder materials and bulk samples are shown in Figure 3. The data obtained for the powder and bulk samples are qualitatively similar. They clearly show (111), (200), (220), (311) and (222) aluminum peaks. At the same time, no diffraction peaks of the reinforcing phases were observed. In general, this is typical for aluminum alloys reinforced with carbon nanostructures even at higher concentrations of reinforcing additives.

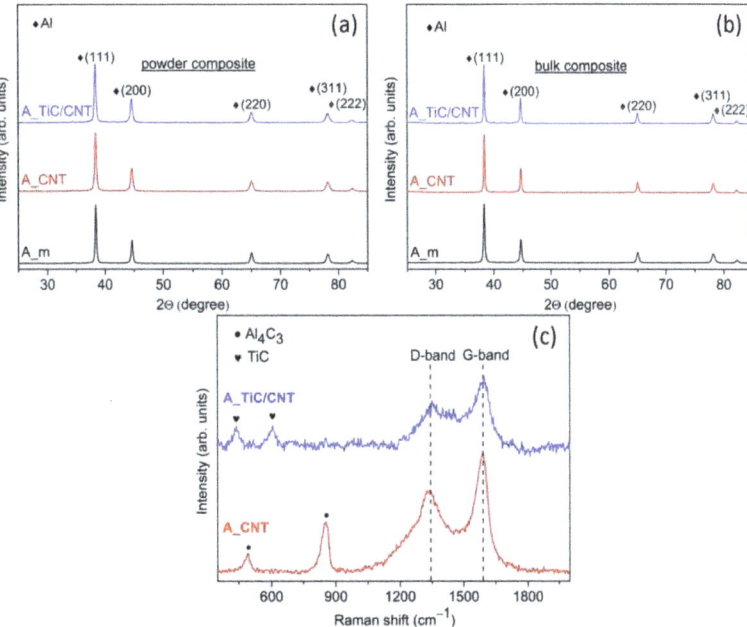

Figure 3. XRD data of powder materials (**a**), bulk samples (**b**) and Raman spectroscopy results of bulk samples (**c**).

In addition, it can be noted that the FWHM for bulk samples decreases compared to powder samples. This indicates an increase in crystallite size during isothermal holding during consolidation. For composite samples A_CNT and A_TiC/CNT containing reinforcing nanostructures, a smaller increase of crystallite size was observed in comparison to A_m without reinforcing additives. This may indicate the anchoring of grain boundaries in the composite materials due to reinforcing nanoparticles. Moreover, A_TiC/CNT samples are characterized by a smaller crystallite size both after ball milling and after consolidation, as compared to A_CNT samples. This may be due to the better distribution of CNT-hybrid nanostructures during ball milling because of their lower volume fraction. Since the XRD data do not contain information on the reinforcing particles and ceramic coating, the Raman spectroscopy of bulk samples was performed (see Figure 3c). The Raman spectroscopy of bulk A_TiC/CNT composites shows the presence of D- and G-lines at ~1350 cm^{-1} and ~1600 cm^{-1} belonging to CNT. In addition, lines at 439 cm^{-1} and 606 cm^{-1} belonging to the TiC-coating are identified. At the same time, the Al_4C_3 lines are absent. At the same time, the Al_4C_3 lines at ~490 cm^{-1} and ~855 cm^{-1} are clearly detected for the A_CNT composites in addition to the D- and G-lines. This shows that the presence of the TiC coating on the CNT surface prevents the formation of in situ Al_4C_3 and acts as a barrier interphase layer between the matrix alloy and CNT.

Figure 4 shows the optical microscopy microstructure of the bulk samples and the surface porosity data. A comparative analysis of the microimages shows that all the bulk samples have low residual porosity (see Figure 4d). This could be an indication that the consolidation mode was chosen correctly.

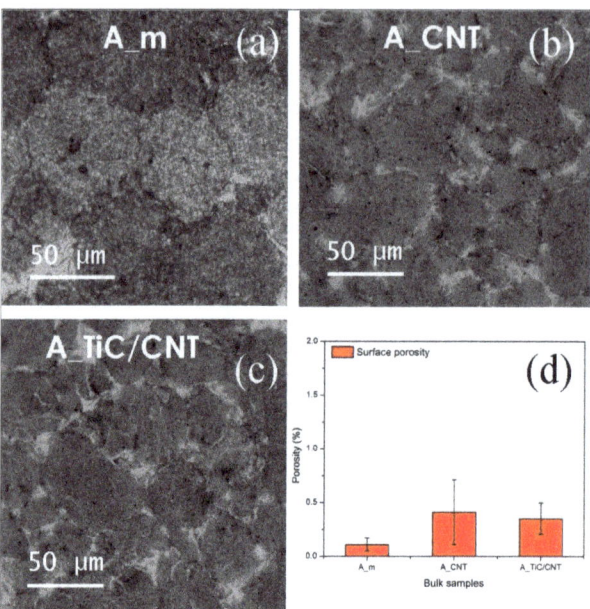

Figure 4. Optical images of the microstructure of bulk samples of matrix material (**a**), CNT-reinforced composite (**b**), TiC/CNT-reinforced composite (**c**) and surface porosity data of these samples (**d**).

Figure 5 shows the engineering curves for compression tests on bulk specimens at 300–500 °C. The flow stress curves are characterized by the presence of sections corresponding to the processes of the strengthening and softening in the deformed material. At 300–400 °C for all the materials tested at the initial stage, the strengthening processes are dominant, which is expressed in the presence of a "hill". This is due to the necessity of energy accumulation for the realization of thermally activated softening mechanisms. In this case, at 500 °C, strengthening and softening processes are balanced. This is due to the higher mobility of the boundaries at this temperature, which contributes to more intensive nucleation and growth of dynamically recrystallized grains as well as more intensive destruction of dislocations. The increased flow stress of composites as compared to the reference matrix material sample is explained, on the one hand, by the inherent submicron structure formed at the HEBM stage and retained during consolidation and, on the other hand, by the presence of reinforcing nanoparticles on the matrix alloy grain boundaries, preventing both the dislocation motion and the boundaries themselves during recrystallization.

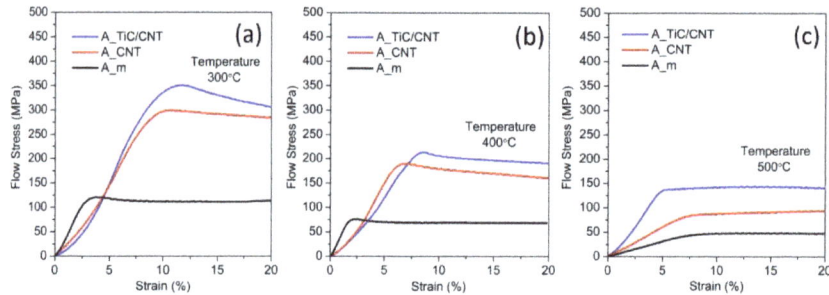

Figure 5. Compression flow stress curves at 300 °C (**a**), 400 °C (**b**), 500 °C (**c**) and strain rate 0.1 s^{-1}.

As expected, an increase in temperature led to a decrease in the flow stress of the bulk samples. For example, an increase in temperature from 300 to 500 °C resulted in a decrease of flow stress from 300 to 86 MPa and from 350 to 137 MPa for A_CNT and A_TiC/CNT, respectively. This clearly demonstrates the positive effect of TiC coating on the CNTs surface on the increase of flow stress during the high-temperature deformation of aluminum matrix composites. Besides, composites reinforced with both TiC/CNT and CNT showed greater flow stress during high-temperature deformation compared to the reference sample A_m of the matrix material obtained by the same process route. Thus, CNT-reinforced composites have on average 1.8–2.5 times higher flow stress at the specified temperatures compared to A_m. Moreover, with increasing temperature, the strengthening effect for A_m CNT decreased. This is consistent with the data reported in [22]. In contrast, the strengthening effect for A_TiC/CNT was constant over the temperature range in question and the TiC/CNT-reinforced composites had 2.9 times higher flow stress than the A_m reference sample. This is due to the fact that in situ Al_4C_3 formation occurs in CNT-reinforced composites, which compromises the structural integrity of CNTs and reduces their bearing capacity. The TiC barrier coating inhibits this reaction, which leads to a more efficient use of the CNT's load-bearing potential. Considering this, it can be assumed that A_TiC/CNT can also operate at temperatures higher than 500 °C. However, this requires further investigations.

To correlate the obtained results with the published data on aluminum alloys and aluminum matrix composites reinforced with other types of micro- and nanoparticles, summary graphs (see Figure 6) of the change in flow stress depending on temperature at a deformation rate of $0.1\ \text{s}^{-1}$ are presented.

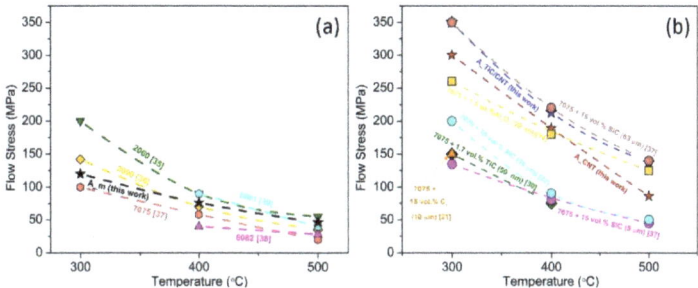

Figure 6. Flow stress of aluminum alloys (**a**) and composites based on them (**b**) at temperature 300–500 °C and strain rate $0.1\ \text{s}^{-1}$.

Figure 6a shows that, for the aluminum alloy systems considered, the flow stress at 300 °C ranges from ~100 to ~200 MPa [35–39]. The highest values correspond to Al-Li alloys [35,36]. An increase in temperature up to 500 °C significantly reduces the above parameter, while reducing its range of variation to ~20 to 55 MPa. This is likely to make it impossible to operate products made from these materials under these conditions. However, AA2060 still appears preferable to alloys of other systems. A slightly different picture was obtained with aluminum alloy composites (see Figure 6b). The addition of 15 vol.% carbon fibers, as mentioned above, provides a flow stress value of 150 MPa at 300 °C, which is five times lower than the flow stress at a normal temperature [21]. This severe drop in strength properties is attributed by the authors to a weak interfacial interaction at elevated temperatures. This leads to a change in the composite fracture mechanism from fiber delamination to fracture at the interface. Approximately the same ~150 MPa value of flow stress was demonstrated by a composite based on the Al7075 alloy reinforced with 1.7 vol.% TiC particles of 50 nm average size [30]. A further increase in the test temperature to 400 °C led to a 2-fold decrease in flow stress to ~75 MPa. According to the authors, this was due to the slippage of the spherical nanoparticles with respect to the grain boundaries under load.

Paper [14] reported that the use of 1.5 wt.% Al_2O_3 particles with an average size of 20 nm to reinforce the alloy Al7075 provides softening reduction with increasing temperature. So, if at 300 °C, the flow stress was ~260 MPa, which is 13% lower than for A_CNT obtained in this work, then at 400 °C the difference does not exceed 5%. At 500 °C, the flow stress of the Al7075-based composite reinforced with 1.5 wt.% Al_2O_3 nanoparticles was already ~125 MPa, which is 30% higher than that of A_CNT. Data on the effect of SiC microparticle size on the heat resistance of Al7075 alloy are given in [37]. The authors show that increasing the average size of SiC microparticles from 5 to 63 μm at 15 vol.% reinforcement led to an increase in the flow stress of the composite material. For example, at 500 °C, the strain strength was 45 MPa and 140 MPa for 5 and 63 μm SiC-reinforced composites, respectively. The authors attribute this effect to the expansion of the Al–SiC interface and a change in the composite fracture mechanism from an adhesive one (along the matrix-reinforcing particle interface) to a transgranular one, in which cracks propagate through the reinforcing particles. The A_TiC/CNT composite material obtained in this work has approximately the same properties as the Al7075-alloy-based composite reinforced with 15 vol.% SiC with an average size of 63 μm. However, the reinforcement volume fraction in the case of A_TiC/CNT is ~23 times lower than that of the Al7075-based composite.

Thus, the presented results show that the surface modification of CNTs with nanoparticles, such as TiC, is a promising structure design strategy to increase the heat resistance of aluminum matrix composites.

4. Conclusions

Two types of bulk composites based on aluminum alloy reinforced with 1 wt.% of CNTs and 1 wt.% of TiC-coated CNTs were produced by powder metallurgy. Compression tests on these three bulk composites at 300–500 °C and a strain rate of $0.1\ s^{-1}$ were performed. It was found that the composites obtained have 1.8–2.9 times higher flow stress at the above temperatures than a reference sample of matrix material without reinforcing particles, obtained by the same technological route. CNT-reinforced composites with TiC coating exhibited better properties than CNT-reinforced composites. For example, the presence of a TiC coating on the CNTs surface resulted in a 14–37% increase in the high-temperature flow stress at 300–500 °C compared with CNT-reinforced composites.

The results show that the use of TiC-coated CNTs as reinforcing particles expands the potential range of application of aluminum matrix composites as a structural material for operation at elevated temperatures. The surface modification of CNTs with ceramic nanoparticles is a promising structure design strategy to enhance the heat resistance of aluminum matrix composites.

Author Contributions: Conceptualization, A.V.A.; methodology, A.V.A.; formal analysis, D.V.B.; investigation, S.A.P.; data curation, D.V.B.; writing—original draft preparation, A.I.E.; writing—review and editing, A.V.A.; supervision, A.V.A.; funding acquisition, A.V.A. All authors have read and agreed to the published version of the manuscript.

Funding: This research was funded by the Russian Science Foundation (project No 18-79-10227).

Institutional Review Board Statement: Not applicable.

Informed Consent Statement: Not applicable.

Data Availability Statement: The data presented in this study are available in this article.

Acknowledgments: The study was carried out using the equipment of the interregional multispecialty and interdisciplinary center for the collective usage of promising and competitive technologies in the areas of development and application in industry/mechanical engineering of domestic achievements in the field of nanotechnology (Agreement No. 075-15-2021-692 of 5 August 2021).

Conflicts of Interest: The authors declare no conflict of interest.

References

1. Poletti, C.; Balog, M.; Simancik, F.; Degischer, H.P. High-temperature strength of compacted sub-micrometer aluminium powder. *Acta Mater.* **2010**, *58*, 3781–3789. [CrossRef]
2. Samal, P.; Vundavilli, P.R.; Meher, A.; Mahapatra, M.M. Recent progress in aluminum metal matrix composites: A review on processing, mechanical and wear properties. *J. Manuf. Process.* **2020**, *59*, 131–152. [CrossRef]
3. Deev, V.B.; Prusov, E.S.; Ri, E.H. Physical Methods of Processing the Melts of Metal Matrix Composites: Current State and Prospects. *Russ. J. Non-Ferr. Met.* **2022**, *63*, 292–304. [CrossRef]
4. Grilo, J.; Carneiro, V.H.; Teixeira, J.C.; Puga, H. Manufacturing Methodology on Casting-Based Aluminium Matrix Composites: Systematic Review. *Metals* **2021**, *11*, 436. [CrossRef]
5. Jendrzejewski, R.; Łubiński, J.; Śliwiński, G. Wear Resistance Enhancement of Al6061 Alloy Surface Layer by Laser Dispersed Carbide Powders. *Materials* **2020**, *13*, 3683. [CrossRef]
6. Li, K.; Liu, X.; Zhao, Y. Research Status and Prospect of Friction Stir Processing Technology. *Coatings* **2019**, *9*, 129. [CrossRef]
7. Jendrzejewski, R.; Van Acker, K.; Vanhoyweghen, D.; Śliwiński, G. Metal matrix composite production by means of laser dispersing of SiC and WC powder in Al alloy. *Appl. Surf. Sci.* **2009**, *255*, 5584–5587. [CrossRef]
8. Li, W.; Assadi, H.; Gaertner, F.; Yin, S. A review of advanced composite and nanostructured coatings by solid-state cold spraying process. *Crit. Rev. Solid State Mater. Sci.* **2019**, *44*, 109–156. [CrossRef]
9. Aborkin, A.V.; Alymov, M.I.; Arkhipov, V.E.; Khrenov, D.S. Formation of heterogeneous powder coatings with a two-level micro-and nanocomposite structure under gas-dynamic spraying conditions. *Dokl. Phys.* **2018**, *63*, 50–54. [CrossRef]
10. Moridi, A.; Hassani-Gangaraj, S.M.; Guagliano, M.; Dao, M. Cold Spray Coating: Review of Material Systems and Future Perspectives. *Surf. Eng.* **2014**, *36*, 369–395. [CrossRef]
11. Liu, J.; Huang, X.; Zhao, K.; Zhu, Z.; Zhu, X.; An, L. Effect of reinforcement particle size on quasistatic and dynamic mechanical properties of Al-Al$_2$O$_3$ composites. *J. Alloy. Compd.* **2019**, *797*, 1367–1371. [CrossRef]
12. Kang, Y.C.; Chan, S.L.I. Tensile properties of nanometric Al$_2$O$_3$ particulate-reinforced aluminum matrix composites. *Mater. Chem. Phys.* **2004**, *85*, 438–443. [CrossRef]
13. Prabhu, B.; Suryanarayana, C.; An, L.; Vaidyanathan, R. Synthesis and characterization of high volume fraction Al-Al$_2$O$_3$ nanocomposite powders by high-energy milling. *Mater. Sci. Eng. A* **2006**, *425*, 192–200. [CrossRef]
14. Saravanan, L.; Senthilvelan, T. Investigations on the hot workability characteristics and deformation mechanisms of aluminium alloy-Al2O3 nanocomposite. *Mater. Des.* **2015**, *79*, 6–14. [CrossRef]
15. Kashyap, K.T.; Patil, R.G. On Young's modulus of multi-walled carbon nanotubes. *Bull. Mater. Sci.* **2008**, *31*, 185–187. [CrossRef]
16. Wong, E.W.; Sheehan, P.E.; Lieber, C.M. Nanobeam mechanics: Elasticity, strength, and toughness of nanorods and nanotubes. *Science* **1997**, *277*, 1971–1975. [CrossRef]
17. Lourie, O.; Wagner, H.D. Evaluation of Young's modulus of carbon nanotubes by micro-Raman spectroscopy. *J. Mater. Res.* **1998**, *13*, 2418–2422. [CrossRef]
18. Yu, M.F.; Lourie, O.; Dyer, M.J.; Moloni, K.; Kelly, T.F.; Ruoff, R.S. Strength and breaking mechanism of multiwalled carbon nanotubes under tensile load. *Science* **2000**, *287*, 637–640. [CrossRef]
19. Peng, B.; Locascio, M.; Zapol, P.; Li, S.; Mielke, S.L.; Schatz, G.C.; Espinosa, H.D. Measurements of near-ultimate strength for multiwalled carbon nanotubes and irradiation-induced crosslinking improvements. *Nat. Nanotechnol.* **2008**, *3*, 626–631. [CrossRef]
20. Liu, Z.Y.; Xiao, B.L.; Wang, W.G.; Ma, Z.Y. Elevated temperature tensile properties and thermal expansion of CNT/2009Al composites. *Compos. Sci. Technol.* **2012**, *72*, 1826–1833. [CrossRef]
21. Lee, W.S.; Sue, W.C.; Lin, C.F. The effects of temperature and strain rate on the properties of carbon-fiber-reinforced 7075 aluminum alloy metal-matrix composite. *Compos. Sci. Technol.* **2000**, *60*, 1975–1983. [CrossRef]
22. Cao, L.; Chen, B.; Wan, J.; Kondoh, K.; Guo, B.; Shen, J.; Li, J.S. Superior high-temperature tensile properties of aluminum matrix composites reinforced with carbon nanotubes. *Carbon* **2022**, *191*, 403–414. [CrossRef]
23. Aborkin, A.V.; Khor'kov, K.S.; Ob'edkov, A.M.; Kremlev, K.V.; Izobello, A.Y.; Volochko, A.T.; Alymov, M.I. Evolution of Multiwalled Carbon Nanotubes and Related Nanostructures during the Formation of Alumomatrix Composite Materials. *Tech. Phys. Lett.* **2019**, *45*, 20–23. [CrossRef]
24. Aborkin, A.V.; Babin, D.M.; Zalesnov, A.I.; Prusov, E.S.; Ob'edkov, A.M.; Alymov, M.I. Effect of ceramic coating on carbon nanotubes interaction with matrix material and mechanical properties of aluminum matrix nanocomposite. *Ceram. Int.* **2020**, *46*, 19256–19263. [CrossRef]
25. Aborkin, A.V.; Khorkov, K.S.; Prusov, E.S.; Ob'edkov, A.M.; Kremlev, K.V.; Perezhogin, I.A.; Alymov, M.I. Effect of Increasing the Strength of Aluminum Matrix Nanocomposites Reinforced with Microadditions of Multiwalled Carbon Nanotubes Coated with TiC Nanoparticles. *Nanomaterials* **2019**, *9*, 1596. [CrossRef] [PubMed]
26. Kremlev, K.V.; Ob'edkov, A.M.; Semenov, N.M.; Kaverin, B.S.; Ketkov, S.Y.; Vilkov, I.V.; Andreev, P.V.; Gusev, S.A.; Aborkin, A.V. Synthesis of Hybrid Materials Based on Multiwalled Carbon Nanotubes Decorated with WC$_{1-x}$ Nanocoatings of Various Morphologies. *Tech. Phys. Lett.* **2019**, *45*, 348–351. [CrossRef]
27. Aborkin, A.V.; Elkin, A.I.; Reshetniak, V.V.; Ob'edkov, A.M.; Sytschev, A.E.; Leontiev, V.G.; Titov, D.D.; Alymov, M.I. Thermal expansion of aluminum matrix composites reinforced by carbon nanotubes with in-situ and ex-situ designed interfaces ceramics layers. *J. Alloy. Compd.* **2021**, *872*, 159593. [CrossRef]

28. Guo, B.; Luo, S.; Wu, Y.; Song, M.; Chen, B.; Yu, Z.; Li, W. Regulating the interfacial reaction between carbon nanotubes and aluminum via copper nano decoration. *Mater. Sci. Eng. A* **2021**, *821*, 141576. [CrossRef]
29. Contreras, A. Wetting of TiC by Al-Cu alloys and interfacial characterization. *J. Colloid Interface Sci.* **2007**, *311*, 159–170. [CrossRef]
30. Huang, C.C.; Qi, L.; Chen, J.; Guan, R.; Ojo, O.A.; Wang, Z.G. Effect of TiC nanoparticles on the hot deformation behavior of AA7075 aluminum alloy. *Mater. Charact.* **2021**, *181*, 111508. [CrossRef]
31. Jagannatham, M.; Chandran, P.; Sankaran, S.; Haridoss, P.; Nayan, N.; Bakshi, S.R. Tensile properties of carbon nanotubes reinforced aluminum matrix composites: A review. *Carbon* **2020**, *160*, 14–44. [CrossRef]
32. Obiedkov, A.M.; Kaverin, B.S.; Egorov, V.A.; Semenov, N.M.; Ketkov, S.Y.; Domrachev, G.A.; Kremlev, K.V.; Gusev, S.A.; Perevezentsev, V.N.; Moskvichev, A.N.; et al. Macroscopic cylinders on the basis of radial-oriented multiwall carbon nanotubes. *Lett. Mater.* **2012**, *3*, 152–156. [CrossRef]
33. Vilkov, I.V.; Kaverin, B.S.; Ob'edkov, A.M.; Semenov, N.M.; Ketkov, S.Y.; Rychagova, E.A.; Gusev, S.A.; Tatarskiy, D.A.; Andreev, P.V.; Aborkin, A.V. Single-step synthesis of tic mesocrystals on the mwcnts surface by the pyrolysis of Cp_2TiCl_2. *Mater. Today Chem.* **2022**, *24*, 100830. [CrossRef]
34. Aborkin, A.V.; Alymov, M.I.; Sobol'kov, A.V.; Khor'kov, K.S.; Babin, D.M. Effect of the Thermomechanical Treatment Conditions on the Consolidation, the Structure, and the Mechanical Properties of Bulk Al–Mg–C Nanocomposites. *Russ. Metall.* **2018**, *2018*, 625–632. [CrossRef]
35. Ou, L.; Zheng, Z.; Nie, Y.; Jian, H. Hot deformation behavior of 2060 alloy. *J. Alloys Compd.* **2015**, *648*, 681–689. [CrossRef]
36. Avramovic-Cingara, G.; McQueen, H.J.; Perovic, D.D. Comparison of torsion and compression constitutive analyses for elevated temperature deformation of Al–Li–Cu–Mn alloy. *Mater. Sci. Technol.* **2014**, *19*, 11–19. [CrossRef]
37. Rajamuthamilselvan, M.; Rajakumar, S.; Kavitha, S. Effect of Different SiCp Particle Sizes on the Behavior of AA 7075 Hot Deformation Composites Using Processing Maps. *Springer Proc. Mater.* **2021**, *5*, 1233–1244.
38. Lin, H.B. Dynamic recrystallization behavior of 6082 aluminum alloy during hot deformation. *Adv. Mech. Eng.* **2021**, *13*, 11. [CrossRef]
39. Ding, S.; Khan, S.A.; Yanagimoto, J. Constitutive descriptions and microstructure evolution of extruded A5083 aluminum alloy during hot compression. *Mater. Sci. Eng. A* **2018**, *728*, 133–143. [CrossRef]

Disclaimer/Publisher's Note: The statements, opinions and data contained in all publications are solely those of the individual author(s) and contributor(s) and not of MDPI and/or the editor(s). MDPI and/or the editor(s) disclaim responsibility for any injury to people or property resulting from any ideas, methods, instructions or products referred to in the content.

Article

Tensile Adhesion Strength of Atmospheric Plasma Sprayed MgAl$_2$O$_4$, Al$_2$O$_3$ Coatings

Andrey Zayatzev [1,*], Albina Lukianova [2], Dmitry Demoretsky [2] and Yulia Alexandrova [1]

[1] Department of Aircraft, Rocket Engines and Power Plants, Moscow Aviation Institute (National Research University), Volokolamskoe Shosse 4, 125993 Moscow, Russia
[2] Department of Machine Building, Samara State Technical University (Samara Polytech), Molodogvardeyskaya St. 244, 443100 Samara, Russia
* Correspondence: skadi221@gmail.com; Tel.: +7-(926)-471-69-24

Citation: Zayatzev, A.; Lukianova, A.; Demoretsky, D.; Alexandrova, Y. Tensile Adhesion Strength of Atmospheric Plasma Sprayed MgAl$_2$O$_4$, Al$_2$O$_3$ Coatings. *Ceramics* **2022**, *5*, 1242–1254. https://doi.org/10.3390/ceramics5040088

Academic Editors: Amirhossein Pakseresht and Kamalan Kirubaharan Amirtharaj Mosas

Received: 14 November 2022
Accepted: 7 December 2022
Published: 9 December 2022

Publisher's Note: MDPI stays neutral with regard to jurisdictional claims in published maps and institutional affiliations.

Copyright: © 2022 by the authors. Licensee MDPI, Basel, Switzerland. This article is an open access article distributed under the terms and conditions of the Creative Commons Attribution (CC BY) license (https://creativecommons.org/licenses/by/4.0/).

Abstract: This study analyses the distribution of stress during the testing of glued cylindrical specimens with thermally sprayed MgAl$_2$O$_4$, Al$_2$O$_3$ oxide coatings in order to evaluate the tensile adhesion strength. The set of studies that make up this work were conducted in order to evaluate the influence of the geometric parameters of cylindrical test specimens, 25 mm in diameter by 16–38.1 mm in height, on the measured tensile adhesion strength of the specimens. The stress and strain states inside the coating and at the coating-substrate interface were determined using the finite element modelling method. The debonding mechanisms, failure mode and influence of the coating microstructure on bond strength are also discussed. The finite element stress analysis shows a significant level of non-uniform stress distribution in the test specimens. The analysis of the results of the modelling stresses and strains using the finite element method for six types of cylindrical specimens, as well as the values obtained for the adhesion testing of MgAl$_2$O$_4$, Al$_2$O$_3$ coatings, show a need to increase the height of the standard cylindrical specimen (according to ASTM C633-13 (2021), GOST 9.304-87). The height should be increased by no less than 1.5–2.0 times to reduce the level of a non-uniform stress distribution in the separation area.

Keywords: tensile adhesion strength; plasma spray coating; MgAl$_2$O$_4$; Al$_2$O$_3$; FEA

1. Introduction

Thermally sprayed coatings of MgAl$_2$O$_4$, Al$_2$O$_3$ are commonly used as a protective layer for parts that require strong insulating properties ($\rho = 10^{-14} - 10^{-15}$ Ohm cm, T = 20 °C) under the conditions of high temperature, vacuums and radiation. Magnesium aluminate spinel compositions using aluminium oxide MgAl$_2$O$_4$ + 30%Al$_2$O$_3$, MgAl$_2$O$_4$ + 50%Al$_2$O$_3$, with a coating thickness of between 0.35 and 0.50 mm, offers satisfactory electrical insulation via high intensity neutron and gamma fluxes in vacuum, at temperatures ranging between 20 and 700 °C. The MgAl$_2$O$_4$, Al$_2$O$_3$ coatings are widely used as an electrical insulation layer for high-voltage electrical devices, channels and parts of magnetohydrodynamic (MHD) generators, sensors, the pads of reactors and International Thermonuclear Experimental Reactor (ITER) blanket modules. These types of coatings are predominantly deposited using atmospheric plasma spraying (APS) [1–3]. The performance of oxide ceramics that are applied to reactor facilities will be determined not only by a significant difference in the thermal expansion coefficients of the substrate and the coating material, but also by phase transformations at high temperatures and through exposure to radiation [4,5]. Compared to aluminium oxide, the cubic crystal lattice of spinel provides greater stability at high fluxes of gamma-neutron irradiation at high temperatures.

The problem of estimating the degree of a non-uniform stress-strain state at the interfacial debonding region of coating and over the thickness has practical applications that will allow us to better interpret the results and optimise the geometry parameters of standard cylindrical specimens, according to the international standards EN ISO 14916:2017,

ASTM C633-13 (2021) and GOST 9.304-87, JIS H 8666:1994 [6–9]. In practical terms, the standards and methods used to estimate the bond strength of thermal spray coatings are reduced to two main approaches for assessing cylindrical and conical specimens under tensile loads (uniaxial stress state). Most methods used to estimate the adhesion strength of a coating to a substrate can also be divided into two main groups; namely, qualitative testing methods and quantitative testing methods [10–15].

The tensile adhesion strength of thermal spray coatings will depend on different mechanical characteristics, including crack resistance (fracture toughness), Young's module, microhardness/hardness, compressive and tensile strength, types of residual stresses (quenching, peening, compressive [16,17]), differences in the coefficients of thermal expansion between the coating and the substrate as well as other features [10,11,18–20]. Parameters such as the roughness of the substrate after grit blasting, coating thickness, the composition of the coating, the tendency of the substrate/coating system to form chemical bonds and spraying methods (APS, HVOF, suspension-HVOF, DGS, LPCS [21–23], etc.) are critically important in terms of the adhesion strength. As such, there is no straightforward method for determining the adhesion strength of the part if it is operating under a complex stress state [24–27].

Meanwhile, test data obtained using different methods are often difficult to compare and are therefore rarely used for analytical description. There are some basic disadvantages that make it unfeasible to apply standard tensile strength test methods to a cylindrical specimen for high strength thermally sprayed ceramic coatings (bond strength more than 100 MPa). These are, firstly, the insufficient adhesive strength of glue at room and high temperatures, and secondly, the penetration of the adhesive to the substrate when the coatings are thin.

In cases where it impossible to perform the glue test, it is possible to apply an alternative pin method (Figure 1). However, this also has some drawbacks due to the lack of precise interpretation of the data due to the small diameter of the conical pin separation area (d_{pin} = 2–2.5 mm) and the incomplete separation of the brittle and highly porous coating.

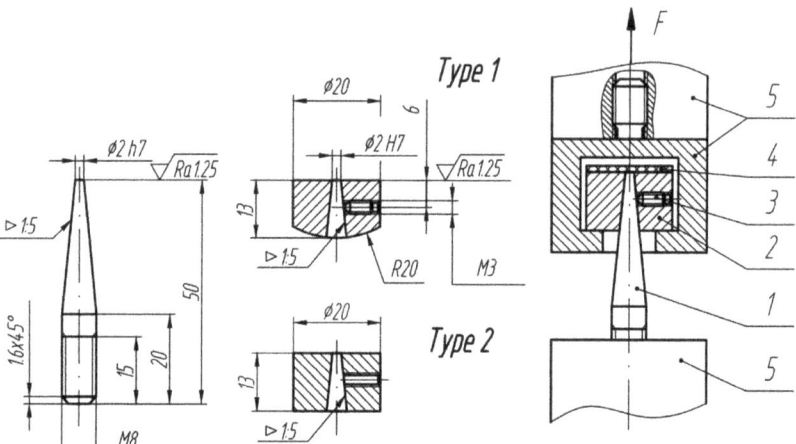

Figure 1. A conical pin specimen and self-aligned test fixture for tensile adhesion test: (**1**) Conical pin; (**2**) Matrix; (**3**) Centring screw; (**4**) Coating; (**5**) Equipment for tensile testing machine: F—ultimate tensile force.

The application of a tensile adhesion test to a conical pin specimen may eliminate the influence of friction forces between the pin and the mating conical pair matrix (Figure 1). The end surfaces of the assembled pair must be carefully grounded before grit blasting and thermal spraying, and the absence of glue then makes it possible to investigate the adhesive strength of the thermally sprayed coatings at high temperatures.

Comparative studies [10,14,24,28] show a significant impact on stress distribution at the bonding interface of the main geometric parameters based on the ratio of height (length)—h_{spec} to depth of the threaded closed hole—$h_{thr.holl}$ and the nominal diameter—d_{spec} of a cylindrical specimen. The studies focus on investigating the influence of specimen height on tensile adhesion strength for both standard and elongated glued specimens with thermal spray coatings assumed by ASTM C633-13 (2021), GOST 9.304-87. The effect of various specimen heights on the tensile adhesion test results and the interface stress distribution at the debonding area will be presented and discussed in the following sections.

2. Experimental Procedures

2.1. Plasma Spraying Conditions

Magnesium aluminum spinel (magnesium aluminate) and aluminium oxide powders ($MgAl_2O_4$ GTV 40.70.1, GTV GmbH, Luckenbach, Westerwald, Germany, particle size 20–45 µm; Al_2O_3 Amperit 740.001, H.C. Starck, particle size 22–45 µm) were deposited onto austenitic stainless steel 316L(N)-IG substrates using atmospheric plasma spraying, through the YPY-8M (JSC "Electromekhanika", Rzhev, Tver region, Russia [29]) technique, with a mounted low-capacity plasma torch. The spraying parameters for two types of oxide coatings and bond coatings are listed in Table 1. Argon-nitrogen was used as plasma-forming mixture, and the constant flow rates are also listed in Table 1. The standard plasma spray conditions of the oxide coating, such as $MgAl_2O_4$, Al_2O_3, are described in various articles [22,24,30,31]. Each set of specimens was deposited by specially designed equipment to set specimens at the same spray distance (Table 1). The spraying of ceramic electrically insulating $MgAl_2O_4$, Al_2O_3 oxide coatings on an austenitic steel substrate of a 316L(N)-IG alloy, standard EN 10088-1:2005 was performed using atmospheric plasma spraying equipment by feeding powdered material into the column of the compressed arc discharge (the anode spot of plasma torch between cathode and anode).

Table 1. Plasma spraying operating parameters.

Parameters	$MgAl_2O_4$, Coating	Al_2O_3, Coating	NiAl, Bond Coat
Current (A) × voltage [V] = power [kW]	330 × 50 = 16.50	320 × 50 = 16.00	330 × 48 = 15.84
Spray distance, [mm]	105		90
Feed powder, [g/min]	15–18		30
Velocity of travel plasma torch, [s/min]	2800		1600
Carrier gas type and flow rates, [L/min]	Ar: 25–30; N_2: 2.0–2.4		Ar: 20–25; N_2: 1.5–2.0
Transporting gas type and flow rate, [L/min]	Ar: 2.5–3.0		
Number of passes	9–10		1–2

For testing purposes, the coatings were deposited onto six cylindrical specimen types (d_{spec} = 25/25.4 mm in diameter; 3 standard and 3 elongated—heights h_{spec} = 28–38.1 mm) by thickness h_c from 0.4 to 0.5 mm with a bond coating of NiAl (powder: Metco 40NS, Oerlicon Metco, Wohlen, Switzerland, particle size 45–90 µm) (Figure 2). In order to isolate the uncoated surfaces of a cylindrical specimen from a stream of spray particles, a protective mask was applied. The first and third types of cylindrical specimen were manufactured in accordance with the standards ASTM C633-13 (2021), GOST 9.304-87, (Figure 2).

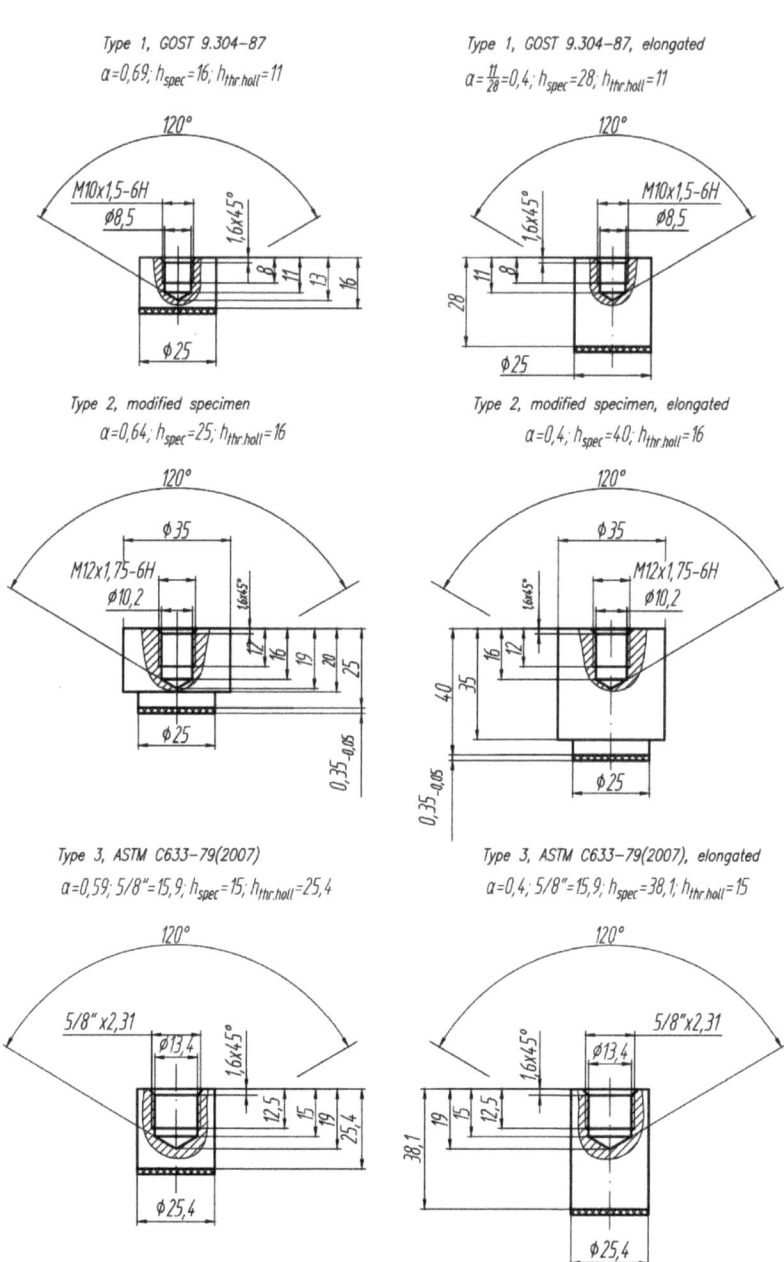

Figure 2. Six types of cylindrical specimens used to evaluate the bond strength of the thermal spray coating.

The geometry of the second type of specimen (Figure 2) was altered to improve the ability of the process to remove the adhesive (glue) surge after curing and to save the possibility of further centring the specimen in the precision bushings (Figure 3). The design of the elongated specimen was changed in accordance with the ratio of the specimen's height h_{spec} to depth of the threaded closed hole $h_{thr.holl}$ using the following equation:

$$\alpha = \frac{h_{thr.holl}}{h_{spec}} = 0.4 \tag{1}$$

The coefficient ratio of the specimen heights $\alpha = 0.4$ was chosen based on the results of similar studies [14].

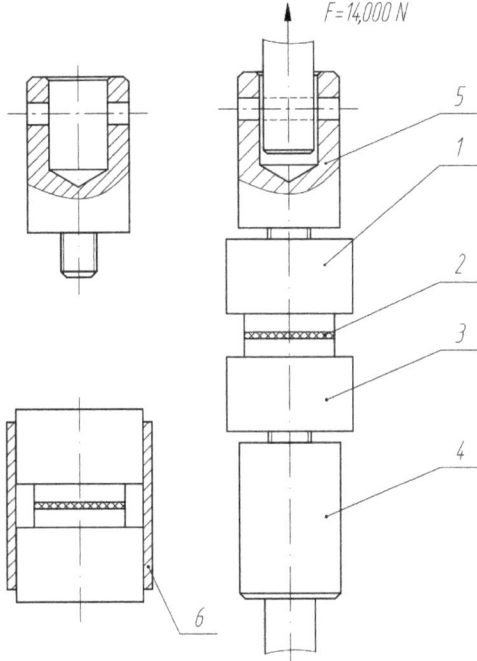

Figure 3. The geometry of a cylindrical specimen (type 2) assembled by tensile force F = 14,000 N: (**1**) Specimen with coating; (**2**) Epoxy; (**3**) Specimen without coating; (**4,5**) Equipment for tensile testing machine; (**6**) Bushing.

Before the substrates were coat-sprayed, they were grit-blasted with 1 mm alumina grit at an air pressure of up to 4–5 kg/cm^2. The mean roughness of the substrate was approximately Rz 45 ± 5 µm.

2.2. Finite Element Analysis, Boundary Conditions

Considerable research has been performed to investigate the stress-strain distribution in oxide coating and substrate interface to develop six finite element models using the ANSYS R16.2 programme, (Ansys Inc., Canonsburg, PA, USA). The axisymmetric models included two identical cylindrical halves, bonded together with a ceramic coating.

In order to facilitate calculations, the adhesive layer (glue) was excluded from the models, and the primary objective of this study was to estimate the level of uniformity in the interface stress distribution. It was therefore possible to simplify the finite element models of the specimens.

The following boundary conditions and assumptions were made for the finite-element models:

- The contact between the substrate and coating was assumed as a rigid connection with a restriction on all degrees of freedom

- Axial symmetry was provided by a cyclic region in the form of a 1/4 model by imposing boundary conditions to the symmetry regions along the XOY and YOZ planes
- The material of the substrate and oxide coating possessed only elastic strain.

The finite element grid near the coating-substrate interface was more refined than in other areas of the model—approximately 0.1 mm (Figure 4). The results of a finite element analysis (FEA) of the models were compared with each other and with the results of the bond strength tests. The finite element models were loaded with a tensile load of F = 14,000 N to the threaded hole, as shown in Figure 4. The mechanical properties of the Al_2O_3 coating and substrate were taken from the material library of the ANSYS R16.2 programme.

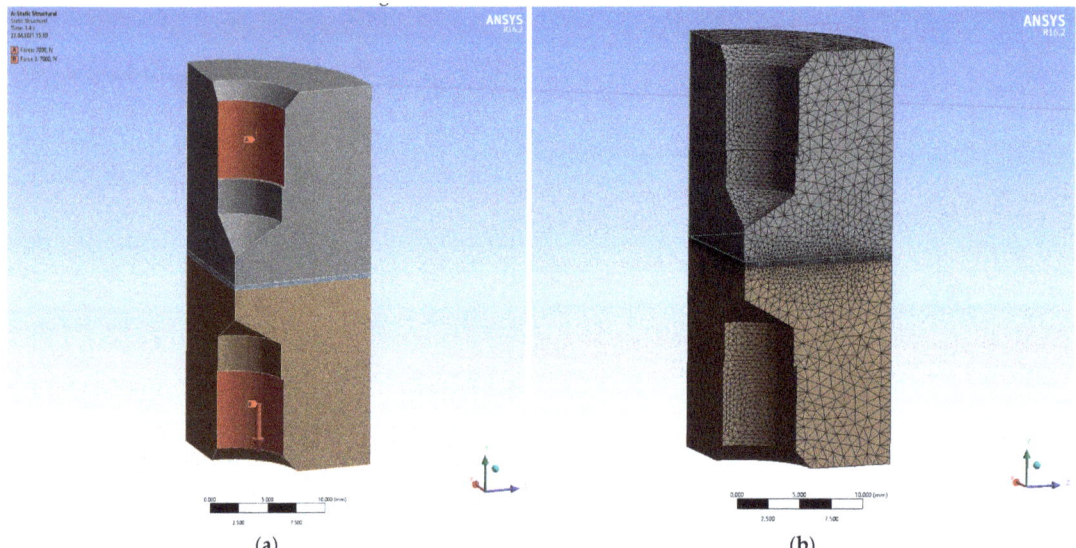

(a) (b)

Figure 4. The axisymmetric finite element model for the standard specimen (type 1): (**a**) the finite element of two identical cylindrical halves bonded together with ceramic coating, GOST 9.304-87; (**b**) the 3-D finite element model grid.

Previous tests [31–33] revealed very similar mechanical and physical properties for the $MgAl_2O_4$, Al_2O_3 oxide coatings, which made it possible to perform FEA only for the Al_2O_3 coating.

2.3. Tensile Adhesion Strength Testing

In this investigation, the test method of estimation tensile adhesion strength was based on the ASTM C633-13 (2021), GOST 9.304-87 [6,9] standards. Six specimens (standard and elongated) were bonded to the same uncoated specimens (counterparts) using a single-part epoxy resin Permobond ES 550 (maximum tensile strength σ_{ten} = 60 ± 3 GPa, maximum shear strength τ_{sh} = 41 ± 2 GPa by T = 20 °C). The alignment of the specimens was provided by centring bushings. The coatings produced were compared in terms of phase composition and microstructure, and the bond strength was also analysed at a temperature of T = 20 °C.

After thermal curing and cooling, the specimens were loaded with axial breaking force until debonding occurred at a traverse speed of 0.5 mm/min in the IR5143-200-11 tensile testing machine (JSC "Tochpribor", Ivanovo, Russia [34]). The tensile adhesion strength of σ_{ten} was calculated as the ratio between ultimate tensile force F divided by cross-sectional area A (A = 491 mm^2 for all testing results). Three identical tests were performed under the same conditions for each type of cylindrical specimen, using the $MgAl_2O_4$, Al_2O_3 plasma spray coating.

3. Results and Discussion

3.1. Results of Tensile Bond Strength of Standard and Elongated Specimens

Verification of the FEA data was performed by comparing them with the summary results of the tensile adhesion testing. The bond strengths of the APS ceramic coatings—$MgAl_2O_4$, Al_2O_3 for both standard and elongated specimens—are shown in Table 2. Before testing the tensile strength of the coating thickness, the h_c and roughness Ra were estimated (Table 2). In the present investigation, the adhesive failure mode occurred at the bond coating-substrate interface [30] for all of the tensile tests of the $MgAl_2O_4$, Al_2O_3 coatings, without the visible penetration of epoxy resin into the coating-substrate interface. The trend of decreased adhesion strength for elongated specimens correlated well with the FEA results.

Table 2. Summary results of tensile bond strength measurements for APS—$MgAl_2O_4$, Al_2O_3 and finite element surface stresses on standard and elongated specimens.

Type of Specimen/Test Method	Type of Coating	Thickness h_c [μm]	Roughness Ra [μm]	Tensile Bond Strength $\overline{\sigma}_{ten} \pm \delta$ [MPa]	Relative Error of Tensile Bond Strength [%]	Finite Element Average Equivalent Stress $\overline{\sigma}_{eq}$ [MPa]	β
Type 1, standard/GOST 9.304-87 [9]	Al_2O_3_APS	420 ± 35	38 ± 10	30.1 ± 1.6	5	60	6.7
	$MgAl_2O_4$_APS	406 ± 27	33 ± 6	28.8 ± 1.9	7	-	-
Type 4, elongated/GOST 9.304-87 [9]	Al_2O_3_APS	484 ± 21	29 ± 4	25.0 ± 2.6	10	45	1.4
	$MgAl_2O_4$_APS	446 ± 45	48 ± 5	22.0 ± 4.3	19	-	-
Type 2, modified/an in-house test method	Al_2O_3_APS	424 ± 38	33 ± 8	23.2 ± 3.2	14	45	3.1
	$MgAl_2O_4$_APS	484 ± 18	31 ± 4	26.4 ± 1.6	6	-	-
Type 5, elongated/an in-house test method	Al_2O_3_APS	426 ± 36	31 ± 6	19.4 ± 0.5	2	44	1.8
	$MgAl_2O_4$_APS	440 ± 29	30 ± 2	20.8 ± 2.0	10	-	-
Type 3, standard/ASTM C633-13 (2021) [6]	Al_2O_3_APS	460 ± 32	37 ± 8	27.8 ± 0.6	2	43	4.9
	$MgAl_2O_4$_APS	430 ± 30	38 ± 4	25.1 ± 3.2	13	-	-
Type 6, elongated/ASTM C633-13 (2021) [6]	Al_2O_3_APS	472 ± 12	33 ± 3	20.7 ± 1.7	8	44	1.7
	$MgAl_2O_4$_APS	450 ± 31	37 ± 3	18.4 ± 1.5	8	-	-

The results of the tensile adhesion strength tests showed a low level of bond strength for applied APS technology for $MgAl_2O_4$, Al_2O_3 coatings [3,10,22,24]. This may be related to the spraying equipment used, the construction of the plasma torch, the residual stress distribution, the fatigue crack propagation, the pore size and distribution [35], or the rate of particle velocity and temperature.

The high level of stress discontinuity at the coating-substrate interface might help explain the increased value of the adhesion strength (Table 2). As the surface morphology images in Figure 5 show, the low level of bond strength conforms well to the results of the metallographic analysis of the scanning electron microscope (SEM) images of the cross sections and surface morphologies of the coatings (large, moulded agglomerates, a large number of pores and other defects).

3.2. Assessment of Stress Distribution during Tensile Testing

The equivalent stress distribution and total strain showed in the three standard and three elongated specimens are shown in Figures 6–8. A calculation of the stress distribution at the interface of three standard specimens showed a high level of non-uniformity over the debonding area. For example, the maximum value of equivalent stress came at the outer edge of the ASTM C633-13 (2021) standard specimen (type 3, Figure 2)—σ_{eq}^{max} = 71.5 MPa and the minimum value occurs at the centre—σ_{eq}^{min} = 14.5 MPa (Figure 8). The international standards procedure assumes that the stress is uniform based on testing a cylindrical specimen. The most uneven distribution of the equivalent stress was found at the interface of the GOST 9.304-87 standard specimen (type 1, Figure 2)—the maximum σ_{eq}^{max} = 59.6 MPa occurs between the centre and the outer edge of the specimen, while the minimum σ_{eq}^{min} = 9 MPa occurs at the centre of the specimen (Figure 6). The maximum value of the equivalent stress can therefore be taken as the average equivalent of the interface for the standard specimen type 1—$\sigma_{eq}^{max} = \overline{\sigma_{eq}}$.

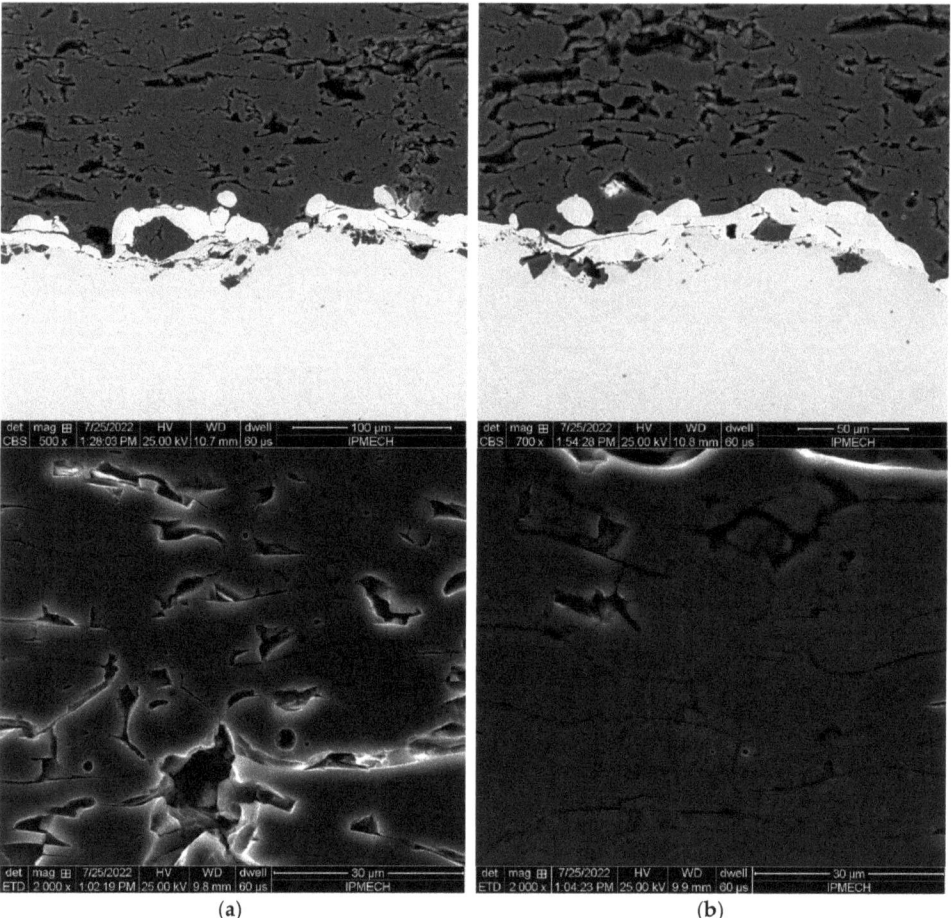

Figure 5. Cross-section SEM images of microstructures of the coating: (**a**) Al$_2$O$_3$; (**b**) MgAl$_2$O$_4$. For spray parameters, refer to Table 2.

The average equivalent stress $\overline{\sigma_{eq}}$ = 43 MPa (Figure 8) showed by FEA is 155% higher than the average tensile adhesion strength σ_{ten} = 27.8 ± 0.6 MPa for Al$_2$O$_4$ (type 3, see Table 2) assumed by the ASTM C633-13 (2021) procedure.

The level of non-uniformity stress distribution in the test specimen was estimated using the following equation:

$$\beta = \frac{\sigma_{eq}^{max}}{\sigma_{eq}^{min}}, \qquad (2)$$

The non-uniform stress distribution will lead to the misinterpretation of the adhesion tensile testing results [14]. Significant excess of peak stress values at the outer edge, compared with average stress distribution, means that the inherent determination of high or low values of adhesion strength can be assumed by the ASTM C633-13 (2021), GOST 9.304-87 test procedure. The result implies that the bond strength depends not only on the geometry of a specimen [14,16], but also on the degree of plasticity and brittleness of the coating (brittle or ductile fracture).

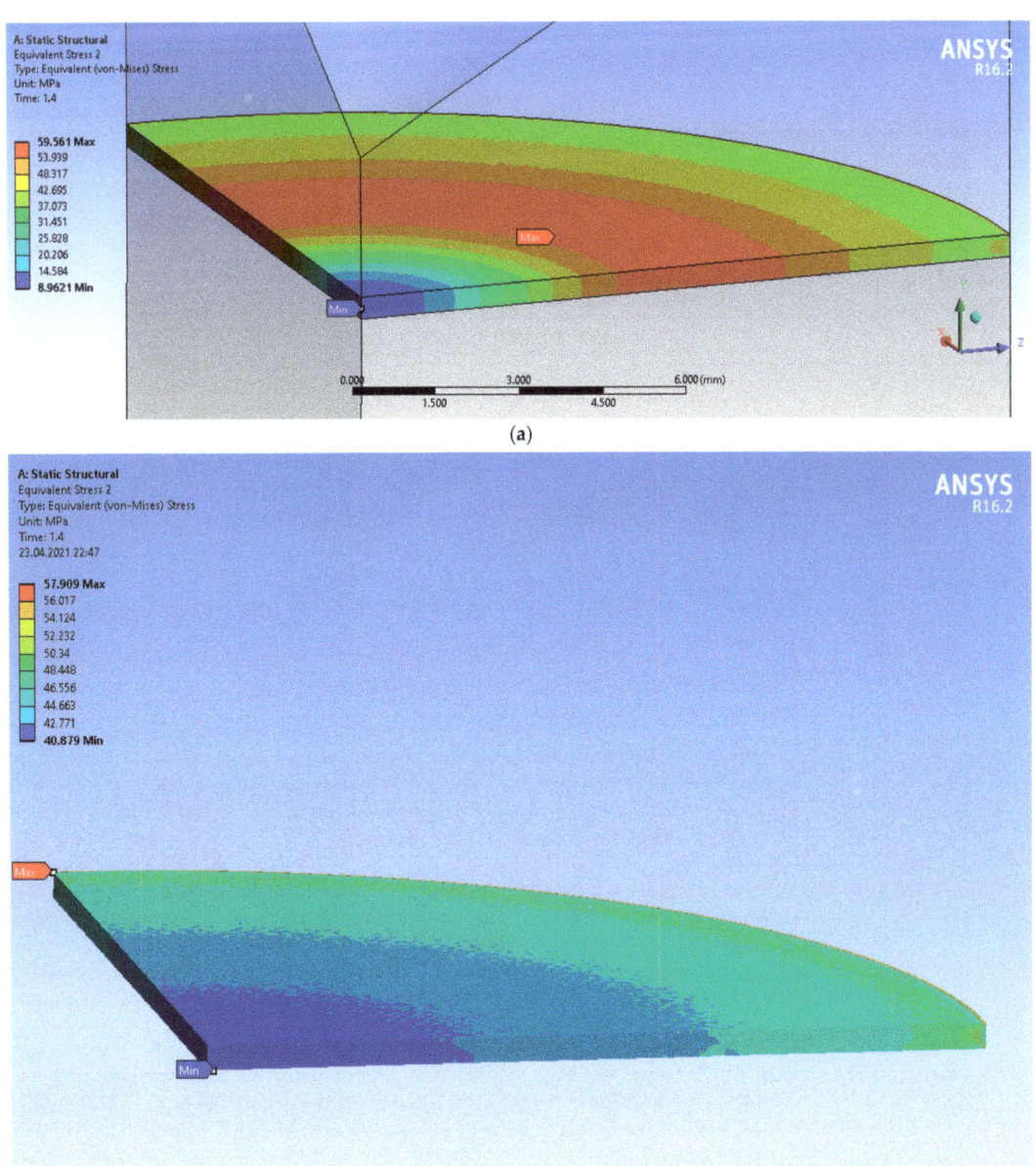

Figure 6. Equivalent stress distribution at the debonding area of the Al_2O_3 coating with a tensile force of F = 14,000 N: (**a**) GOST 9.304-87 standard specimen, type 1; (**b**) Elongated specimen, type 4. For geometric parameters, see Figure 2.

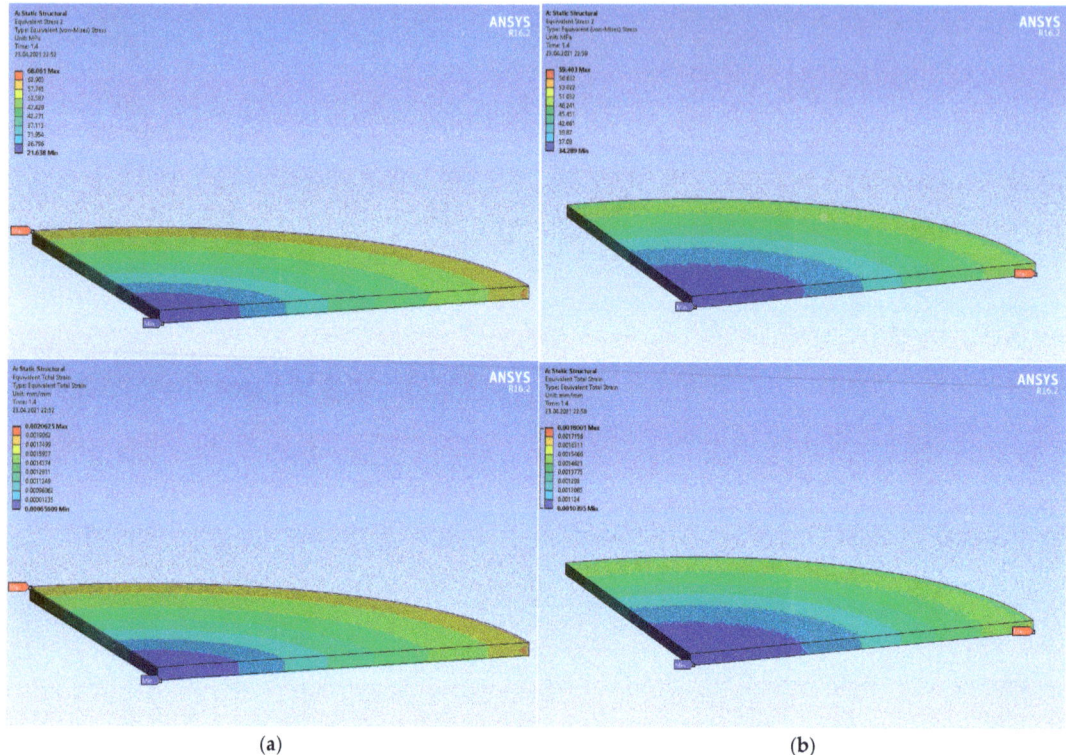

(a)   (b)

Figure 7. The equivalent stress distribution and total strain at the debonding area of the coating Al_2O_3 with tensile force F = 14,000 N: (**a**) Modified specimen, type 2; (**b**) Elongated specimen, type 5. For geometric parameters, refer to Figure 2.

The main reasons for the appearance of significant non-uniform interface stress distribution should also be mentioned. These are the appearance of additional shear stress at the outer edge of the coating (the Edge effect), and the decrease in the stiffness of the specimen due to the close proximity of the threaded hole in relation to the detachment surface of the coating.

The impact of the Edge effect can be minimized by increasing the diameter of the cylindrical specimen, up to d_{spec} = 40–60 mm, as well as the length of a cylindrical specimen. The distance from the threaded closed holes to the plane face of a cylindrical specimen with a diameter of d_{spec} = 25/25.4 mm should be enough to minimise the stiffness reduction. Applying elongated specimens with lengths of h_{spec} = 38–50 mm and more [9] will lead to more uniform stress distribution.

The finite element analysis and laboratory testing of the elongated cylindrical specimens with a diameter of d_{spec} = 25/25.4 mm confirmed that the results of estimating bond strength using a standard cylindrical specimen assumed by ASTM C633-13 (2021), GOST 9.304-87 can be significantly higher than the actual tensile adhesion strength [2]. With the exception of the GOST 9.304-87 standard specimen, the equivalent stress peaks were not identified at the centre of the specimens, but rather, at the outer edge in all models.

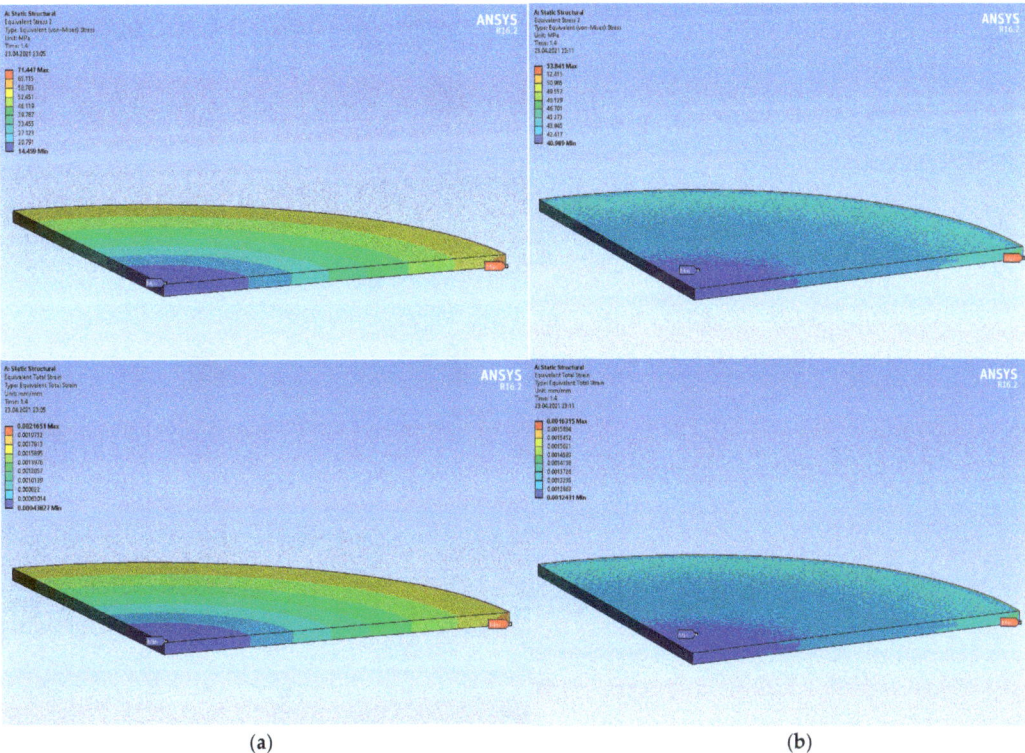

(a)   (b)

Figure 8. Equivalent stress distribution and total strain at the debonding area of the Al_2O_3 coating with a tensile force of F = 14,000 N: (**a**) ASTM C633-79 (2021) standard specimen, type 3; (**b**) Elongated specimen, type 6. For geometric parameters, see Figure 2.

4. Conclusions

According to the results of this scientific study and the analysis of the existing literature, we were able to make three conclusions. Firstly, the significant peak stress at the outer edge of the specimen, compared to the average stress calculated by tensile testing, implies that the bond strength should be estimated using elongated specimens, according to the ASTM C633-13 (2021), GOST 9.304-87 test procedure. Secondly, compared to conventional tests (EN ISO 14916:2017, ASTM C633-13 (2021), GOST 9.304-87), the accuracy of tensile bond strength estimation may be higher when the new geometry of a cylindrical specimen is applied. Thirdly, to provide a more accurate method, researchers should carry out finite element analysis and an epoxy tensile adhesion test using elongated cylindrical specimens using different diameters, d_{spec} = 20–60 mm (using the procedure defined in ASTM C633-13 (2021), GOST 9.304-87).

Author Contributions: A.Z. developed and planned the concept and methodology for conducting and reporting this science work. A.L., D.D. and A.Z. carried out the simulations of stress-strain stated at the coating-substrate interface and analysed the data on the tensile tests. D.D. undertook the spraying of coatings on specimens. A.Z. wrote the draft document. Y.A. organised SEM analysis and contributed to the discussion and interpretation of the results. All authors have read and agreed to the published version of the manuscript.

Funding: The study was carried out with the financial support of the Ministry of Science and Higher Education of the Russian Federation within the framework of the state task theme no. AAAAA12-2110800012-0.

Institutional Review Board Statement: Not applicable.

Informed Consent Statement: Not applicable.

Data Availability Statement: Not applicable.

Conflicts of Interest: The authors declare no conflict of interest.

References

1. Niemi, K.; Hakalahti, J.; Hyvärinen, L.; Laurila, J.; Vuoristo, P.; Berger, L.-M.; Toma, F.-L.; Shakhverdova, I. Influence of chromia alloying on the characteristics of APS and HVOF sprayed alumina coatings. In Proceedings of The ITSC 2011, International Thermal Spray Conference & Exposition, Düsseldorf, Germany, 27–29 September 2011; Volume 276.
2. Sang, P.; Chen, L.-Y.; Zhao, C.; Wang, Z.-X.; Wang, H.; Lu, S.; Song, D.; Xu, J.-H.; Zhang, L.-C. Particle size-dependent microstructure, hardness and electrochemical corrosion behavior of atmospheric plasma sprayed NiCrBSi coatings. *Metals* **2019**, *9*, 1342. [CrossRef]
3. Junge, P.; Rupprecht, C.; Greinacher, M.; Kober, D.; Stargardt, P. Thermally Sprayed Al2O3 Ceramic Coatin–s for Electrical Insulation Applications. In Proceedings of The ITSC, International Thermal Spray Conference, Vienna, Austria, 4–6 May 2022; pp. 72–81.
4. Popov, A.I.; Lushchik, A.; Shablonin, E.; Vasil'chenko, E.; Kotomin, E.A.; Moskina, A.M.; Kuzovkov, V.N. Comparison of the F-type center thermal annealing in heavy-ion and neutron irradiated Al$_2$O$_3$ single crystals. *Nucl. Instrum. Methods Phys. Res. Sect. B Beam Interact. Mater. At.* **2018**, *433*, 93–97. [CrossRef]
5. Lushchik, A.; Feldbach, E.; Kotomin, E.A.; Kudryavtseva, I.; Kuzovkov, V.N.; Popov, A.I.; Seeman, V.; Shablonin, E. Distinctive features of diffusion-controlled radiation defect recombination in stoichiometric magnesium aluminate spinel single crystals and transparent polycrystalline ceramics. *Sci. Rep.* **2020**, *10*, 7810. [CrossRef]
6. *ASTM C633-13*; Standard Test Method for Adhesion or Cohesion Strength of Thermal Spray Coatings. ASTM International: West Conshohocken, PA, USA, 2021. Available online: https://www.astm.org/c0633-13r21.html (accessed on 7 November 2022).
7. *EN ISO 14916*; Thermal spraying—Determination of Tensile Adhesive Strength. ISO: Geneva, Switzerland, 2017. Available online: https://www.iso.org/standard/60995.html (accessed on 7 November 2022).
8. *JIS H 8666*; Test Methods for Ceramic Sprayed Coatings. Japanese Standards Association: Tokyo, Japan, 1994. Available online: https://jis.eomec.com/abolished/jish86661994?abolishedid=10886661994#gsc.tab=0 (accessed on 7 November 2022).
9. *GOST 9.304-87*; Unified System of Corrosion and Ageing Protection. Thermal Sprayed Coatings. General Requirements and Methods of Control. GOST—USSR Ministry of Chemical and Petroleum Engineering: Moscow, Russia, 1987. Available online: https://docs.cntd.ru/document/1200014731 (accessed on 7 November 2022).
10. Hadad, M.; Marot, G.; Démarécaux, P.; Chicot, D.; Lesage, J.; Rohr, L.; Siegmann, S. Adhesion tests for thermal spray coatings: Correlation of bond strength and interfacial toughness. *J. Surf. Eng.* **2007**, *23*, 279–283. [CrossRef]
11. Vencl, A.; Arostegui, S.; Favaro, G.; Zivic, F.; Mrdak, M.; Mitrović, S.; Popovic, V. Evaluation of adhesion/cohesion bond strength of the thick plasma spray coatings by scratch testing on coatings cross-sections. *Tribol. Int.* **2011**, *44*, 1281–1288. [CrossRef]
12. Chen, Z.; Zhou, K.; Lu, X.; Yee, C.L. A review of the mechanical methods for evaluating coating adhesion. *Acta Mech.* **2014**, *255*, 431–452. [CrossRef]
13. Lorenzo-Bañuelos, M.; Díaz, A.; Rodríguez, D.; Cuesta, I.I.; Fernández, A.; Alegre, J.M. Influence of Atmospheric Plasma Spray Parameters (APS) on the Mechanical Properties of Ni-Al Coatings on Aluminum Alloy Substrate. *Metals* **2021**, *11*, 612. [CrossRef]
14. Han, W.; Rybicki, E.F.; Shadley, J.R. An improved specimen geometry for ASTM C633-79 to estimate bond strengths of thermal spray coatings. *J. Therm. Spray Technol.* **1993**, *2*, 145–150. [CrossRef]
15. Lindner, T.; Saborowski, E.; Scholze, M.; Zillmann, B.; Lampke, T. Thermal Spray Coatings as an Adhesion Promoter in Metal/FRP Joints. *Metals* **2018**, *8*, 769. [CrossRef]
16. Lyphout, C.; Nylen, P.; Östergren, L.G. Adhesion Strength of HVOF Sprayed IN718 Coatings. *J. Therm. Spray Technol.* **2011**, *21*, 145–150. [CrossRef]
17. Stokes, J.; Looney, L. FEA of residual stress during HVOF thermal spraying. *J. Mat. Eng. Perf.* **2009**, *18*, 21–25. [CrossRef]
18. Boruah, D.; Robinson, B.; London, T.; Wu, H.; de Villiers-Lovelock, H.; McNutt, P.; Doré, M.; Zhang, X. Experimental evaluation of interfacial adhesion strength of cold sprayed Ti-6Al-4V thick coatings using an adhesive-free test method. *Surf. Coat. Technol.* **2020**, *381*, 125–130. [CrossRef]
19. Marot, G.; Démarécaux, P.; Lesage, J.; Hadad, M.; Siegmann, S.T.; Staia, M.H. The interfacial indentation test to determine adhesion and residual stresses in NiCr VPS coatings. *Surf. Coat. Technol.* **2018**, *202*, 4411–4416. [CrossRef]
20. Ang, A.S.M.; Sanpo, N.; Sesso, M.L.; Kim, S.Y.; Berndt, C.C. Thermal Spray Maps: Material Genomics of Processing Technologies. *J. Therm. Spray Technol.* **2013**, *22*, 1170–1183. [CrossRef]
21. Kiilakoski, J.; Trache, R.; Björklund, S.; Joshi, S.; Vuoristo, P. Process parameter impact on suspension-HVOF-sprayed Cr$_2$O$_3$ coatings. *J. Therm. Spray Technol.* **2019**, *28*, 1933–1944. [CrossRef]
22. Yilmaz, Ş. An evaluation of plasma-sprayed coatings based on Al2O3 and Al2O3–13 wt.% TiO2 with bond coat on pure titanium substrate. *J. Therm. Spray Technol.* **2009**, *35*, 2017–2022.

23. Winnicki, M. Advanced functional metal-ceramic and ceramic coatings deposited by low-pressure cold spraying: A review. *Coatings* **2021**, *11*, 1044. [CrossRef]
24. Toma, F.-L.; Berger, L.-M.; Scheitz, S.; Langner, S.; Rödel, C.; Potthoff, A.; Sauchuk, V.; Kusnezoff, M. Comparison of the microstructural characteristics and electrical properties of thermally sprayed Al2O3 coatings from aqueous suspensions and feedstock powders. *J. Therm. Spray Technol.* **2012**, *21*, 480–488. [CrossRef]
25. Hadad, M.; Marot, G.; Démarécaux, P.; Lesage, J.; Michler, J.; Siegmann, S.T. Adhesion tests tor thermal spray coatings: Application range of tensile, shear and interfacial indentation methods. In Proceedings of The ITSC 2005: Thermal Spray Connects: Explore Its Surfacing Potential (DVS-ASM), Basel, Switzerland, 2–4 May 2005; pp. 759–764.
26. Yamazaki, Y.; Arai, M.; Miyashita, Y.; Waki, H.; Suzuki, M. Determination of interfacial fracture toughness of thermal spray coatings by indentation. *J. Therm. Spray Technol.* **2013**, *22*, 1358–1365. [CrossRef]
27. Mueller, E. Stress peening—A sophisticated way of normal shot peening. *J. Mater. Sci. Eng.* **2019**, *A9*, 56–63.
28. Liu, Y.; Fu, S.; Wang, Z.; Yan, X.; Xi, N.; Wu, Y.; Chen, H. Tensile properties, shear strength calculation and cracking behavior of bulk composite comprised of thick HVOF sprayed coating and steel substrate. *Surface and Coating Technol.* **2019**, *374*, 807–814. [CrossRef]
29. Public Joint Stock Company. Electromekhanika. Available online: http://www.el-mech.ru/products/pokryt/pokryt_67.html (accessed on 7 November 2022).
30. Zaytsev, A.N.; Yagopol'skiy, A.G.; Aleksandrova, Y.P. Assessing the impact of the structure and chemical composition of plasma-sprayed coatings on their adhesion and tribological properties. *BMSTU J. Mech. Eng.* **2018**, *5*, 48–59.
31. Toma, F.-L.; Scheitz, S.; Berger, L.-M.; Sauchuk, V.; Kusnezoff, M.; Thiele, S.V. Comparative study of the electrical properties and characteristics of thermally sprayed alumina and spinel coatings. *J. Therm. Spray Technol.* **2011**, *20*, 195–204. [CrossRef]
32. Zaytsev, A.N.; Lukianova, A.N.; Demoretsky, D.A. Assessment of Shear Bond Strength of Thermal Spray Coatings by Applying Prismatic Samples. *Solid State Phenom.* **2022**, *337*, 35–41. [CrossRef]
33. Pogrebnjak, A.D.; Tyurin, Y.N. Modification of material properties and coating deposition using plasma jets. *UFN* **2005**, *175*, 515–544. [CrossRef]
34. Public Joint Stock Company. Tochpribor. Available online: https://www.tochpribor-kb.ru (accessed on 7 November 2022).
35. Siao, M.A.A.; Berndt, C.C. A review of testing methods for thermal spray coatings. *Int. Mat. Rev.* **2014**, *59*, 179–223.

Article

Electron-Beam Deposition of Metal and Ceramic-Based Composite Coatings in the Fore-Vacuum Pressure Range

A. V. Tyunkov [1], A. S. Klimov [1], K. P. Savkin [2], Y. G. Yushkov [1] and D. B. Zolotukhin [1,*]

[1] Department of Physics, Tomsk State University of Control Systems and Radioelectronics, Tomsk 634050, Russia
[2] Institute of High Current Electronics, Tomsk 634050, Russia
* Correspondence: denis.b.zolotukhin@tusur.ru

Abstract: We present the experimental results on the fabrication of metal-ceramic coatings by electron-beam evaporation of alumina ceramic and copper powder composites with different fractions of the components (with Cu powder fraction from 0.1 to 20%) pre-sintered by an electron beam. The mass-to-charge composition of the multi-component plasma, generated in the electron beam transport region, was measured, demonstrating that the fraction of target ions in plasma grows with the electron beam power density. The morphology and electrical conductivity of fabricated coatings were investigated; it was found that the increase in Cu fraction in the deposited coating from 0 to 20% decreases both the volumetric and surface resistance of the coatings in around 8 orders of magnitude, thereby being a convenient tool to control the coating properties.

Keywords: electron-beam evaporation; fore-vacuum pressure range; beam plasma; coating deposition; metal-ceramic coatings

1. Introduction

Many requirements on industrial products are determined by the properties of the surface layers of the material from which the product is made. The use of expensive and scarce materials in industrial production is often not feasible. In practice, the required properties of a product are achieved by using materials with special coatings that endow such properties [1–3].

Ceramic structures, owing to their high strength properties, are of interest as a reinforcing material in the composites [4,5]. They are refractory, have high specific strength and high mechanical characteristics; they preserve their properties at elevated temperatures. Alumina ceramics is a dielectric, which can be used as an insulating coating in microelectronics [6]. Such coatings can be used for reinforcement, wear-resistance, refractory, optical, biocompatible, and decorative purposes [7–9].

As of today, among the many methods of coating deposition [10–12], beam-plasma methods have a special place. Multifunctional dielectric and metal coatings produced by magnetron or vacuum-arc sputtering and laser ablation in plasma chemical reactors, as well as by electron-beam evaporation, are applied to handle a wide range of practical problems related to surface modification of various materials. In the long list of beam-plasma coating technologies, the electron-beam evaporation technique is characterized by higher rates of deposition and, hence, a higher process performance [13]. The electron-beam evaporation method, when applied to dielectric targets, is hampered by the electron-beam charging of the target surface. To avoid this, it is required to apply, at least at in the initial stages of the technological process, special methods and approaches [14]. It complicates the production equipment and makes the entire process less manageable and efficient.

Fore-vacuum plasma-cathode electron sources [13] can generate electron beams in a previously inaccessible range of elevated fore-vacuum pressures (1–100 Pa). They have all the known advantages of conventional plasma sources of electrons, such as high current density and reliability and are not demanding in terms of operation in harsh vacuum conditions and in the presence of aggressive gases. The beam plasma generated in the beam transport region at elevated fore-vacuum pressures effectively neutralizes the electron-beam charging of the surface of electrically non-conductive materials [15]. This feature allows one to create plasma that contains evaporation products of a dielectric (ceramics) and deposit its vapor onto the substrate, thus forming a dielectric coating. It is additionally possible, using an electron beam, to create a metal-ceramic coating by evaporating composite targets of compound composition based on ceramic and the addition of practically any metal. The properties of such coatings can be readily controlled by varying the ratio of the components used in the preparation of the target to be evaporated.

In this article, we summarize our studies of the process of obtaining metal-ceramic coatings by electron-beam evaporation of aluminum ceramic and copper composites. The importance of thin films deposited from an evaporated composite target is justified by the fact that the properties of such coatings can be adjusted in a wide range by varying the composition and the fractions of elements in the composite targets.

2. Sample Preparation

The targets for electron beam evaporation were made in our lab from an Al_2O_3 fine dispersive powder with a particle size of ≈30–60 μm and a Cu powder with a particle size of ≈30–50 μm. The targets had different weight ratios of metal to ceramics. A total of four targets with different contents of ceramics and metal were used in the experiments. The component composition of the targets is given in Table 1.

Table 1. Component composition of Al_2O_3-Cu targets.

Component	Component Percentage, Mass %			
	Target 1	Target 2	Target 3	Target 4
Al_2O_3	99.9	99	90	80
Cu	0.1	1	10	20

To ensure an even distribution of the material over a target volume, the powder components were mixed for 30 min. Powder was mixed using a conventional ball mill with the addition of ceramic granules. Then, the resulting mixture was poured into a mold and pressed in a pallet at a pressure of 200 MPa. The diameter of the produced target was 10 ± 0.1 mm with a height of 3 ± 0.15 mm. Micrographs of aluminum oxide and copper powders used in the experiment are shown in Figure 1.

 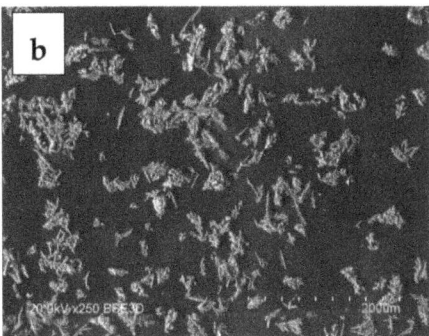

Figure 1. Micrographs of original Al_2O_3 (**a**) and Cu (**b**) powders used to fabricate the compacts.

The coatings were deposited on titanium samples (disks, 20 mm in diameter and 3 mm thick), which were located on a stainless holder; it was possible to place up to four samples per cycle. The sample surfaces were prepared using a Saphire 320 (Germany) grinder and polisher with an automatic Rubin 500 head, with working wheels 200 mm in diameter, the grinder speed 50–600 rev/min and the option to load up to 5 samples per automatic cycle.

The elemental composition of coatings was studied using a Hitachi S3400N SEM, equipped with a BrukerX'Flash 5010 energy-dispersive microanalyzer.

The electrical resistance of the synthesized metal-containing ceramic coatings was studied using an E6-13A teraohmmeter of «Punane-Rat», Estonia. The experimental stand for studying the electrical resistance is discussed in detail in [16].

3. Experimental Setup

Experiments on electron-beam synthesis of coatings were conducted using a fore-vacuum plasma-cathode electron source based on a hollow-cathode glow discharge, operating in a continuous mode [17] (Figure 2). The beam was transported towards the evaporated target at a residual atmosphere of 5 Pa. Under the electron beam action, the target material evaporated and partially ionized, thereby providing the deposition of coating on the sample surface. In one experiment, the coatings were synthesized simultaneously on three samples. The temperatures of the sample surface and of the evaporated target were monitored by a rapid Raytek optical pyrometer; they were 150 and 2100 degrees celsius, respectively. The chamber vacuum was maintained by an ISP-500C spiral fore-vacuum pump. The secondary plasma created during the target material evaporation, consisting of the "products" of the gaseous residual atmosphere and the evaporated target, was analyzed using a modified RGA–300 quadrupole mass-spectrometer, operating as a mass analyzer of the beam plasma ions [18].

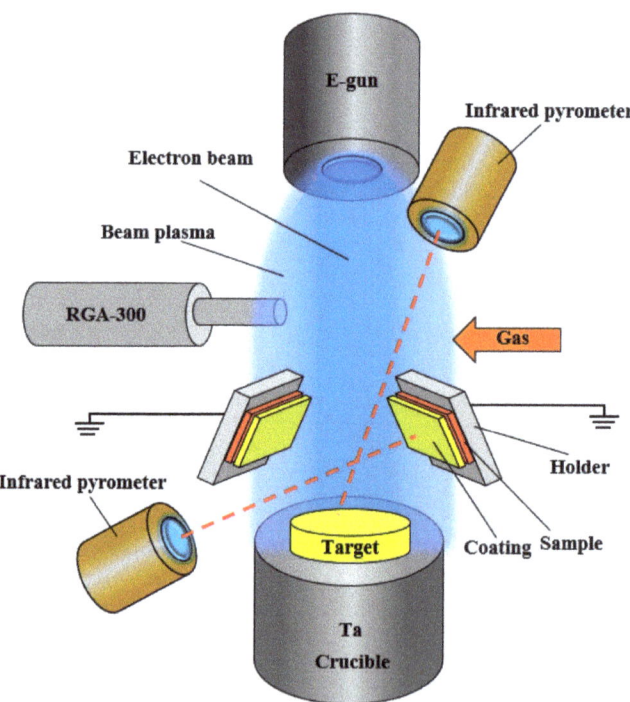

Figure 2. Schematic diagram of the coating deposition experiment.

4. Results

The ion mass-to-charge spectrum of the plasma vapor generated during the electron beam evaporation of target 2 with 1% copper content is shown in Figure 3. This spectrum was recorded during the electron beam evaporation at a power density of 1 kW/cm^2. As seen, when the evaporation is not intensive, the spectrum includes predominantly ions of the residual atmosphere. The spectrum also registers atomic ions of the ceramic material: lithium, sodium, aluminum, as well as copper. It should be noted that for the indicated copper fraction, the beam power density of about 1 kW/cm^2 is a threshold value. It is from this level of the power density that the mass analyzer begins to register the copper ions. With a further increase in the electron beam power density, the amplitude of copper ion peaks falls off abruptly against the background of the growing amplitude of aluminum ions.

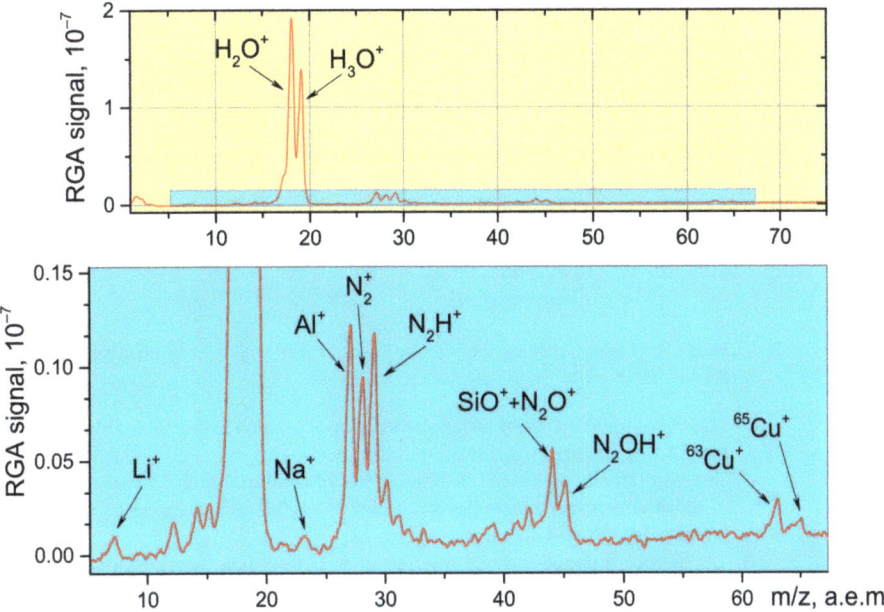

Figure 3. Mass spectrum of the beam plasma ions at a power density of 1 kW/cm^2.

A similar situation is observed in the plasma generated during the evaporation of target 3 with 10% copper content (Figure 4). At a beam power density of 0.75 kW/cm^2, the spectrum exhibits exclusively ions of the residual atmosphere. At a beam power density of 1 kW/cm^2, the spectrum shows the peaks of copper and sodium ions. In contrast to the evaporation of 1% copper target, in this case, the copper peaks are distinctly identified in the spectrum due to the larger fraction of copper atoms. No destruction of the sample surface is observed in this case. As the beam power density increases, the proportion of impurity ions in the spectrum decreases, and the amplitude of ceramic base atoms (aluminum) increases. This fact can be explained as follows. With an increase in the beam power density, the temperature of the target being processed increases. As the temperature increases, fusible (compared to aluminum oxide) impurities (sodium, lithium, copper) melt and, because of differences in the density with the base, move to the surface where they evaporate.

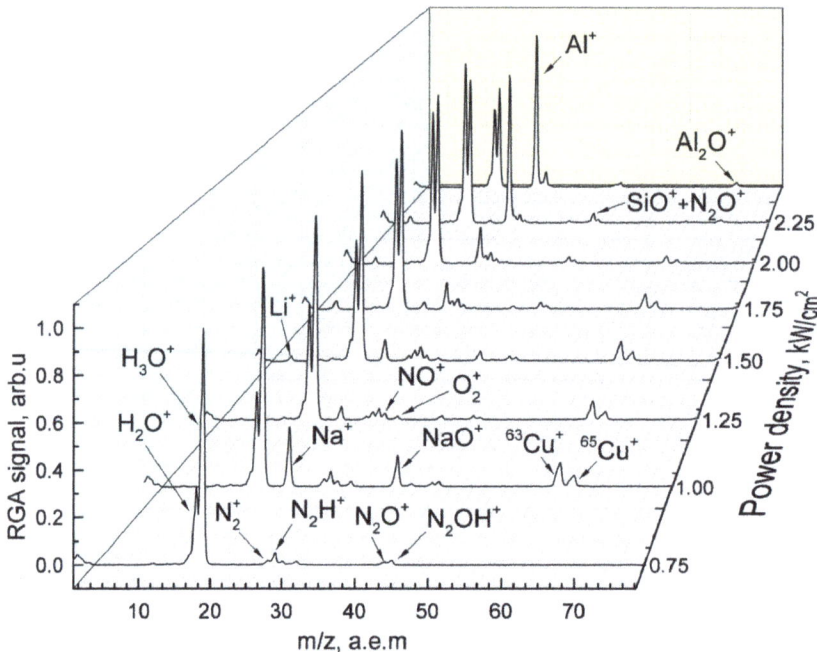

Figure 4. Dependence of the mass-to-charge composition of beam plasma ions on the beam power density.

Since for the case of target 2, the content of such impurities did not exceed 1%, they evaporated within a short time, and the spectra registered only ions of the target material and residual atmosphere. The 10% content of copper in target 3 ensured a longer presence of these ions in the mass-spectra; however, in the end, the ions of the ceramic material remained dominant.

Figure 5 shows typical micrographs of the fabricated coatings depending on the percentage of the components in the target material (see Table 1). As seen, the coating surface is fairly uniform and almost free of defects (Figure 5a–c). Nevertheless, for target 4 with 20% copper (Figure 5d), there are condensed copper droplets on the surface with the size ranging from 500 nm to 2 µm. The coating thickness varies from 1.5 to 2 µm. The deposition rate was greater than 1 µm per minute.

The formation of these droplets on the surface is caused by the explosive boiling of copper in the melting pool of target 4. As a result, along with the vapor or plasma, the micro droplets form, with the velocity comparable to that of copper vapor, and bring about the micro defects in the deposited coating. To reduce this negative effect, one can use high-speed scanning of the target or indirect heating of the target similar to what we used earlier in [19].

Figure 6 shows the elemental composition and its spatial distribution for the obtained coatings. One can conclude from Figure 6 that the components of the evaporated target distributed rather uniformly in the deposited coatings.

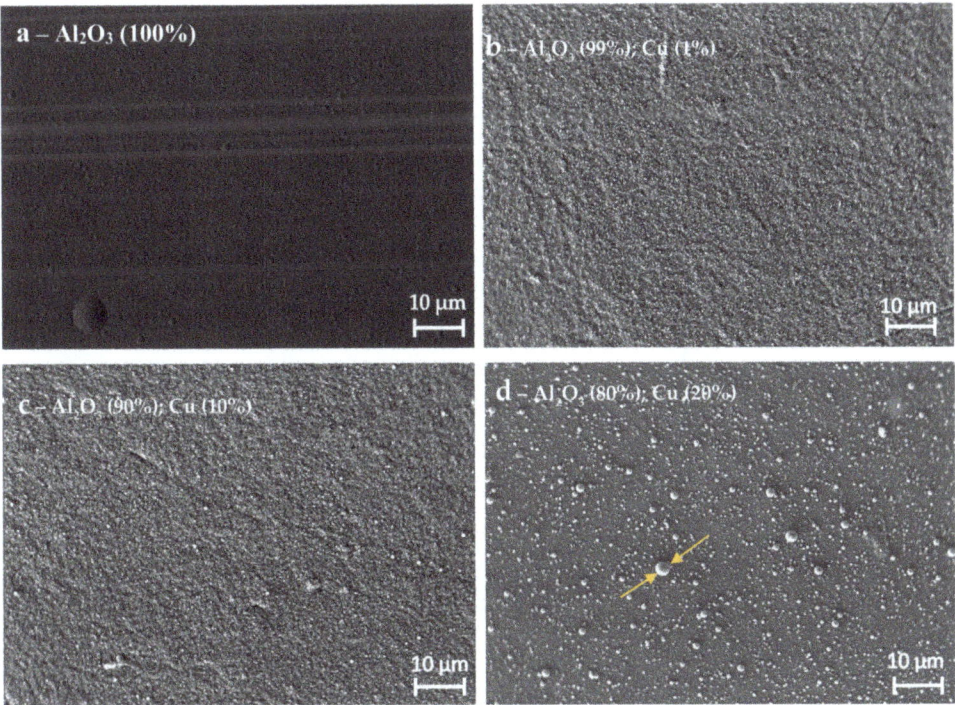

Figure 5. Micrographs of the fabricated coatings.

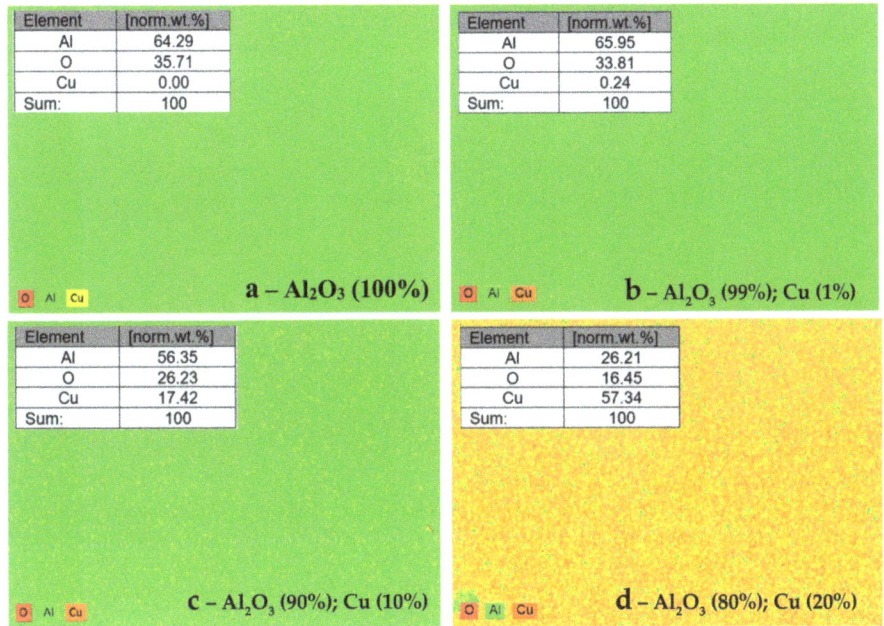

Figure 6. Surface distribution of the coating components.

It should be noted that the copper content in the coatings obtained by the evaporation of targets 3 and 4 (Figure 6c,d) considerably exceeds the copper content in these targets. This is due to a significant difference in melting T_m and boiling T_b points of the components of the target material (T_m = 1083 °C, T_b = 2567 °C for copper; T_m = 2047 °C, T_b = 2980 °C for alumina ceramics). Atoms of relatively fusible copper are quicker to reach the target surface layers during heating, and, hence, they are likely the first to evaporate. Similar to electron beam power density, the density of copper vapor exceeds the density of alumina ceramic vapor, which results in an imbalanced content of elements in the coating and in the target.

The adhesion of metal-ceramic coatings has not been carried out specifically; however, the expression "scratch method", which consists of applying a linearly-increasing force to the diamond indenter with its simultaneous uniform displacement along the film surface [20], showed that the adhesion of the coatings ranges from 3.2 to 4.1 J/m^2.

Figure 7 shows the volume and surface resistance of the obtained coatings depending on the fraction of the copper impurity in the target material. For pure ceramic target 1, practically without copper addition, the measured values of the volume and surface resistance of coatings condensed on the surface of titanium samples were at a level of 10^{10} Ω and $5 \cdot 10^9$ Ω/sq (Ohms per square). An increase in the mass fraction of copper in the evaporated target to 1% resulted in a sharp decrease in the volume and surface resistance down to $2 \cdot 10^3$ Ω and $5 \cdot 10^5$ Ω/sq, respectively. A further increase in the proportion of copper in the composite targets to 20% promoted a smoother decrease in electrical resistivity of the deposited coatings down to 10^3 Ω/sq for the surface resistance, and to less than 100 Ω for volume resistance.

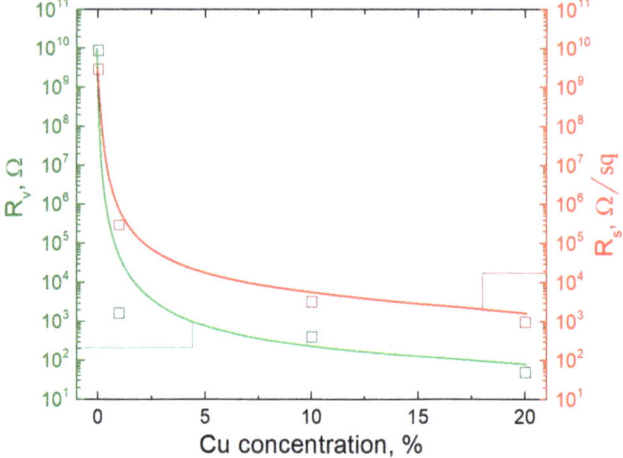

Figure 7. Dependences of the volume and surface resistance on the proportion of copper content in the coating.

We suppose that the main reason for the decrease in the volume resistance is the conductivity matrix, which is created as a result of mutual overlapping of the synthesized copper particles submerged in the bulk of condensed ceramic coating. The surface resistance also decreases with increasing copper proportion in the composite metal-ceramic target as a result of the copper particles residing on the surface of the synthesized coating, and, at the same time, having direct contact with the bulk conductivity matrix consisting of similar particles. Thus, the experimental curves shown in Figure 7 indirectly confirm the correspondence between the stoichiometric composition of the materials of the composite target, subjected to electron-beam evaporation, and the coatings synthesized as a result of condensation of evaporation products.

5. Conclusions

In this article, by example of aluminum oxide ceramics and copper, we present the results of experimental studies that convincingly demonstrate the possibility of fabricating metal-ceramic coatings by electron-beam evaporation of a sintered target, containing powder components of ceramic and metal, in the fore-vacuum pressure range. The mass-to-charge composition of the multicomponent plasma generated in the beam transport region and the morphology and electrical conductivity of obtained coatings were investigated for different ratios of ceramics and copper in the evaporated target. It was experimentally shown that by simply varying the fraction of the metal (copper) within dielectric (alumina) powder, one can predictably control the functional properties (such as volume and surface resistance) of the thin coating deposited on a substrate as a result of electron beam evaporation of the mixed-powder target. These results justify the further search of possibilities to control other properties (such as porosity, hardness, Young modulus, corrosion resistance, etc.) by the choice of proper materials and their fractions in the composite target. The results of the studies carried out expand the field of the possible applications of electron-beam synthesis of functional coatings using fore-vacuum plasma-cathode electron sources.

Author Contributions: Conceptualization, A.V.T. and Y.G.Y.; methodology, A.S.K. and K.P.S.; validation, A.V.T., K.P.S. and Y.G.Y.; investigation, A.V.T.; data curation, K.P.S.; writing—original draft preparation, A.V.T.; writing—review and editing, D.B.Z.; visualization, A.V.T.; supervision, Y.G.Y.; project administration, Y.G.Y.; funding acquisition, Y.G.Y. All authors have read and agreed to the published version of the manuscript.

Funding: The research of plasma parameters were supported by the program of the Ministry of Science and High Education of Russian Federation for youth laboratories, Project No. FEWM-2021–0013. The coating fabrication and studies were supported by the Russian Science Foundation (Grant No. 21-79-1003 5), https://rscf.ru/project/21-79-10035/.

Institutional Review Board Statement: Not applicable.

Informed Consent Statement: Not applicable.

Data Availability Statement: Data are available upon reasonable request.

Conflicts of Interest: The authors declare no conflict of interest.

References

1. Tejero-Martin, D.; Bennett, C.; Hussain, T. A review on environmental barrier coatings: History, current state of the art and future developments. *J. Eur. Ceram. Soc.* **2021**, *41*, 1747–1768. [CrossRef]
2. Lee, K.N.; Miller, R.A.; Jacobson, N.S.; Opila, E.J. Environmental durability of mullite coating/SiC and mullite-YSZ coating/SiC systems. *Ceram. Eng. Sci. Proc.* **1995**, *16*, 1037–1044. [CrossRef]
3. Lee, K.N. Current status of environmental barrier coatings for Si-Based ceramics Surf. *Coat. Technol.* **2000**, *133*, 1–7. [CrossRef]
4. Lee, K.N.; Fox, D.S.; Robinson, R.C.; Bansal, N.P. Environmental barrier coatings for silicon-based ceramics. In *High Temp. Ceram. Matrix Compos*; Wiley-VCH Verlag GmbH & Co. KGaA: Weinheim, Germany, 2006; pp. 224–229. [CrossRef]
5. Krause, A.R.; Garces, H.F.; Senturk, B.S.; Padture, N.P. 2ZrO$_2 \cdot$Y$_2$O$_3$ thermal barrier coatings resistant to degradation by molten CMAS: Part II, interactions with sand and fly ash. *J. Am. Ceram. Soc.* **2014**, *97*, 3950–3957. [CrossRef]
6. Yushkov, Y.G.; Oks, E.M.; Tyunkov, A.V.; Yushenko, A.Y.; Zolotukhin, D.B. Electron-Beam Deposition of Aluminum Nitride and Oxide Ceramic Coatings for Microelectronic Devices. *Coatings* **2021**, *11*, 645. [CrossRef]
7. Padture, N.P. Advanced structural ceramics in aerospace propulsion. *Nat. Mater.* **2016**, *15*, 804–809. [CrossRef] [PubMed]
8. Sundaram, K.B.; Alizadeh, J. Deposition and optical studies of silicon carbide nitride thin films. *Thin Solid Films* **2000**, *370*, 151–154. [CrossRef]
9. Hou, N.Y.; Perinpanayagam, H.; Mozumder, M.S.; Zhu, J. Novel Development of Biocompatible Coatings for Bone Implants. *Coatings* **2015**, *5*, 737–757. [CrossRef]
10. Yushkov, Y.G.; Oks, E.M.; Tyunkov, A.V.; Zolotukhin, D.B. Electron-Beam Synthesis of Dielectric Coatings Using Forevacuum Plasma Electron Sources (Review). *Coatings* **2022**, *12*, 82. [CrossRef]
11. Anders, A. Physics of arcing, and implications to sputter deposition. *Thin Solid Films* **2006**, *502*, 22–28. [CrossRef]
12. Kelly, P.J.; Arnell, R.D. Magnetron sputtering: A review of recent developments and applications. *Vacuum* **2000**, *56*, 159–172. [CrossRef]
13. Yushkov, Y.G.; Oks, E.M.; Tyunkov, A.V.; Zolotukhin, D.B. Alumina Coating Deposition by Electron-Beam Evaporation of Ceramic Using a Forevacuum Plasma-Cathode Electron Source. *Ceram. Int.* **2019**, *45*, 9782–9787. [CrossRef]

14. Burdovitsin, V.A.; Medovnik, A.V.; Oks, E.M.; Skrobov, E.V.; Yushkov, Y.G. Potential of a Dielectric Target during Its Irradiation by a Pulsed Electron Beam in the Forevacuum Pressure Range. *Tech. Phys.* **2012**, *57*, 1424–1429. [CrossRef]
15. Yushkov, Y.G.; Oks, E.M.; Tyunkov, A.V. Deposition of Boron-Containing Coatings by Electron-Beam Evaporation of Boron-Containing Targets. *Ceram. Int.* **2020**, *46*, 4519–4525. [CrossRef]
16. Savkin, K.P.; Bugaev, A.S.; Nikolaev, A.G.; Oks, E.M.; Kurzina, I.A.; Shandrikov, M.V.; Yushkov, G.Y.; Brown, I.G. Decrease of ceramic surface resistance by implantation using a vacuum arc metal ion source. In Proceedings of the 2012 25th International Symposium on Discharges and Electrical Insulation in Vacuum (ISDEIV), Tomsk, Russia, 2–7 September 2012; pp. 554–557. [CrossRef]
17. Zolotukhin, D.B.; Oks, E.M.; Tyunkov, A.V.; Yushkov, Y.G. Deposition of Dielectric Films on Silicon Using a Fore-Vacuum Plasma Electron Source. *Rev. Sci. Instrum.* **2016**, *87*, 063302. [CrossRef] [PubMed]
18. Tyunkov, A.V.; Oks, E.M.; Yushkov, Y.G.; Zolotukhin, D.B. Ion Composition of the Beam Plasma Generated by Electron-Beam Evaporation of Metals and Ceramic in the Forevacuum Range of Pressure. *Catalysts* **2022**, *12*, 574. [CrossRef]
19. Tyunkov, A.V.; Yushkov, Y.G.; Zolotukhin, D.B. Generation of Metal Ions in the Beam Plasma Produced by a Forevacuum-Pressure Electron Beam Source. *Phys. Plasmas* **2014**, *21*, 123115. [CrossRef]
20. Yushkov, Y.G.; Oks, E.M.; Tyunkov, A.V.; Zolotukhin, D.B. Dielectric Coating Deposition Regimes during Electron-Beam Evaporation of Ceramics in the Fore-Vacuum Pressure Range. *Coatings* **2022**, *12*, 130. [CrossRef]

Article

Electron-Beam Sintering of Al_2O_3-Cr-Based Composites Using a Forevacuum Electron Source

Aleksandr Klimov [1,*], Ilya Bakeev [1], Anna Dolgova [1], Efim Oks [1,2], Van Tu Tran [1] and Aleksey Zenin [1]

1. Laboratory of Plasma Electronics, Tomsk State University of Control Systems and Radioelectronics, 634050 Tomsk, Russia
2. Laboratory of Plasma Sources, Institute of High Current Electronics SB RAS, 634034 Tomsk, Russia
* Correspondence: klimov@main.tusur.ru; Tel.: +7-905-990-5241

Abstract: We describe our studies of the influence of Cr content in an Al_2O_3-Cr composite on its thermal and electrical conductivity properties during and after electron-beam sintering in the forevacuum range of pressure. Sintering was carried out using a plasma-cathode forevacuum-pressure electron source of an original design, capable of processing non-conducting materials directly. It is shown that the chromium content affects the efficiency of the beam power transfer to the irradiated composite. The efficiency decreases with increasing chromium content. Measurement of the composite's coefficient of thermal conductivity, in the temperature range 50–400 °C, shows that it varies almost linearly from 25 W/(m·K) to 68 W/(m·K) as the Cr content in the composite increases from 25% to 75% wt. The electrical conductivity properties after sintering exhibit a non-linear behavior. The conduction activation energy E_a, measured via the dependence of the current through composites of different compositions, is slightly lower than the Al_2O_3 band-gap. The addition of metallic Cr results in a disproportionate decrease in E_a, almost by an order of magnitude, from 6.9 eV to 0.68 eV. By varying the chromium content, it is possible to form a material with thermal and electrical conductivities controllable over a wide range.

Keywords: pressureless sintering; composite ceramics; electron beam; electron-beam irradiation; sintering; conduction activation energy; Al_2O_3-Cr; thermal conductivity; electrical conductivity; forevacuum pressure region

Citation: Klimov, A.; Bakeev, I.; Dolgova, A.; Oks, E.; Tran, V.T.; Zenin, A. Electron-Beam Sintering of Al_2O_3-Cr Based Composites Using a Forevacuum Electron Source. *Ceramics* 2022, 5, 748–760. https://doi.org/10.3390/ceramics5040054

Academic Editor: Amirhossein Pakseresht

Received: 12 September 2022
Accepted: 11 October 2022
Published: 14 October 2022

Publisher's Note: MDPI stays neutral with regard to jurisdictional claims in published maps and institutional affiliations.

Copyright: © 2022 by the authors. Licensee MDPI, Basel, Switzerland. This article is an open access article distributed under the terms and conditions of the Creative Commons Attribution (CC BY) license (https://creativecommons.org/licenses/by/4.0/).

1. Introduction

Metal–ceramic composites offer numerous advantages. They are materials of high hardness and mechanical strength that can operate at temperatures above 1000 °C [1]. Wide commercial use of composites is possible, contingent upon the availability of rapid production methods and a good understanding of their thermal and electrical characteristics. Of the various kinds of ceramic materials, Al_2O_3-based ceramic is one of the most widely used, possessing high strength, high hardness, and excellent thermal resistance [2–5]. However, its high brittleness restricts its possible applications. The introduction of malleable metal phases to ceramics is an effective method of reducing its brittleness. Metal–ceramic composites obtained in this way acquire not only low brittleness but also often new electrical, optical, magnetic, and thermal properties [6–11]. In the work described here, for the metallic component we select Cr, which has a high melting point (1863 °C [12]), high-temperature oxidation resistance, and high-temperature plasticity [13,14]. These properties may be useful in creating thermally, chemically, and mechanically strong metal–ceramic materials. Among the main areas of application of such materials are the coating of jet nozzles and the protective coating of gas furnaces, crucibles, heat shields, etc.

The main methods used for forming composite materials can be divided into two types. The first is selective sintering by an electron or laser beam [15,16], and the second includes hot pressing, [17,18], cold pressing, microwave, thermal [19], and spark plasma sintering [20–22]. The sintering of powders involves issues with controlling the heating rate and the uniformity of target heating. Furnace heating and sintering require time-consuming exposure at high temperatures and have low energy efficiency. When a target is irradiated by laser radiation, heating of the material starts from the surface and proceeds to the core unevenly [23]. This temperature gradient leads to uneven distribution of grain size and of density with depth [24]. These problems can be resolved by lowering the heating rate, thereby increasing the overall processing time. Additionally, the efficiency of the laser energy transfer to the target depends on its optical properties, which restricts the range of materials that can be used for sintering.

In the case of target heating by an electron beam, the target optical properties are not important. However, a new problem arises, which is removing the electrical charge carried by the electron beam onto the dielectric target. The charged surface causes electron beam defocusing and diminishes the efficiency of the beam power transfer to the irradiated target [25,26]. This issue can be resolved by use of a forevacuum-pressure plasma-cathode electron source for the e-beam irradiation. Such sources generate electron beams at a pressure of 1–100 Pa, and the negative surface charge is compensated by the positive ion flux from the beam plasma created during beam propagation at such elevated pressure values [27]. We have previously shown [28–30] that the use of electron beams generated by forevacuum plasma electron sources is a useful approach for sintering ceramic compacts. In the work described here, we have used this method for electron-beam sintering of Al_2O_3-Cr-based composites and have explored the thermal and electrical properties of the Al_2O_3-Cr composites produced.

2. Materials and Methods

We used commercially available Al_2O_3 powders with particle size 10–30 µm, and Cr powder with particle size 50 µm.

The main parameters pertaining to the experiment are shown in Table 1 [31–34].

Table 1. Material parameters at 20 °C.

Material	Theoretical Density, g/cm^3	Thermal Conductivity, W/(m·K)	Melting Point, °C	Specific Electrical Conductivity, S/m	Band Gap, eV
Al_2O_3	3.97	27–34	2054	10^{-9}–10^{-11}	5.1–8.7
Cr	7.19	93.9	1907	5.104×10^6	-

The composites for sintering were produced by mixing ceramic and metal powders in various mass proportions. The sample and the mixture used to produce it were assigned the same nomenclature label. The proportions are shown in Table 2.

Table 2. Mixture contents used to prepare Al_2O_3-Cr composite.

Mixture Label	Al_2O_3 Content, % wt.	Cr Content, % wt.
100A	100	0
75A	75	25
50A	50	50
25A	25	75

Pellets 3 ± 0.1 mm thick and 10 ± 0.1 mm in diameter were formed from the mixtures by uniaxial pressing. The composites were processed in a vacuum chamber equipped with necessary pumping equipment and manipulators; see Figure 1a. A forevacuum plasma-cathode electron source [35] was used for composite heating. A beam-focusing system of a special design allowed an electron beam of 0.6 mm in diameter to be generated in the vacuum chamber, under forevacuum conditions at a pressure of 30 Pa (helium). For sintering, the composite of given composition was placed in the vacuum chamber on a graphite holder of special design. The holder assembly consisted of a graphite crucible with mounting and bracing that minimized heat loss to its fastening elements. The composite heating efficiency was improved by placing a heat-reflecting shield around the graphite holder to reduce heat transfer to the chamber walls by thermal radiation. After installing the composite and evacuating the chamber to a working pressure of 3.0 Pa, the electron source was turned on. Sintering was performed as follows. First, a smooth heating of the composite at constant electron beam energy of 15 keV, by slowly increasing the beam current from 10 to 100 mA, depending of the composite composition; then, exposure to the beam at a constant temperature of 1400 °C for 10 minutes; next, cooling by reducing the beam current, followed by turning off the electron source and further cooling in the vacuum chamber for 20 minutes. Since the area of the irradiated surface was much greater than the cross-sectional area of the electron beam at its point of incidence on the composite, the electron beam was rapidly scanned over the composite surface. The scanning was performed using a magnetic deflecting system controlled by a sweep circuit. The scanning frequency was 100 Hz, and the scanned area was 1×1 cm^2.

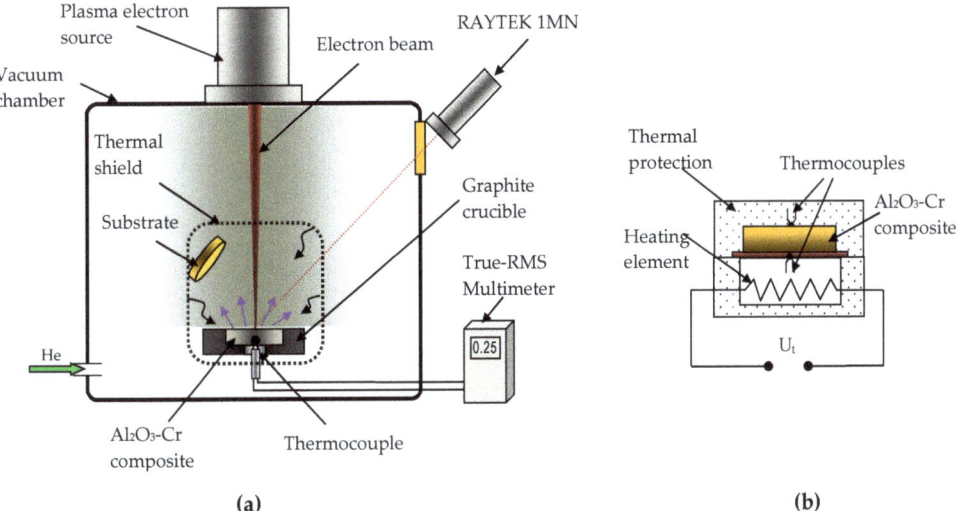

Figure 1. Schematic diagram of the experimental setup (**a**) and electrical heater (**b**).

The composite surface during sintering was monitored remotely by a RAYTEK 1MH (Raytek Corp., Santa Cruz, CA, USA) infrared pyrometer with measurement range 550–3000 °C. A tungsten–rhenium thermocouple was used to measure the temperature of the composite non-irradiated side. The thermocouple and the composite surface were in tight contact with each other. The thermocouple and the pyrometer were calibrated by heating of a thin copper plate. The difference in readings did not exceed 10 °C.

Current through the composite during its irradiation was measured by replacing the thermocouple with a flat metal electrode (not shown in Figure 1). The sample was placed on this electrode on the irradiated side. The current through the sample was measured by a True-RMS Multimeter 289 (Fluke Corp., Everett, WA, USA), with one sensing wire connected to the electrode and the other grounded.

In order to investigate the thermal properties of the composites after sintering, a model of an electrical heater was made; see Figure 1b. The heater could heat composite materials up to 400 °C in a controlled manner. In addition, it was possible to measure the heat flow through the samples and to measure the temperature of the irradiated and non-irradiated surfaces. The temperature was measured using standard chromel–copel thermocouples.

Sample microstructure and elemental composition were studied using a JEOL JSM-7500FA (JEOL Ltd., Freising, Germany) scanning electron microscope, equipped with a set of add-on units for energy dispersive elemental analysis (EDS) and electron backscatter diffraction (EBSD) (Bruker Nano GmbH, Berlin, Germany). We used the facilities at the TPU Center for Sharing Use, "Nanomaterials and Nanotechnologies", supported by the Ministry of Education and Science of Russia under grant number 075-15-2021-710.

3. Experimental Results and Analysis

3.1. Electron-Beam Sintering

Focused electron-beam heating in the forevacuum medium enabled us to heat the composite surfaces quite easily to temperatures of 800–900 °C. The temperature growth rate at constant beam power of 40 W/min was 80 °C/min; see Figure 2.

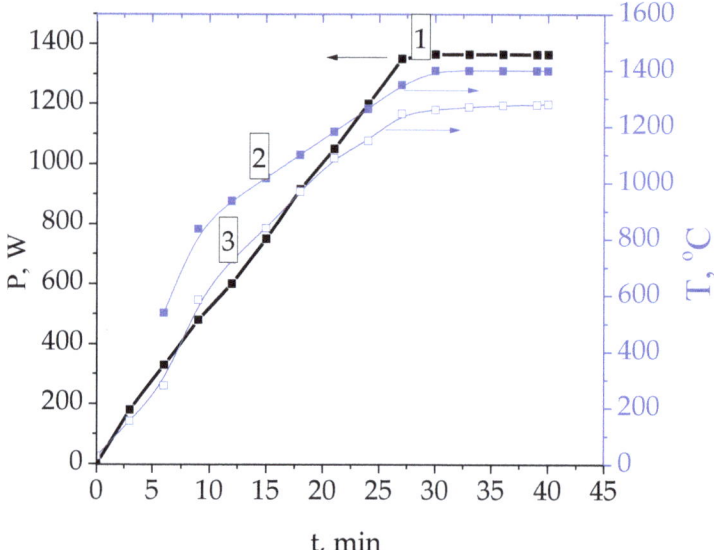

Figure 2. Dependencies of the electron beam power 1 and the temperature of irradiated 2 and non-irradiated 3 surfaces of the 75A composite, as a function of time.

The sample temperature growth rate decreased to 26–30 °C/min, depending on the chromium content in the composite. The temperature-increase rate for zero chromium content is 30 °C/min, and the addition of chromium decreases the temperature growth rate to 26 °C/min. This change in the rate of temperature rise is not significant, but is still noticeable. Clearly the thermal parameters, especially the coefficient of thermal conductivity, affect the composite heating. Since chromium has a greater coefficient of thermal conductivity than aluminum oxide, 94 W/(m·K) and, on average, 30 W/(m·K), respectively, the addition of chromium results in greater heat transfer to the graphite crucible and the holder elements, thereby reducing the composite heating. This circumstance should be taken into account when automating and optimizing the process.

3.2. Microstructure and Parameters

After sintering, the samples were cut in half along a diameter and were polished. In order to remove the products of grinding, they were then rinsed in alcohol and distilled water in an ultrasonic bath for 20 minutes. The microstructures of 75A, 50A, and 25A composites are shown in Figure 3. The samples display homogeneous areas with good compaction, as well as pores of various sizes formed at grain boundaries. EDX analysis revealed that the gray areas correspond to Al_2O_3 ceramic matrix and the lighter areas are Cr. The size of Al_2O_3 grains is 20–40 μm, and the size of Cr grains is 20 to 80 μm. With the increase in Cr content, the Cr grain size increases due to smaller grains amalgamating into bigger grains.

Figure 3. SEM image of composite microstructure for samples (**a**) 75A, (**b**) 50A, and (**c**) 25A.

The elemental composition of the selected areas, obtained by energy-dispersive analysis, is shown in Figure 4. The increase in Cr content corresponds to its content in the initial mixture of powders used for sintering.

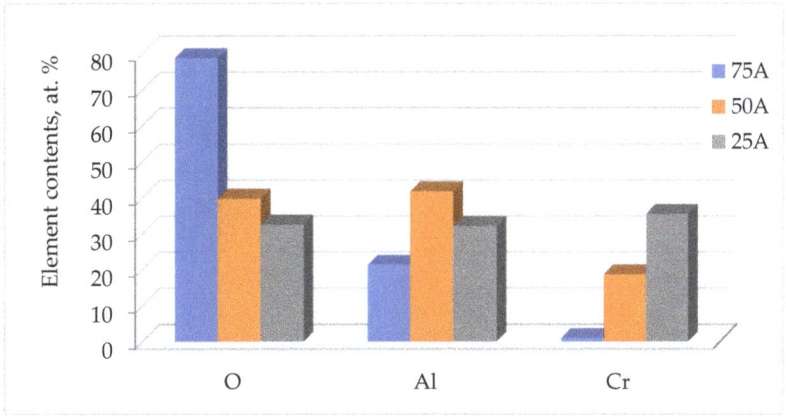

Figure 4. Contents of O, Al, and Cr in the cross sections of 75A, 50A, and 25A composites.

Composite parameters before and after electron-beam sintering are shown in Table 3.

Table 3. Composite parameters before and after sintering.

Sample		100A	75A	50A	25A
Mass m, mg	before	395	420	493	610
	after	356	401	479	580
Thickness h, mm	before	2.78	2.83	2.82	2.73
	after	2.6	2.63	2.68	2.61
Diameter d, mm	before	10.3	10.32	10.23	10.30
	after	9.32	9.54	9.69	9.86
Density ρ, g/cm^3	before	1.7	1.77	2.13	2.68
	after	2.02	2.14	2.43	2.92

The maximum increase in density after irradiation, 21%, was for the composite containing 75% aluminum oxide. The minimum increase in density, 9%, was for the 25A sample, with the lowest content of Al_2O_3. It is apparent that since aluminum oxide has greater shrinkage in the course of sintering, the corresponding samples with greater content of it must have smaller geometric dimensions and, hence, a higher density after sintering. Compared with the sintering of similar composites using the hot-pressing method [36], the porosity value in this work turned out to be higher. This difference may be due to the peculiarity of the electron beam method—sintering without applying pressure.

The mass of all composites changes (decreases) after sintering; see Table 3. A possible reason could be mass evaporation during sintering. However, the sintering temperature of 1400 °C is not high enough for melting and evaporating the composite components; see Table 1. To further explore the possibility of mass loss by evaporation, we conducted experiments to study the composition of coatings on witness substrates. Substrates in the form of flat steel disks were placed at a distance of 5 cm from the sintered composite, as shown in Figure 1. After sintering, the sample surface elemental composition was studied by energy-dispersive analysis. According to the composition measurements for three compacts, the substrate coatings contained elemental constituents of the composites, namely chromium and aluminum; see Figure 5.

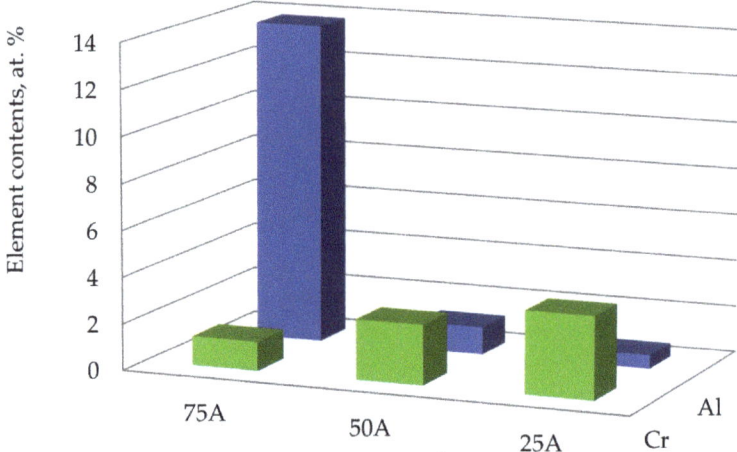

Figure 5. Contents of Al and Cr in the witness substrate coatings: 75A, 50A, and 25A.

As shown in Figure 5, the coatings contain a significant amount of the composite elements. Thus, for the composite with 75% aluminum oxide (75A), the substrate coating contains over 13% at. aluminum. The witness substrates used for the other composites with a lower aluminum oxide content demonstrate a considerable decrease in aluminum in the coating. For the 50A composite, the aluminum content on the substrate is less than for 75A, by almost a factor of 10. At the same time, the chromium content in the substrate coating increases almost proportionally to the increase in chromium content in the sintered composite—from 1.26% for composite 75A with 25% Cr to 3.56% for composite 25A with 75% Cr. Evidently, these elements appear on the substrate due to evaporation from the composite surface. Despite the rather low temperature, according to pyrometer readings, for the evaporation of these elements to occur, such an effect can occur during electron-beam sintering of ceramics employing an electron beam deflecting system. When scanning the ceramic surface, the high power-density electron beam can cause local heating of the surface at the impact point; the beam cross-section at the impact point is less than 1 mm^2. Since the pyrometer measures the mean value of the composite surface temperature over an area of about 2.5 cm^2, the local temperature increase at the beam impact point is not readily registered. Additionally, mass loss from the composite surface can occur due to evaporation of low-melting-point impurities with a content in the aluminum oxide powder used in the experiments that can be as high as 5%.

Another possible mechanism for heating to the evaporation temperatures of Al_2O_3 and Cr could be heating due to the current through the composite bulk, as is the case for the flash-sintering technique. In this technique, a constant electric field of 100–150 V/m is established between the two opposite surfaces of the sintered sample, which is simultaneously heated to a high temperature. The electrical conduction, arising as a result of the temperature increase, leads to Joule heating of the sample. This significantly reduces the ceramic sintering time.

Measurement of the current through the composite show that it can reach several milliamperes; see Figure 6.

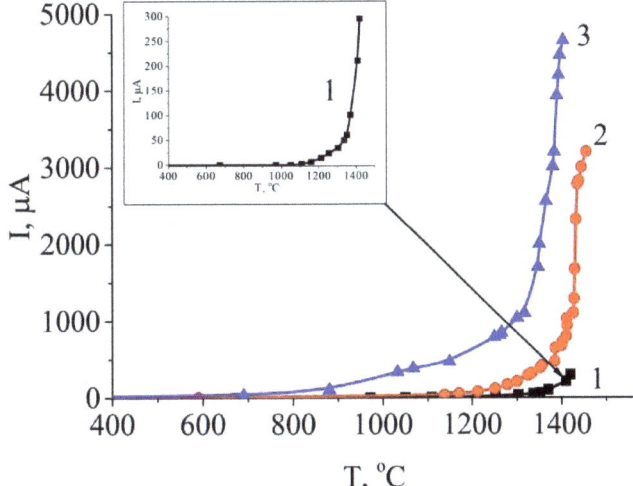

Figure 6. Temperature dependence of current through composites during electron-beam sintering: (1) composite 75A; (2) composite 50A; (3) composite 25A.

The current increases with increasing chromium content, which is directly related to the increase in the composite electrical conductivity. The addition of a metal, as a more electrically conducting material, increases the overall electrical conductivity of the composite. The dependence of the coefficient of electrical conductivity is well-known to be exponential with temperature:

$$\gamma = \gamma_0 \cdot \exp\left(-\frac{\Delta E_a}{kT}\right) \tag{1}$$

where

γ_0 is an electrical conductivity of the conductor/dielectric, S/m;

γ_0 is a temperature-independent coefficient determined by the properties of the conductor/dielectric, S/m;

k is Boltzmann's constant, J/K;

T is the temperature of the irradiated composite surface, K; and

E_u is the conduction activation energy, eV.

The conduction current in semiconductors, of which aluminum oxide may be referred to as a particular kind, depends on the coefficient of electrical conductivity, as well as on the field strength in the semiconductor. Assuming that the field strength is determined by the difference of potentials between the irradiated and non-irradiated surfaces, one can write an equation for evaluating the value of the conduction current flowing through the composite as a function of temperature during electron-beam irradiation:

$$I_\gamma = \frac{\Delta\varphi}{h} \cdot S \cdot \gamma_0 \cdot \exp\left(-\frac{E_a}{kT}\right) \qquad (2)$$

where

S is the composite base area, m^2;

$\Delta\varphi$ is the difference of potentials between the irradiated and non-irradiates surfaces of the composite, V; and

h is the composite thickness.

Expression (2) allows the conduction activation energy to be estimated for composites with different elemental contents. Assuming that the temperature dependence of the potential of the composite surface is not as strong as that of the coefficient of electrical conductivity and plotting the graphs $\ln(I) = f\left(\frac{1}{T}\right)$ using the experimental data of Figure 4, one can determine the activation energy from the slope of the straight lines obtained. The dependencies $\ln(I) = f\left(\frac{1}{T}\right)$, plotted for the three composites over the temperature range 1000–1400 °C, where a noticeable increase in current is observed, are shown in Figure 7. As can be seen, the experimental points of the logarithm of the current from the inverse temperature fit into a linear dependence. From one perspective, this serves as an argument in favor of the chosen mechanism for increasing electrical conductivity with increasing temperature and the correctness of choosing Formula (1) for theoretical estimates.

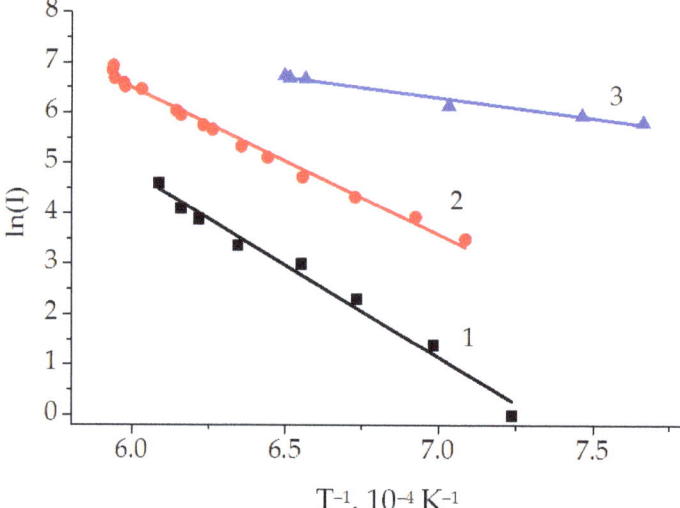

Figure 7. Temperature dependencies $\ln(I) = f\left(\frac{1}{T}\right)$ for composites of different compositions: (1) composite 75A; (2) composite 50A; (3) composite 25A.

The fact that the experimental data fit well to straight lines indirectly indicates that the assumption not to take into account the change in potential on the composite surface over the given temperature range is correct.

The calculated values of the conduction activation energy E_a are given in Table 4.

Table 4. Conduction activation energy for composites of different compositions.

Sample	75A	50A	25A
E_a, eV	3.1 ± 0.4	2.5 ± 0.3	0.68 ± 0.08

The obtained values of E_a for the Cr-containing composites are slightly lower than the Al_2O_3 band gap, which is predictable, since it was a metal that was added to the composite. The addition of metallic Cr results in a disproportional decrease in E_a. Thus, for the composite with 25% chromium content, the conduction activation energy, or band gap, decreases compared to pure Al_2O_3, from an average value of 6.9 eV (see Table 2) to 3.1 eV, i.e., by more than a factor of two. The addition of 75% Cr leads to a further decrease in E_a, down to 0.68 eV, which is almost by an order of magnitude.

As shown in [37], the electrical conductivity of aluminum oxide-based composites can be adjusted, when reinforced with conductive or semi-conductive phases (such as silicon carbide, for example), added in an amount at which they penetrate into an insulating aluminum oxide matrix. After sintering, such a composite can be used in many industries. The main factors affecting the electrical properties of composites with reinforced semiconductor phases are the volume fraction of SiC and the content of other impurities. The addition of SiC improves the electrical conductivity, which increases with an increase in the volume fraction of SiC [37]. Thus, in a composite with 20 vol.% SiC, the conductivity of 4.05×10^{-2} S·m^{-1} was measured, which is an increase of four orders of magnitude compared to the reference monolithic alumina (7.80×10^{-6} S·m^{-1}).

A rather strong dependence of the composite electrical conductivity on the content of metal phase has been observed [38], when adding Ti to an Al_2O_3 ceramic matrix. Specific electrical resistance, with the addition of 20% vol. Ti, decreases from 10^{12} Ohm·m to 10^{-2}–10^{-3} Ohm·m, and the fall is rather abrupt. The authors have explained this by the formation of conducting paths through the composite bulk, due to the melting of fine grains of Ti. In the present work, the changes are not so severe, which may be related to pressureless sintering. Chromium grains combine without forming conducting paths throughout the composite volume; see Figure 3.

The electrical parameters of the Al_2O_3-Cr composite can be controlled over a fairly wide range.

3.3. Composite Thermal Conductivity

Thermal conductivity is an important parameter in such applications of Al_2O_3 ceramics as high-temperature structural components, refractories, gas burners, wear parts, and cutting tools. To reduce thermal shock, the thermal conductivity of the composite in all these applications should be as high as possible. It can be expected that Cr particles improve the thermal conductivity of Al_2O_3-based composites due to the inherent high thermal conductivity of Cr.

To measure the thermal conductivity, the sintered composites were placed in a heating device with a fixed heater temperature T_1 and a temperature T_2 on the composite side not subject to irradiation; see Figure 1b. The coefficient of thermal conductivity λ was determined using the expression:

$$\lambda = \frac{Q \cdot h}{\Delta T \cdot S} \quad (3)$$

where λ is the coefficient of thermal conductivity, W/(m·K);
Q is the heat flux through the composite, W/m^2;
ΔT is the difference of temperatures: $T_1 - T_2$, °C; and
S is the composite surface area, m^2.

The obtained value of the coefficient of thermal conductivity corresponds to the average temperature $\Delta T/2$.

The measured thermal conductivity over the temperature range 50–400 °C is shown in Figure 8. As seen, the coefficient of thermal conductivity decreases with increasing temperature for composites of any composition, as reported in the literature [39]. The general pattern here is as follows: the thermal conductivity of ceramics of a crystalline structure, especially an oxide, with an increase in temperature, as a rule, drops significantly [40]. This is based on the idea of heat transfer in solid non-metallic bodies by thermal elastic waves—phonons. The thermal conductivity of the composite is closely related to their microstructure and depends on the free path length of the phonons: the degree of disturbance of the harmonic oscillations of heat waves during their passage through a given substance. Phonons are also known to interact with lattice defects, grain boundaries, and other microstructure defects. The presence of a metallic phase in the form of chromium inclusions leads to a higher porosity value, characteristic of composites and, as a result, negatively affect thermal conductivity. The resulting internal stresses in composites also lead to a decrease in thermal conductivity [41]. However, despite these negative factors, the thermal conductivity of the composite increases with an increase in the chromium content and is still higher than that of pure aluminum oxide.

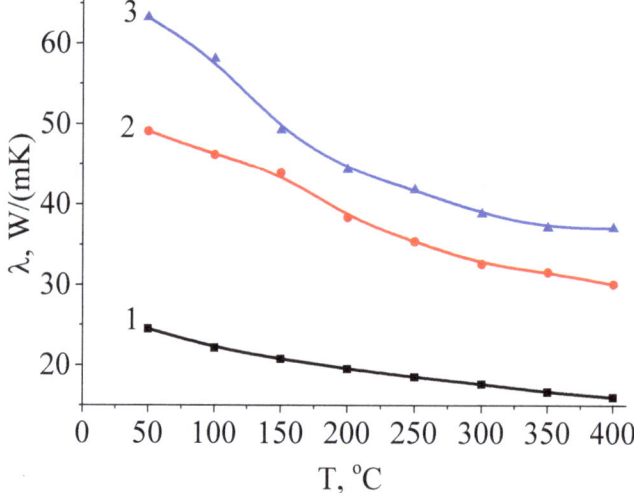

Figure 8. Temperature dependence of thermal conductivity for composites of different compositions: (1) composite 75A; (2) composite 50A; (3) composite 25A.

The coefficient of thermal conductivity of the composite with 75% content of Al_2O_3 is 25 W/m·K and does not differ significantly from that of pure Al_2O_3 at the same temperature [42]. The values of thermal conductivity measured at room temperature are somewhat lower than the data given in the literature [43] and are measured for mono-cast Al_2O_3 (28–30 W/m·K). A possible reason is the greater porosity of the materials obtained in this work. With the addition of Cr, the composite conductivity rises almost proportionally to the content of Cr. Thus, for the composite with 75% content of Cr, i.e., three times as much compared to that in the 25% Cr composite, the conductivity increases from 25 to 68 W/m·K. For both Al_2O_3 and Cr, the thermal conductivity decreases with temperature [44], as is reflected in Figure 8. Compared with the data of [45], the thermal conductivity of the composite remains at a high level and does not decrease to values below 15 W/m·K.

4. Conclusions

Electron-beam irradiation allows Al_2O_3-Cr-based composites to be sintered at a temperature of 1400 °C. The complete sintering cycle, including heating and cooling, is no longer than 50 minutes. By varying the Cr content, one can change the electrical and thermal conductivity properties of the composite. In this case, the thermal conductivity in the temperature range of 20–400 °C varies directly proportionally to the Cr content and inversely proportionally to temperature. The thermal conductivity increases from 25 to 68 W/m·K, when the Cr content increases from 25% to 75%, and decreases with increasing temperature, especially for composites with higher Al_2O_3 content.

The electrical conductivity properties, illustrated by the current through the composite and the conduction activation energy, depend on the Cr content nonlinearly. The addition of 75% Cr to the composite decreases the Al_2O_3-Cr activation by an order of magnitude compared to that of Al_2O_3, from 6.9 to 0.68 eV. At the same time, the addition of 50% Cr reduces this energy only by a factor of two, to 2.5 eV. In general, by varying the chromium content, it is possible to produce materials with values of electrical conductivity controllable over orders of magnitude and thermal conductivity controllable within range limits differing by almost a factor of two.

Author Contributions: Conceptualization, A.K. and E.O.; methodology, A.Z.; software, A.D.; validation, A.K., A.Z., and I.B.; formal analysis, A.K.; investigation, A.Z. and V.T.T.; resources, E.O.; data curation, A.K.; writing—original draft preparation, A.K.; writing—review and editing, E.O.; visualization, I.B.; supervision, E.O.; project administration, A.K.; funding acquisition, A.K. and V.T.T. All authors have read and agreed to the published version of the manuscript.

Funding: The work was supported by grant FEWM-2020-0038 from the Ministry of Science and Higher Education of the Russian Federation, and a study of coatings composition was funded by research project 20-38-90184 from RFBR.

Institutional Review Board Statement: Not applicable.

Informed Consent Statement: Not applicable.

Data Availability Statement: Not applicable.

Acknowledgments: Special thanks to Ian Brown (Berkeley Lab) for the English correction and helpful discussion.

Conflicts of Interest: The authors declare no conflict of interest. The funders had no role in the design of the study; in the collection, analyses, or interpretation of data; in the writing of the manuscript; or in the decision to publish the results.

References

1. Foltz, J.V.; Blackmon, C.M. *Metal Matrix Composites in Metals Handbook*, 10th ed.; Second Print, Properties and Selection; ASM: Metals Park, OH, USA, 1992; Volume 2, p. 2536.
2. Fan, R.; Liu, B.; Zhang, J.; Bi, J.; Yin, Y. Kinetic evaluation of combustion synthesis $3TiO_2 + 7Al \rightarrow 3TiAl + 2Al_2O_3$ using non-isothermal DSC method Mater. *Chem. Phys.* **2005**, *91*, 140–145.
3. Niu, F.Y.; Wu, D.J.; Zhou, S.Y.; Ma, G.Y. Power prediction for laser engineered net shaping of Al_2O_3 ceramic parts. *J. Eur. Ceram. Soc.* **2014**, *34*, 3811–3817. [CrossRef]
4. Reddy, M.P.; Ubaid, F.; Shakoor, R.A.; Parande, G. Effect of reinforcement concentration on the properties of hot extruded Al-Al_2O_3 composites synthesized through microwave sintering process. *Mater. Sci. Eng. A* **2017**, *696*, 60–69. [CrossRef]
5. Ramesh, M.; Marimuthu, K.; Karuppuswamy, P.; Rajeshkumar, L. Microstructure and properties of YSZ-Al_2O_3 functional ceramic thermal barrier coatings for military applications. *Bol. Soc. Esp. Cerám. Vidr.* **2021**, in press. [CrossRef]
6. Pietrzak, K.; Chmielewski, M.; Wlosinski, W. Sintering Al_2O_3–Cr composites made from micro-and nanopowders. *Sci. Sinter.* **2004**, *36*, 171–177. [CrossRef]
7. Oh, S.T.; Sekino, T.; Niihara, K. Fabrication and mechanical properties of 5 vol% copper dispersed alumina nanocomposites. *J. Eur. Ceram. Soc.* **1998**, *18*, 31–37. [CrossRef]
8. Sekino, T.; Nakahira, A.; Nawa, M.; Niihara, K. Fabrication of Al2O3/W nanocomposites. *J. Powder Metal.* **1991**, *38*, 326–330. [CrossRef]
9. Pan, Y.; Xiao, S.; Lu, X.; Zhou, C.; Li, Y.; Liu, Z.; Qu, X. Fabrication, mechanical properties and electrical conductivity of Al_2O_3 reinforced Cu/CNTs composites. *J. Alloys Compd.* **2019**, *782*, 1015–1023. [CrossRef]
10. Hossain, S.; Rahman, M.M.; Chawla, D.; Kumar, A.; Seth, P.P.; Gupta, P.; Jamwal, A. Fabrication, microstructural and mechanical behavior of Al-Al_2O_3-SiC hybrid metal matrix composites. *Mater. Today Proc.* **2020**, *21*, 1458–1461. [CrossRef]
11. Farvizi, M.; Javan, M.K.; Akbarpour, M.R.; Kim, H.S. Fabrication of NiTi and NiTi-nano Al_2O_3 composites by powder metallurgy methods: Comparison of hot isostatic pressing and spark plasma sintering techniques. *Ceram. Int.* **2018**, *44*, 15981–15988. [CrossRef]
12. Okamoto, H. The Cr-O (chromium-oxygen) system. *J. Phase Equilb.* **1997**, *18*, 402. [CrossRef]

13. Gulbransen, E.A.; Andrew, K.F. Kinetics of the oxidation of chromium. *J. Electrochem. Soc.* **1957**, *104*, 334–338. [CrossRef]
14. Holzwarth, U.; Stamm, H. Mechanical and thermomechanical properties of commercially pure chromium and chromium alloys. *J. Nucl. Mater.* **2002**, *300*, 161–177. [CrossRef]
15. Pfeiffer, S.; Florio, K.; Puccio, D.; Grasso, M.; Colosimo, B.M.; Aneziris, C.G.; Graule, T. Direct laser additive manufacturing of high performance oxide ceramics: A state-of-the-art review. *J. Eur. Ceram. Soc.* **2021**, *41*, 6087–6114. [CrossRef]
16. Vailes, J.; Hagedorn, Y.C.; Wilhelm, M.; Konrad, W. Additive manufacturing of ZrO_2-Al_2O_3 ceramic components by selective laser melting. *Rapid Prototyp. J.* **2013**, *19*, 51–57.
17. Nguyen, T.D.; Caccia, M.; McCormack, C.K.; Itskos, G.; Kenneth, H.S. Corrosion of Al_2O_3/Cr and Ti_2O_3/Cr composites in flowing air and CO_2 at 750 °C. *Corros. Sci.* **2021**, *179*, 109115. [CrossRef]
18. Zhang, X.; Zhang, Y.; Tian, B.; Jia, Y.; Liu, Y.; Song, K.; Volinsky, A.A.; Xue, H. Cr effects on the electrical contact properties of the Al_2O_3-Cu/15W composites. *Nanotechnol. Rev.* **2019**, *8*, 128–135. [CrossRef]
19. Betül, K.Y.; Hüseyin, Y.; Yahya, K.T. Evaluation of mechanical properties of Al_2O_3–Cr_2O_3 ceramic system prepared in different Cr_2O_3 ratios for ceramic armour components. *Ceram. Int.* **2019**, *45*, 20575–20582.
20. Pulgarín, H.L.C.; Albano, M.P. Sintering and Microstructure of Al_2O_3 and Al_2O_3-ZrO_2. *Ceramics. Procedia Mater. Sci.* **2015**, *8*, 180–189. [CrossRef]
21. Daguano, J.K.M.F.; Santos, C.; Souza, R.C.; Balestra, R.M.; Strecker, K.; Elias, C.N. Properties of ZrO_2–Al_2O_3 composite as a function of isothermal holding time. *Int. J. Refract. Met. Hard Mater.* **2007**, *25*, 374–379. [CrossRef]
22. Maca, K.; Pouchly, V.; Shen, Z. Two-step sintering and spark plasma sintering of Al_2O_3, ZrO_2 and $SrTiO_3$ ceramics. *Integr. Ferroelectr.* **2008**, *99*, 114–124. [CrossRef]
23. Liu, X.; Zou, B.; Xing, H.; Huang, C. The preparation of ZrO2-Al2O3 composite ceramic by SLA-3D printing and sintering processing. *Ceram. Int.* **2019**, *46*, 937–944. [CrossRef]
24. Hu, K.; Li, X.; Qu, S.; Li, Y. Effect of Heating Rate on Densification and Grain Growth During Spark Plasma Sintering of 93W-5.6Ni-1.4Fe Heavy Alloys. *Metall. Mater. Trans.* **2013**, *44*, 4323–4336. [CrossRef]
25. Luo, G.N.; Yamaguchi, K.; Terai, T.; Yamawaki, M. Charging effect on work function measurements of lithium ceramics under irradiation. *J. Alloys Compd.* **2003**, *349*, 211–216. [CrossRef]
26. Mekni, O.; Goeuriot, D.; Damamme, G.; Raouadi, K.; Sao, S.J.; Meunier, C.; Aoufi, A. Dynamic investigation of charging kinetics in sintered yttria stabilized zirconia and α-alumina polycrystalline ceramics under electron beam irradiation. *Ceram. Int.* **2016**, *42*, 8729–8737. [CrossRef]
27. Burdovitsin, V.A.; Klimov, A.S.; Medovnik, A.V.; Oks, E.M. Electron beam treatment of non-conducting materials by a fore-pump-pressure plasma-cathode electron beam source. *Plasma Sources Sci. Technol.* **2010**, *19*, 055003. [CrossRef]
28. Klimov, A.S.; Bakeev, I.Y.; Dvilis, E.S.; Oks, E.M.; Zenin, A.A. Electron beam sintering of ceramics for additive manufacturing. *Vacuum* **2019**, *169*, 108933. [CrossRef]
29. Klimov, A.S.; Bakeev, I.Y.; Oks, E.M.; Zenin, A.A. Electron-beam sintering of an Al_2O_3/Ti composite using a forevacuum plasma-cathode electron source. *Ceram. Int.* **2020**, *46*, 22276–22281. [CrossRef]
30. Klimov, A.S.; Zenin, A.A.; Bakeev, I.Y.; Oks, E.M. Formation of gradient metalloceramic materials using electron-beam irradiation in the forevacuum. *Russ. Phys. J.* **2019**, *62*, 1123–1129. [CrossRef]
31. Abyzov, A.M. Aluminum oxide and alumina ceramics (Review). Part 1. Properties of Al_2O_3 and industrial production of dispersed Al_2O_3. *Novye Ogneup.* **2019**, *1*, 16–23. [CrossRef]
32. Jacobs, J.A.; Testa, S.M. Overview of Chromium (VI) in the Environment: Background and History. In *Chromium (VI) Handbook*, 1st ed.; CRC Press: Boca Raton, FL, USA, 2005; pp. 1–21.
33. Martienssen, W.; Warlimont, H. *Springer Handbook of Condensed Matter and Materials*; Springer: Berlin/Heidelberg, Germany, 2005; pp. 431–476.
34. Toyoda, S.; Shinohara, T.; Kumigashira, H.; Oshima, M.; Kato, Y. Significant increase in conduction band discontinuity due to solid phase epitaxy of Al_2O_3 gate insulator films on GaN semiconductor. *Appl. Phys. Lett.* **2012**, *101*, 231607. [CrossRef]
35. Klimov, A.; Bakeev, I.; Oks, E.; Zenin, A. Forevacuum plasma source of continuous electron beam. *Laser Part. Beams* **2019**, *37*, 203–208. [CrossRef]
36. Chmielewski, M.; Pietrzak, K. Processing, microstructure and mechanical properties of Al2O3–Cr nanocomposites. *J. Eur. Ceram. Soc.* **2007**, *27*, 1273–1279. [CrossRef]
37. Galusek, D.; Galusková, D. Alumina matrix composites with non-oxide nanoparticle addition and enhanced functionalities. *Nanomaterials* **2015**, *5*, 115–143. [CrossRef] [PubMed]
38. Shi, S.; Cho, S.; Goto, T.; Sekino, T. The effects of sintering temperature on mechanical and electrical properties of Al_2O_3/Ti composites. *Mater. Today Commun.* **2020**, *25*, 101522. [CrossRef]
39. Grimvall, G. *Thermophysical Properties of Materials*, 2nd ed.; Elsevier Science: Amsterdam, The Netherlands, 1999.
40. Parchovianský, M.; Galusek, D.; Švančárek, P.; Sedláček, J.; Šajgalík, P. Thermal behavior, electrical conductivity and microstructure of hot pressed Al2O3/SiC nanocomposites. *Ceram. Int.* **2014**, *14*, 14421–14429. [CrossRef]
41. Hasselman, D.P.H.; Johnson, L.F. Effective thermal conductivity of composites with interfacial thermal barrier resistance. *J. Compos. Mater.* **1987**, *21*, 508–515. [CrossRef]
42. Hostaša, J.; Pabst, W.; Matějíček, J. Thermal conductivity of Al_2O_3–ZrO_2 composite ceramics. *J. Am. Ceram. Soc.* **2011**, *94*, 4404–4409. [CrossRef]

43. McCluskey, P.H.; Williams, R.K.; Graves, R.S.; Tiegs, T.N. Thermal Diffusivity/Conductivity of Alumina—Silicon Carbide Composites. *J. Am. Ceram. Soc.* **1990**, *73*, 461–464. [CrossRef]
44. Moore, J.P.; Williams, R.K.; Graves, R.S. Thermal conductivity, electrical resistivity, and Seebeck coefficient of high-purity chromium from 280 to 1000 K. *J. Appl. Phys.* **1977**, *48*, 610–617. [CrossRef]
45. Wada, S.; Piempermpoon, B.; Nakorn, P.N.; Wasanapiarnpong, T.; Jinawath, S. Thermal conductivity of Al_2O_3 ceramics: The inconsistency between measured value and calculated value based on analytical models for a composite. *J. Sci. Res. Chula. Univ.* **2005**, *30*, 109–120.

Article

Electron-Beam Synthesis and Modification and Properties of Boron Coatings on Alloy Surfaces

Yury Yushkov [1,2], Efim Oks [1,2], Andrey Kazakov [2], Andrey Tyunkov [1,2] and Denis Zolotukhin [2,*]

1. Institute of High Current Electronics SB RAS, 634055 Tomsk, Russia
2. Department of Physics, Tomsk State University of Control Systems and Radioelectronics, 634050 Tomsk, Russia
* Correspondence: zolotukhinden@gmail.com

Abstract: In this study, fore-vacuum plasma electron beam sources were used to deposit a few micron-thick boron coatings on A284 and ZrNb1 alloys and modify their surfaces. The coating deposition rate with a continuous 1 kW electron beam that evaporated the boron target at a distance of 10 cm was 0.5 μm/min, and the boron coating density was 2.2 g/cm^3. Based on the comparison of data on the mass-to-charge composition, beam plasma density, and coating parameters, the contribution of the plasma phase of the evaporated material to the growth of coatings was greater than that of the vapor phase. Using the scanning electron and atomic force microscopy techniques, surface modification by repeated electron beam pulses with electron energies of 8 and 6 keV and a beam power per pulse of 2 J/cm^2 and 2.25 J/cm^2, respectively, transformed a relatively smooth coating surface into a hilly structure. Based on a structural phase analysis of coatings using synchrotron radiation, it was concluded that the formation of the hilly coating structure was due to surface melting under the repeated action of electron beam pulses. The microhardness, adhesion, and wear resistance of coatings were measured, and their corrosion tests are presented herein. The pure boron coatings obtained and studied are expected to be of use in various applications.

Keywords: boron coatings; electron beam deposition; fore-vacuum electron source; film properties; electron beam modification

Citation: Yushkov, Y.; Oks, E.; Kazakov, A.; Tyunkov, A.; Zolotukhin, D. Electron-Beam Synthesis and Modification and Properties of Boron Coatings on Alloy Surfaces. *Ceramics* **2022**, *5*, 706–720. https://doi.org/10.3390/ceramics5040051

Academic Editors: Amirhossein Pakseresht and Kamalan Kirubaharan Amirtharaj Mosas

Received: 14 September 2022
Accepted: 8 October 2022
Published: 10 October 2022

Publisher's Note: MDPI stays neutral with regard to jurisdictional claims in published maps and institutional affiliations.

Copyright: © 2022 by the authors. Licensee MDPI, Basel, Switzerland. This article is an open access article distributed under the terms and conditions of the Creative Commons Attribution (CC BY) license (https://creativecommons.org/licenses/by/4.0/).

1. Introduction

Boron-based coatings are promising protective surface coatings [1]. They are used to harden the surface of parts and structural materials in the field of mechanical engineering [2]. They are characterized by high hardness, resistance to wear [3], corrosion [4], and high thermal stability [5]. For example, wurtzite boron nitride has comparable hardness to natural diamond [6]. Pure boron thin films are used as materials in electronic [7] and optical devices [8] as well as in protective layers of thermonuclear installations [9]. Moreover, boron coatings of monoisotopic compositions find applications in the nuclear industry; for example, ^{10}B-based coatings are promising as burnable neutron absorbers in nuclear reactors [10] and as absorber coatings of neutron detectors [11]; they are also used for the delivery of ^{10}B atoms deposited on the surface of nanosized particles to the malignant neoplasm during boron neutron capture therapy [12], while ^{11}B-based coatings are used for the ^{11}B aneutronic fusion of protons and ^{11}B atomic nuclei, which may be an alternative to deuterium and tritium fusion [13]. Hence, developing techniques to deposit boron coatings, investigating the properties of boron coatings, and revealing the interrelationship between these properties and conditions that brought about their formation are crucial.

Boriding is an industrial technique that has been used widely for decades to develop boron-containing layers on metal and alloy surfaces. In this process, boride atoms are diffused into the surface of a metal component, resulting in the formation of metal borides in the surface layer, thereby increasing the surface hardness and wear resistance. Conventional boriding can be achieved in a solid, liquid, or gaseous [14,15] medium [16,17].

However, there are disadvantages to using it, including the high energy consumption required for heating parts and electrolysis of boron-containing media, the long duration of surface diffusive saturation with boron, the use of hazardous and toxic substances, and environmental pollution.

Vacuum-plasma methods, such as magnetron sputtering [18,19] or cathodic arc deposition [20–22], can be used as alternatives for creating boron surface layers owing to their eco-friendliness because the equipment of this type generates a boron flux from a consumable boron-containing solid cathode during an electric discharge in a vacuum chamber. Moreover, there is no heating requirement for the surface because the main process involves the deposition of boron atoms or ions onto the surface. Another advantage of these methods over conventional boriding is the short process duration, which is determined by the flux intensity. The maximum flux intensity is limited by thermal stability in the discharge of the consumed boron-containing cathode. At maximum discharge parameters, the deposition rate of boron-containing coatings at a characteristic distance of 10 cm from the cathode can reach approximately 20–30 nm/min [19] for magnetron sputtering and approximately 100 nm/min for arc deposition [22].

A method for depositing boron coating has been proposed based on the electron beam evaporation of a boron solid target using an electron beam at a fore-vacuum pressure [23]. Briefly, a boron or boron-containing ceramic target is heated locally by an electron beam to the melting temperature and the melted surface area undergoes an intense evaporation. The flux of boron vapor, partially ionized by the beam, deposits onto the substrate surface forming a boron-containing coating [24]. The maximum temperature of the boron-containing target is limited in this case not by its thermal resistance, as in magnetron or vacuum arc deposition of boron coatings [18–22] but by the temperature of intense boiling of boron, at which an unwanted flux of droplets occurs because of the splashing of the molten target material. Thus, the boron coating deposition rate under this method is much higher than that under magnetron sputtering or arc deposition and can reach up to 1 μm/min [24], which provides a higher coating production performance. This work aims to further develop this method using an additional action of a wide-aperture electron beam in the fore-vacuum on the deposited boron coating. The characteristics and properties of boron coatings are studied by surface analysis, which includes structural phase analysis with synchrotron radiation generated by the VEEP-2 electron storage ring based in the Siberian Center for Synchrotron and Terahertz Radiation at G. I. Budker Institute of Nuclear Physics, SB of RAS [25].

2. Experimental Setup and Diagnostics

Figure 1 illustrates the setup for depositing boron coatings by electron-beam heating and evaporation of a crystal boron target. The target was evaporated using a fore-vacuum plasma electron source operating in continuous mode [26]. In such a source, electrons are extracted from a hollow cathode glow discharge plasma. The discharge current varies from 100 to 500 mA, while the discharge burning voltage, depending on the discharge current and gas pressure, varies from 200 to 500 V. At the maximum discharge current, the source is capable of generating electron beams with a current of up to 200 mA and energy of up to 20 keV, while the beam can be focused on the target surface by the magnetic field of the focusing system to a diameter of 3 mm. The electron beam current is controlled by the discharge current. The average power density in the electron beam focal spot on the target surface is 500 W/mm^2, which is several times greater than the radiation power density on the sun's surface (approximately 60 W/mm^2). Hence, the source is capable of melting and evaporating the surface of a target composed of any refractory material.

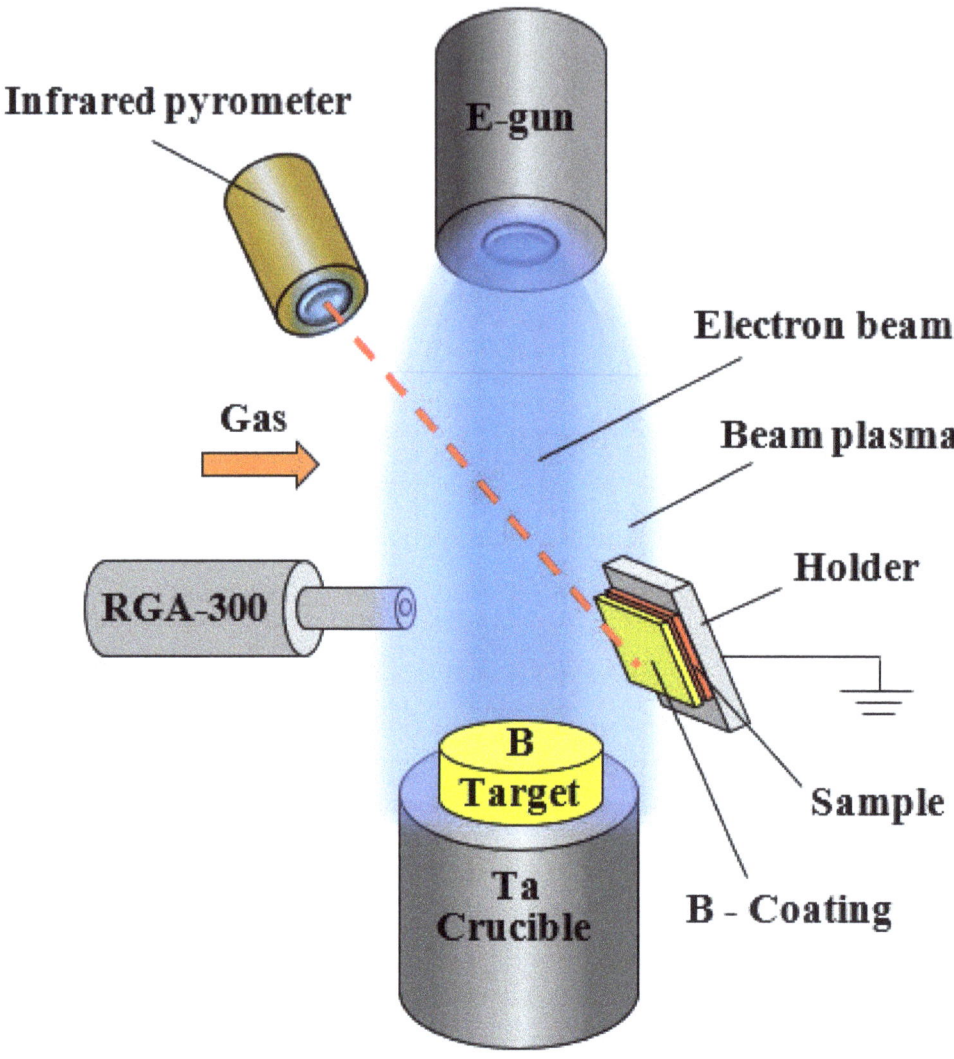

Figure 1. Schematic of the electron-beam synthesis of boron-based coatings.

A 4-mm-thick, 1 × 1 cm² boron plate target fabricated using the hot pressing of 99.6% pure boron crystals 1–10 μm in size was placed at the bottom of the vacuum chamber on a carbon crucible. The average density of the target due to the pores between crystals was 1.2 g/cm³, which is two times less than the density of amorphous boron. Because the pores occupied 50% of the target volume, the effective surface-to-volume ratio for the target material was about 3×10^3 cm^{-1} and the total effective surface area of the target was about 0.1 m². The electron beam toward the boron target was transported through the vacuum chamber filled with 99.9% pure nitrogen at a pressure of 10 Pa. An ISP-500C helical mechanical oil-free fore-vacuum pump with a pumping rate of 500 l/min was used to maintain the vacuum in the chamber. Prior to the experiment, the vacuum chamber was evacuated to a residual pressure of 1 Pa.

Because boron is a wide-gap semiconductor, its specific resistivity is as high as 1 MOhm × cm at a normal temperature of 20 °C; [27], which is insufficient for the complete drain of the electron beam charge. Therefore, at the initial heating of the boron target, its negative surface charge introduced by the electron beam was neutralized by a flux of positive ions from the beam plasma [28]. As the beam heated up the boron target, its specific resistivity decreased to 0.1 Ohm × cm at a temperature of about 750 °C. At this resistivity, the effect of the target surface charging by the beam becomes immaterial because of the electron beam current running through the boron target volume.

To prevent the boron target from thermal shock destruction, the target was heated in two stages, the parameters of which were determined experimentally [24]. Initially, the electron beam with a current of 80 mA at an accelerating voltage of up to 5 kV heated up the target to a temperature of approximately 900 °C for 70 s. Subsequently, the accelerating voltage was increased slowly up to 9 kV for 1 min; in this case, there was an observed local melting of the target surface at the focal spot of the beam. Target evaporation occurred and the boron coating was deposited on the sample surface at a temperature of the melting pool of approximately 2400 °C, and after increasing the accelerating voltage to 10 kV and the beam current to 100 mA.

The samples with polished surfaces on which boron coatings were applied were disks of A284 steel and a special reactor alloy ZrNb1 with a hexagonal crystal lattice; they had a 2 cm diameter and 0.5 cm thickness. The samples were placed at an angle of 30° relative to the electron beam propagation axis. The sample surfaces were normally oriented to the center of the evaporated target. A high-performance Raytek-MM1MH optical pyrometer was used to monitor the temperature of the sample surface during the deposition process; the temperature did not exceed 500 °C. Plasma composition during the coating was controlled with a quadrupole mass analyzer based on an RGA-300 residual gas analyzer. We have produced several dozen samples with boron coating thicknesses of 1–7 μm throughout the course of our research. Herein, for comparison, we analyze coatings fabricated under the same conditions, such as a distance of 10 cm from the beam-heated boron target to the sample surface and an overall coating deposition time of 6 min and 10 s. Considering that the first two heating stages under which the deposition did not occur took 1 min and 10 s, the direct deposition time was 5 min.

The deposition rate of boron coatings by a 1 kW electron beam was measured with an MII-4 interferometer and an MNL-1 interference microscope-profilometer. The thickness of the deposited boron coatings was about 2.1 μm. Thus, under the experimental conditions, the coating growth rate was about 0.5 μm/min. The weight gain because of boron coating deposition, measured with a VL-220M analytical balance with a precision of ±10 μg, was 1.45 mg. Considering a sample surface area of 3.13 cm^2, the density of boron coatings was estimated to be $\rho \approx 2.2$ g/cm^3, which is close to the density of crystalline boron of 2.34 g/cm^3.

Figure 2 shows a setup for the modification of boron coatings obtained with a pulsed electron beam. The pulsed electron beam was generated by a fore-vacuum plasma-cathode electron source based on an arc discharge [29]. The electron source was mounted on the vacuum chamber, which was evacuated with an ISP-500C pump. The working gas was helium at a pressure of 10 Pa. The samples were placed on a movable grounded holder, which allowed several samples to be sequentially irradiated after the chamber evacuation. To prevent the sample surface from beam exposure, a protective stainless-steel screen was applied on top of one of the samples. The treatment of boron-coated samples was performed in two regimes through a series of 300 pulses with a repetition rate of 2 pulses per second (p.p.s.). The length of each pulse was τ_e = 500 μs. The beam diameter was 7 cm. In regime 1, the treatment was performed at a beam current amplitude of 20 A and an electron energy of 8 keV; in regime 2, it was performed at a beam current amplitude of 30 A and an electron energy of 6 keV. The electron current density j_e on the boron coating surface and the energy per pulse P_e were 0.5 A/cm^2 and 2 J/cm^2 for regime 1, and 0.75 A/cm^2 and 2.25 J/cm^2 for regime 2, respectively. The pulsed beam parameters for modifying the boron coatings were selected empirically: on the one hand, they should lead to a noticeable change in the surface morphology; on the other hand, they should be soft enough to prevent the coating from cracking, or its evaporation.

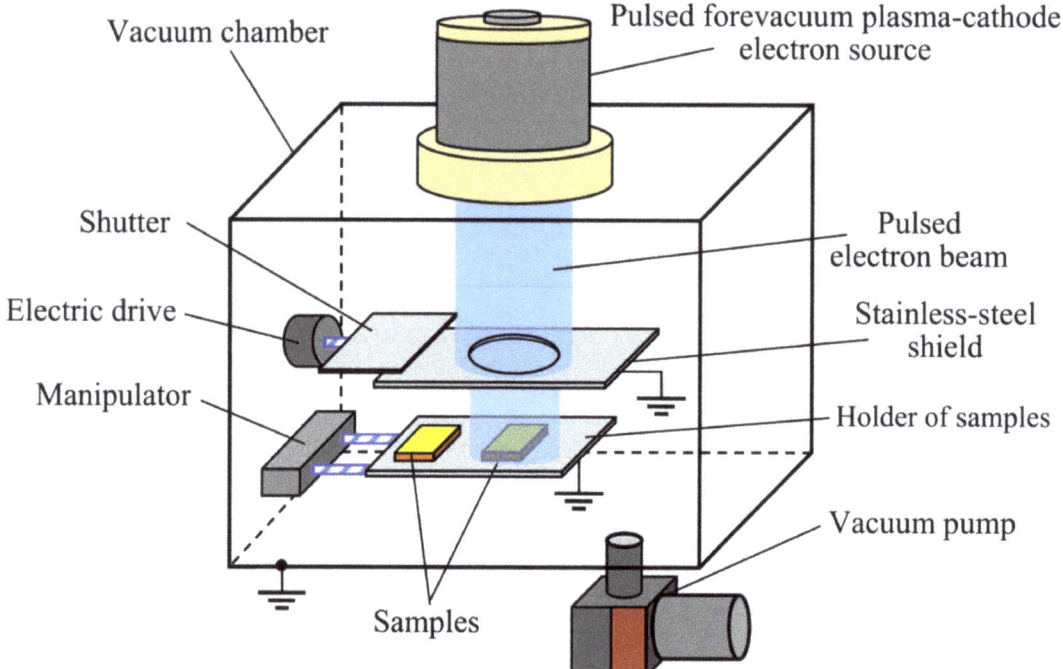

Figure 2. Setup for the modification of boron coatings with a pulsed electron beam.

The morphology of the final coatings was studied using a Hitachi S3400N scanning electron microscope and a Solver P47 atomic force microscope. Elemental composition was analyzed with a Bruker X'Flash 5010 energy dispersive spectrometer. The surface hardness of coatings was determined using the micro-Vickers technique. A square-section diamond indenter with a dihedral angle of 136° acted on the sample surface at various points with a constant load of 100 g while the penetration depth and indentation area were recorded. The phase composition of coatings was measured at the Synchrotron Radiation Station for High-Precision X-ray Diffraction Studies of Materials (also called the "Anomalous Scattering" station) on beamline No. 2 of the VEPP-3M electron storage ring at the Siberian Synchrotron Radiation Center (Budker Institute of Nuclear Physics, SB of RAS).

For the evaluation of the adhesive properties of the coatings, the scratch method was used. The method consisted of applying a force F (linearly growing with time) to the diamond indenter with its simultaneous uniform displacement along the coating surface. At the critical load F_c, the coating begins to break down. The critical load F_c is determined using the sensors of acoustic emission and friction force, indenter immersion depth, indenter loading force, and optical microscopy.

Adhesion can be characterized by a parameter G (specific peel work). The calculation formula connecting the parameter G with the critical lateral load F_c at the beginning of the film detachment from the substrate is as follows:

$$G = (F_c)2d / \pi(r_c)4E_{IT},$$

where d is the film thickness; r_c is the radius of the contact spot at the moment of peeling; E_{IT} is the Young's modulus of the substrate material. The determination of adhesion was carried out using a Micro-Scratch Tester MST-S-AX-0000 device.

The wear resistance of the obtained coatings was measured using a Pinon Disc and Oscillating TRIBO tester (France) using the "ball on disk" method. The sample surface was pressed by a tungsten carbide spherical tip with a load of 2 N. The coefficient of friction was determined by measuring the deflection of the lever. The wear rate was calculated by the formula:

$$V = 2\pi RA/FL,$$

where R is the track radius, μm; A is the cross-sectional area of the wear groove, μm^2; F is the value of the applied load, N; L is the distance traveled by the ball, m.

3. Results and Discussion

During the target heating and coating deposition, the ion composition of the plasma was measured with a quadrupole mass spectrometer based on an RGA-300 residual gas analyzer [30]. In our previous work [31], at an electric beam power of 0.4 kW, which corresponds to the preheating of the target at a beam current of 80 mA and an electron energy of 5 keV, the density of such a plasma is about 3×10^{10} cm^{-3}. Initially, at the target heating, in addition to nitrogen ions, a significant number of ions, up to 50% of the total ions, were produced through water vapor ionization processes: HO$^+$, H$_2$O$^+$, and water dissociation products O$^+$, H$_3^+$, H$_2^+$, and H$^+$. The sources of water molecules were the walls of the vacuum chamber with a surface area of approximately 1 m^2, the overall effective surface of the boron crystalline target with dimensions $1 \times 1 \times 0.4$ cm, and the total area of about 0.1 m^2. The appearance of water molecules on the walls and in the target was due to their exposure to the atmosphere prior to the experiments. With the electron beam power increasing to 0.7 kW, the surface temperature of the boron target at the beam focal spot reached 2100 °C; this started the melting process, and the plasma spectrum recorded traces of ^{10}B$^+$ and ^{11}B$^+$ ions. At this temperature, water molecules apparently evaporated from the target and the heated walls, resulting in the appearance of peaks of water vapor ions and their derivatives.

A further increase in the electron beam power of up to 1 kW led to an increase in the target temperature at the focal spot to 2400 °C. Consequently, a brightly glowing melt area about 4 mm in diameter formed on the target surface, from which an intense evaporation of boron occurred. In this case, the plasma density increased to approximately 1.6×10^{11} cm^{-3}. The fraction of boron ions in this plasma, evaluated by the height of its ion peaks in the spectrometer signals, was approximately 75%, and their total density in the plasma was about 1.2×10^{11} cm^{-3}. The ratio of ^{10}B$^+$ to ^{11}B$^+$ isotopes in the plasma at a beam power of 1 kW was 1:4, which is close to their natural ratio. Thus, the concentrations of ^{10}B$^+$ and ^{11}B$^+$ in the beam plasma were 2.4×10^{10} and 9.6×10^{10} cm^{-3}, respectively.

Because boron coatings are formed from two-phase states, namely, plasma and boron vapor, it is important to know the contribution of each phase in coating formation. Because the thickness, elemental composition, and specific density of coatings were determined using independent methods, one can demonstrate that to form such boron coatings, the flux density of atomic and ion boron onto the sample surface must be approximately 9.6×10^{16} cm^{-2} s^{-1}. The speed of boron isotope ions of mass M_i that leave the beam plasma with plasma electron temperature T_e, in eV units, is determined by the ambipolar speed of sound $v_i = \sqrt{eT_e/M_i}$. At $T_e \approx 3$ eV, this speed is $v_{10B} = 5.4 \times 10^5$ cm/s for ^{10}B$^+$ ions and $v_{11B} = 5.2 \times 10^5$ cm/s for ^{11}B$^+$ ions. Because only singly charged ions were registered in the beam plasma, the flux density of ^{10}B$^+$ ions on the sample surface equaled the product of their density in the plasma and their speed from plasma, which is 1.3×10^{16} cm^{-2} s^{-1}. This value was 5.0×10^{16} cm^{-2} s^{-1} for the ^{11}B$^+$ isotope. Thus, the flux density of all boron ions on the sample surface was around 6.3×10^{16} cm^{-2} s^{-1}. Comparing this value with the above estimate of the total boron particle flux both in ionized and neutral states, it was concluded that the contribution of the plasma phase to the formation of boron coatings is about 65 at.%, which exceeds the 35 at.% contribution of boron vapor.

Figure 3 shows the scanning electron microscopy (SEM) images of the surfaces of the obtained boron coatings. The coatings do not contain any defects, pores, or cracks, indicative of coating uniformity and smoothness. The boron coating deposited without the electron beam comprises small tightly packed segments 30–150 nm in size, while that deposited with a pulsed electron beam (regime 1) has a structure with discernible round hills on the surface. The characteristic side of the hill base in the image field is approximately 1–3 μm.

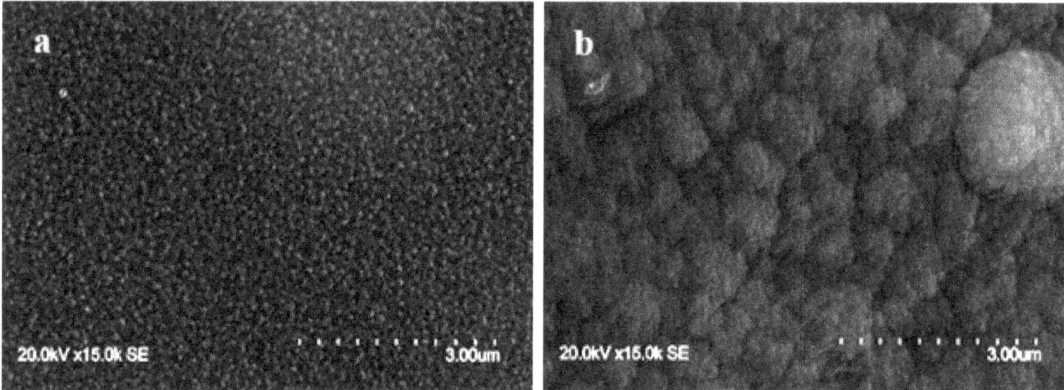

Figure 3. Scanning electron microscopy (SEM) images of boron coatings on sample surfaces (**a**) after deposition on the substrate and (**b**) after processing with electron beam pulses in regime 1 at a peak current of 20 A and an electron energy of 8 keV.

Figure 4a shows the elemental composition of the coating, comprising 95.5 at.% boron with small admixtures of carbon (3.7 at.%) and oxygen (0.9 at.%). Analysis of the elemental composition of coatings modified and unmodified by the electron beam showed that beam treatment did not practically change the coating composition. The carbon content in the coating is caused by the presence of this element in the substrate material (steel A284). Figure 4b,c shows the coating profiles measured using a Solver P47 SEM before and after treatment with a pulsed electron beam in regime 2. For a better visual perception of the beam effect, the scale of the height axes in Figure 4b,c is two orders of magnitude smaller than the scales of width and length, and the columnar structure of the surface in Figure 4c is actually a landscape of gently sloping hills. Hence, the deposition of coating forms a relatively smooth surface with nonuniformities a few fractions of a micrometer in size, while subsequent treatment by a pulsed electron beam leads to the formation of gentle hills with a height of up to 1.5 μm. In contrast to SEM, the use of atomic force microscopy (AFM) to determine the surface relief may screen small nonuniformities adjacent to the larger ones with a larger height. Thus, Figure 4b shows mostly hills with a base diameter of 1–10 μm although their structure qualitatively matches the surface structure in Figure 3b with discernible smaller hills. The change in energy during the pulse action on the coating does not significantly affect the picture of the surface; however, the size and height of the hills in regime 1 are approximately 20% greater than those in regime 2.

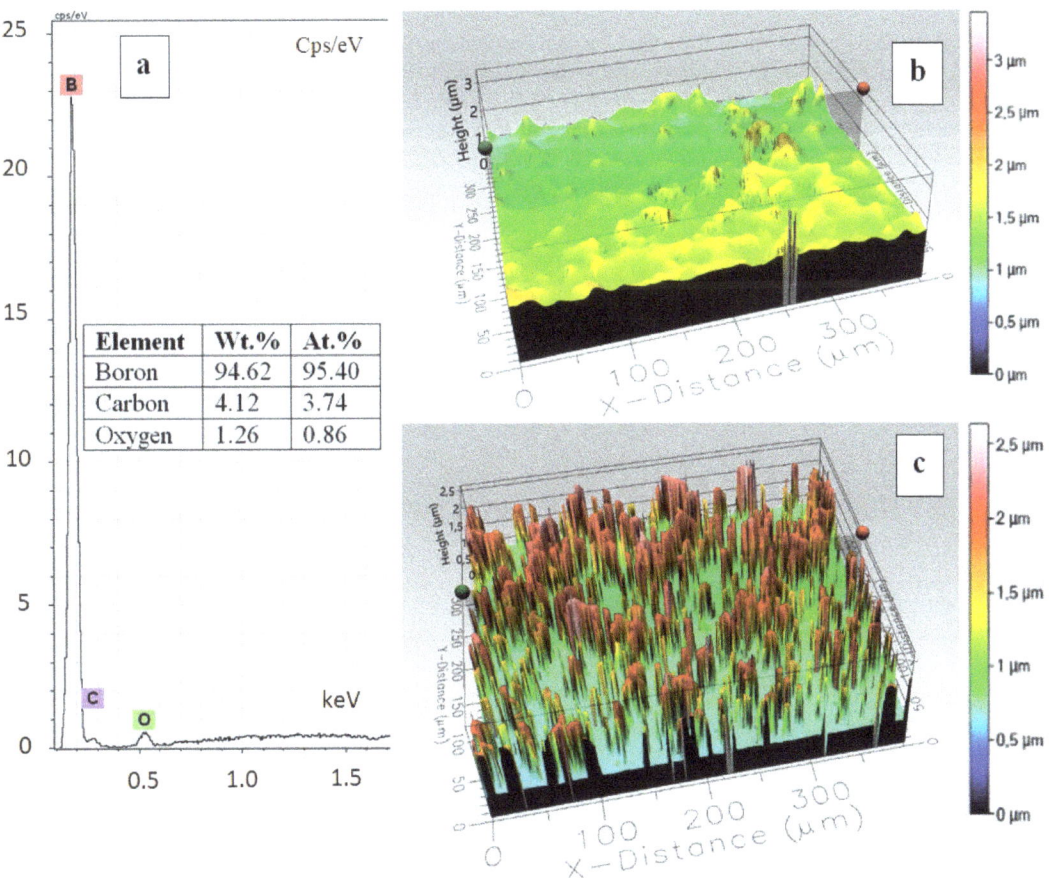

Figure 4. (a) Elemental composition of the obtained coatings. The surface profiles of (b) boron-based coating and (c) boron-based coating after pulse modification in regime 2 measured under a Solver P47 atomic force microscope.

Accelerated electrons in a solid are decelerated in the layer of their maximal range of penetration. Based on previous works [32], this range R_e in micron units at an electron beam energy E_e = 0.5–10 keV in a solid with density ρ (in g/cm³) with an accuracy of better than 15% can be estimated by

$$R_e = 9 \times 10^{-2} \, \rho^{-0.8} E_e^{1.3} \tag{1}$$

For boron coating ρ ≈ 2.2 g/cm³ and electron beam energy E_e = 8 keV, the value of R_e is 0.72 μm for regime 1, while for E_e = 6 keV, the R_e value is 0.49 μm for regime 2. Thus, for both regimes, the electron energy is released in the layer whose thickness is less than that of the boron coating.

The dependence of energy density distribution Q absorbed from the beam of accelerated electrons by the boron coating at depth x can be determined by [33]

$$Q(x) = (E_e/R_e)(1 - x/R_e)^{5/4}(3 - 2\exp(-(Z+8/4) \times (x/R_e)))(j_e \tau_e) \tag{2}$$

where j_e and τ_e are the densities of the electron beam current on the sample surface and the pulse duration, respectively, and Z = 5 is the number of electrons in a boron atom. When evaluating (x), it is convenient to substitute in the first factor in Expression (2), the electron energy E_e in eV and R_e in cm, and the remaining factors should use R_e in μm. In this case, the Q value is expressed in J/cm³ and the depth x is expressed in μm. The dependences of the energy density distribution for the electron beam in both regime 1 and regime 2 are shown in Figure 5.

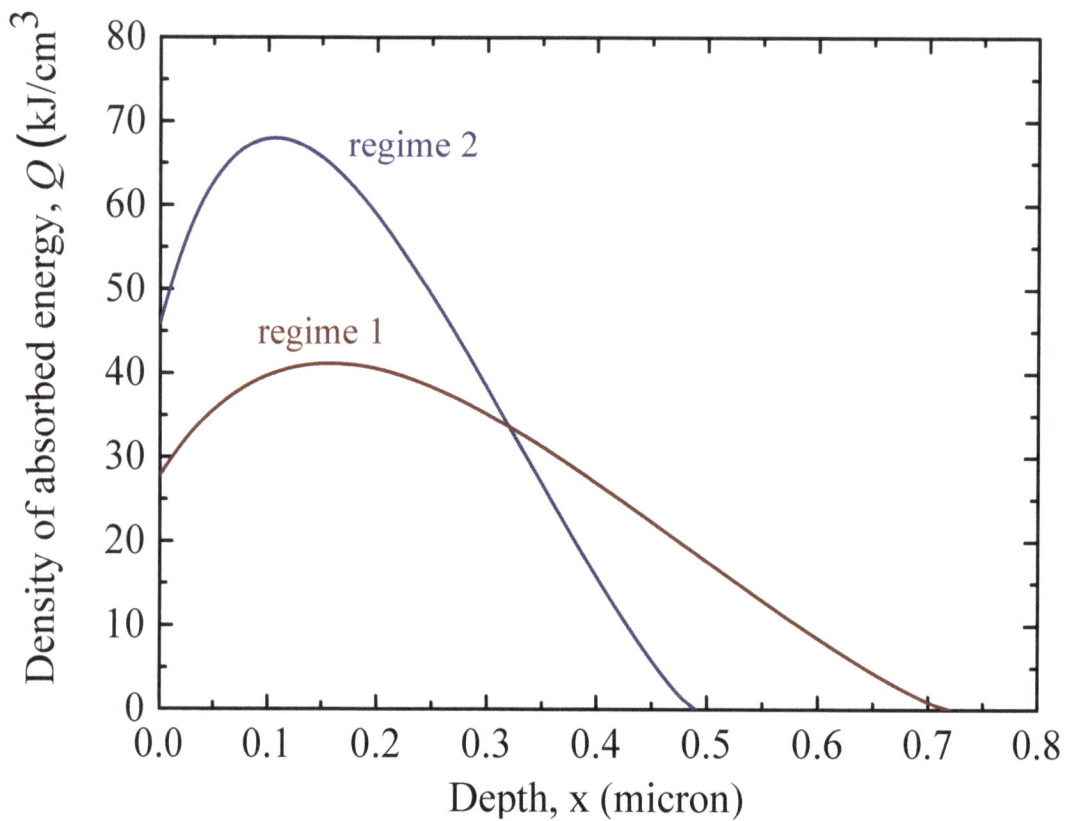

Figure 5. Distribution of the energy density Q absorbed by the boron coating from the beam of accelerated electrons versus depth for two regimes. The beam parameters are τ_e = 500 ms, regime 1: j_e =0.5 A/cm², E_e = 8 keV, regime 2: j_e =0.75 A/cm², E_e = 6 keV.

The boron temperature profile with depth x will approximately match the profile of $Q(x)$, provided that the condition $\sqrt{\alpha \tau_e} \ll R_e$ holds true, where $\alpha \approx 0.1$ cm²/s is the thermal diffusivity of the boron coating. At experimental electron energies, this condition is strictly satisfied for submicrosecond electron beam pulses, when the heat due to the energy imparted from the beam at a depth R_e does not have time to propagate deep into the surface of the solid. However, even though this condition is not satisfied in our experimental conditions due to much longer pulse width (hundreds of microseconds), anyway, as follows from the dependences in Figure 5, about 70% of the released energy of accelerated electrons and, therefore, the most intense heating of the coating material occurs at a depth of $0.5R_e$, which corresponds to 17% of the overall coating thickness at E_e = 8 keV and about 12% at E_e = 6 keV. That means that a significant part of electron beam energy can be deposited even at much longer pulses. Moreover, because the experimental processing of coatings was performed using a series of 300 pulses for 150 s, it is possible that the gradual heating of the coating surface during this period of time took place.

The surface hilly structure appeared due to the beam action, which is apparently related to cyclic temperature effects. The energy density of the pulsed electron beam on the coating surface is about 2 J/cm^2 per pulse, while the thermal conductivity coefficient of boron is an order of magnitude less than those of the majority of metals. Thus, the surface of the boron coating at the depth with the maximum release of the beam energy can heat up to a temperature at which the thin surface layer of the coating begins to melt. A similar relief of the surface of TiNi alloy was observed during repeated treatment by an electron beam with an electron energy of 20 keV and a beam power per pulse of 4 J/cm^2 [34]. Under the beam action, a hilly structure was formed; the number of pulses (128) in this case was comparable to that in our experiments (300 pulses). From a prior study [32], such surface relief may be associated with the development of instabilities on the melt/vapor phase interface under repeated action of the electron beam and cyclic melting and cooling of a thin surface layer. Similar effects seem to take place in our case too. Another thing that can be noted is that the pulsed electron beam treatment may have an effect on the structure of the deposited coating at different depths. Although the study of such effects is an interesting task, it was outside of the scope of the current research focused on the study of the surface properties of the coating.

The structural phase properties of boron coatings were studied using synchrotron radiation on beamline 2 of the VEPP-3 electron storage ring at the Anomalous Scattering station. In the X-ray diffraction patterns of boron coatings treated and untreated with the electron beam, reflections of low intensity and considerable width are observed. Among them, there are two remarkable reflections that correspond to interplanar distances of 0.27 and 0.24 nm (areas 1, 2, 3, and 4 in the inset of Figure 6). These reflections are associated with ultrafine crystals of nonstoichiometric boron nitride. Notably, these reflections decrease for regime 2, which may be indicative of the partial destruction of crystals by the beam with a high power density. However, the X-ray patterns do not show any reflections that can be associated with the crystal structure of boron, as is the case for the hexagonal crystal structure of the zirconium alloy. Thus, the hilly surface of the boron coating after the beam treatment does not exhibit a pronounced inner crystalline structure, and the hills themselves are not specific crystalline formations. This fact again verifies the nature of their formation as a result of repeated melting and cooling of the coating surface at the depth of the maximum beam energy release during the cyclic beam action.

Figure 7 shows the microhardness results of a steel substrate, a crystalline boron target, boron coatings, and boron coatings treated with a pulsed electron beam (regime 1, regime 2). The microhardness of the crystalline boron target is approximately three times less than that of the boron coating. This is apparently due to the density of the target (1.2 g/cm^3) being lower than the measured density of the coating (2.2 g/cm^3). The microhardness of the boron coating is 12 ± 0.35 GPa, while additional surface modification with a pulsed beam further increases the microhardness, up to 15.5 ± 0.45 GPa.

The adhesion measurements of the samples with boron-based coatings showed that the pulsed beam treatment of coatings did not affect the adhesion value between the boron coating and the sample surface. This is because the beam mainly affected the surface layers of the coating as adhesion is the interface property of the coating–substrate boundary on which the effect of beam treatment was weak. Figure 8 shows the typical micrographs of the surface under different loads F exerted on a diamond indenter with a radius of 100 µm. As the pressure on the indenter increases, it begins to submerge into the coating. This is accompanied by an increase in the coefficient of friction, indicating the growing resistance of the sample to the indenter movement. At a load of 6 N exerted on the coating, the coefficient of friction begins to fluctuate, which is indicative of the destruction of the surface structure. In the micrograph of the 6 N load, the start of the local film peeling can be seen. A further increase in the load on the indenter leads to increased fluctuations in the coefficient of friction and in the submergence depth, which is indicative of the film peeling off the substrate. The maximum load on the indenter was 30 N; nevertheless, it sufficed to completely peel off the coating. Traces of the coating remained on the substrate surface. Thus, it can be estimated that the specific peel work of the coating was about 100 J/m^2.

The wear resistance of the A284 steel samples with deposited boron coatings and the samples with the same coatings modified by a pulsed electron beam was also measured. The wear rate of the original sample was 6×10^{-4} mm^3/N·m, while that of the boron-based coating sample was significantly lower and equal to 0.8×10^{-4} mm^3/N·m. The wear rate of the coating modified by the pulsed electron beam was 1.3×10^{-4} mm^3/N·m, lower than that for the uncoated samples but higher than that for the boron coating without beam treatment. Thus, the boron-based coating increases the surface wear resistance by a factor of 7.5, but the coating modification by a pulsed electron beam rolls it back by about 60%. Meanwhile, even a surface with boron coating modified by the beam has a wear resistance 4.5 times higher than that without coating. In our opinion, the reduced wear resistance of the boron coating after modification is associated with an increase in the coefficient of friction due to the formation of the surface hilly structure.

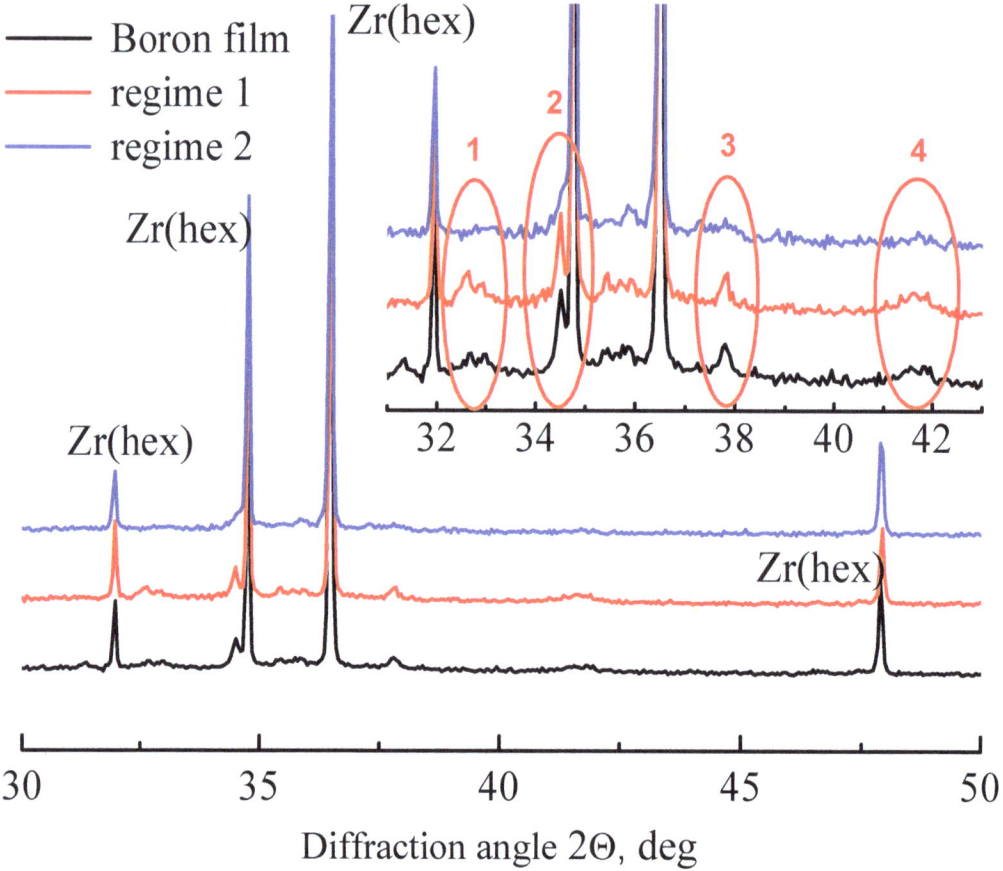

Figure 6. X-ray patterns of the surfaces of boron coatings on samples of the ZrNb1 alloy with a hexagonal crystal lattice. The inset shows the regions where the reflections of crystalline boron appear: regions 1, 2, 3, and 4. The wavelength of synchrotron radiation is 0.154 nm.

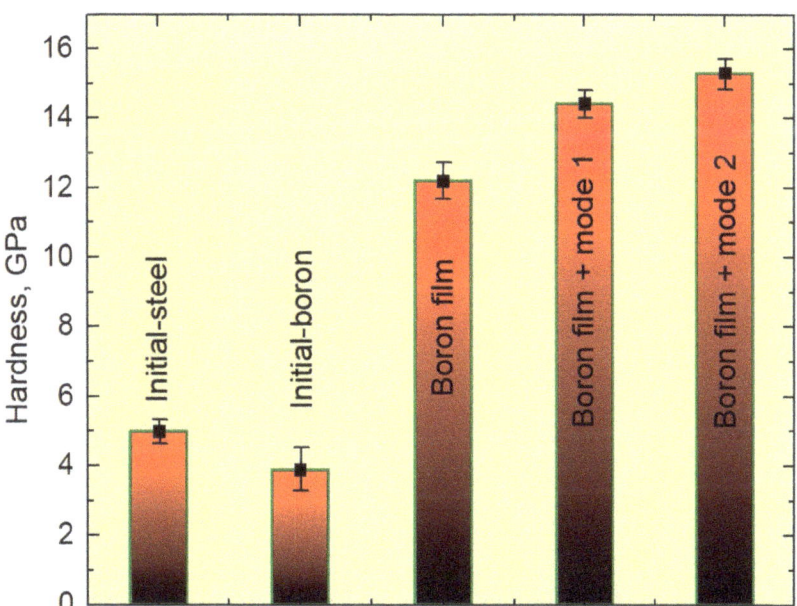

Figure 7. Microhardness of the steel substrate, the boron target, and fabricated boron coatings.

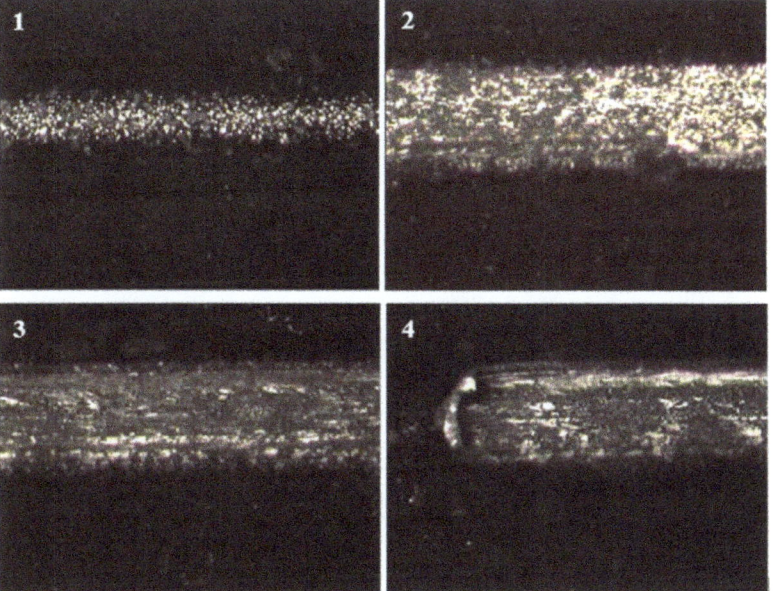

Figure 8. Micrographs of the boron coating surface under different loads exerted on a diamond indenter with a radius of 100 μm (regime 2), taken with an optical microscope; indenter loads: 1–3 N; 2–6 N; 3–9 N; 4–30 N.

Additionally, a corrosion rapid test was performed. Uncoated and boron-coated steel samples were placed in 25 wt.% saturated aqueous NaCl solution and exposed at 70 °C for 200 h. We tested 2 samples without coatings and 10 samples with deposited boron coatings, including those treated with a pulsed electron beam. Uncoated samples bore traces of pitting corrosion, distinctly seen in Figure 9a. Signs of corrosion were not noticeable on the surface of all the boron-coated samples. As an example, we include here pictures of the sample surface with boron coating treated with a pulsed electron beam (regime 2) and the sample kept in the solution. This verifies the high corrosion resistance of boron coatings and the absence of slits and cracks in them.

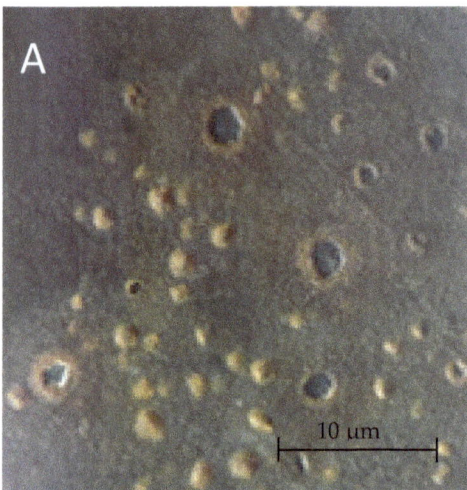
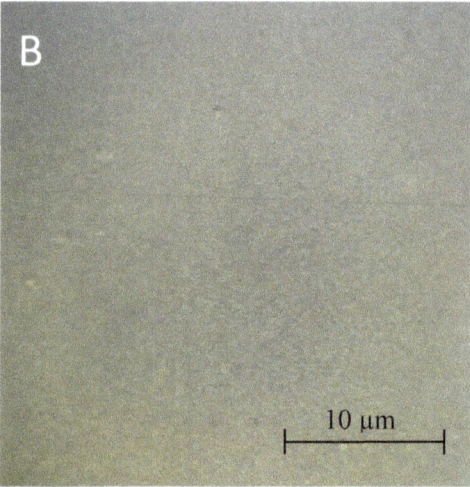

Figure 9. Photographs of A284 steel sample surfaces after a 200-h exposure in 25 wt.% NaCl aqueous solution: (**A**) uncoated sample, (**B**) sample with boron coating.

4. Conclusions

Electron-beam evaporation was used on boron targets to fabricate coatings on the surface of A284 steel and ZrNb1 alloy with a thickness of a few microns and a uniform structure. The deposition rate of boron coating at a power of 1 kW of a continuous beam of a fore-vacuum source was about 0.5 µm/min. The coating density of 2.2 g/cm^3 was close to the density of crystalline boron. On comparing the data on the mass-to-charge composition, beam plasma density, and coating parameters, it was concluded that the contribution of the plasma phase to the growth of the boron coating exceeded that of the vapor phase.

The boron coatings were modified by a pulsed beam with electron energies of 8 and 6 keV and powers of 2 and 2.2 J/cm^2 for a duration of 300 pulses at a repetition rate of 2 p.p.s. Using SEM and AFM, it was determined that such modifications yielded a considerable change in the surface morphology, forming a hilly structure with a characteristic size of the hill base of 1–10 µm and a height of 0.2–1.5 µm. Analysis of the absorbed beam energy showed that at a boron coating thickness of about 2 µm, 70% of the electron energy was released at a depth constituting less than 20% of this thickness. Structural phase analysis of coatings using synchrotron radiation showed that boron, both in modified and unmodified coatings, was present in the form of amorphous or ultrafine phases. Based on the bulk of the data obtained, it was concluded that the formation of the hilly surface structure was not caused by crystallization due to the electron beam but was associated with cyclic melting and cooling of the surface under the action of repetitive electron beam pulses.

Research on mechanical properties of the boron coatings showed that modification of its surface by the electron beam improved the coating hardness from 12 ± 0.35 to 15.5 ± 0.45 GPa. Based on the study of the coating–surface adhesion, it was shown that the specific peel work of the coating was about 100 J/m^2. The measured wear resistance of boron coatings was of the order of 10^{-4} mm^3/N·m, which exceeded the wear resistance of the A284 steel substrate manyfold. Based on the tests conducted in saturated salt solution, we demonstrated that the deposited coatings have anti-corrosive properties.

Author Contributions: Conceptualization, Y.Y. and E.O.; methodology, Y.Y. and A.T.; investigation, Y.Y, A.T. and A.K.; resources, E.O.; data curation, A.K.; writing—original draft preparation, Y.Y.; writing—review and editing, D.Z.; visualization, D.Z.; supervision, E.O.; project administration, Y.Y.; funding acquisition, E.O. All authors have read and agreed to the published version of the manuscript.

Funding: This work was conducted with financial support from the Russian Federation represented by the Ministry of Science and Higher Education under Project No. 075-15-2021-1348 within the framework of event No. 2.1.6.

Institutional Review Board Statement: Not applicable.

Informed Consent Statement: Not applicable.

Data Availability Statement: Data are available upon reasonable request.

Acknowledgments: The authors are grateful to Georgy Yushkov (Institute of High Current Electronics SB RAS, Tomsk), the father of one of the co-authors of this article (Yu. Yushkov), for advice in the analysis of the results and preparing the manuscript as well as to Ian Brown (Berkeley Lab) for useful discussions and English language improvement.

Conflicts of Interest: The authors declare no conflict of interest.

References

1. He, D.; Shang, L.; Lu, Z.; Zhang, G.; Wang, L.; Xue, Q. Tailoring the mechanical and tribological properties of B_4C/a-C coatings by controlling the boron carbide content. *Surf. Coat. Technol.* **2017**, *329*, 11–18. [CrossRef]
2. Bolton, W. *Engineering Materials Technology*, 3rd ed.; Butterworth Heinemann: Oxford, UK, 1998.
3. Nakazawa, H.; Sudoh, A.; Suemitsu, M.; Yasui, K.; Itoh, T.; Endoh, T.; Narita, Y.; Mashit, M. Mechanical and tribological properties of boron, nitrogen-coincorporated diamond like carbon films prepared by reactive radio-frequency magnetron sputtering. *Diam. Relat. Mater.* **2010**, *19*, 503–506. [CrossRef]
4. Baule, N.; Kim, Y.S.; Zeuner, A.T.; Haubold, L.; Kühne, R.; Eryilmaz, O.L.; Erdemir, A.; Hu, Z.; Zimmermann, M.; Schuelke, T.; et al. Boride-carbon hybrid technology for ultra-wear and corrosive conditions. *Coatings* **2021**, *11*, 475. [CrossRef]
5. Zhou, B.; Piliptsov, D.G.; Jiang, X.; Kulesh, E.A.; Rudenkov, A.S. Boron-Carbon Coatings: Methods of Deposition, Structure Features and Mechanical Properties, Problems of Physics, Mathematics and Technics. 2019, Volume 3, pp. 7–12. Available online: https://www.mathnet.ru/links/a5a0d8d29ad7416ca0e9b32191f67544/pfmt647.pdf (accessed on 10 September 2022). (In Russian).
6. Pan, Z.; Sun, H.; Zhang, Y.; Chen, C. Harder than Diamond: Superior Indentation Strength of Wurtzite BN and Lonsdaleite. *Phys. Rev. Lett.* **2009**, *102*, 055503. [CrossRef]
7. Mannix, A.J.; Zhou, X.; Kiraly, B.; Wood, J.D.; Alducin, D.; Myers, B.D.; Liu, X.; Fisher, B.L.; Santiago, U.; Guest, J.R.; et al. Synthesis of borophenes: Anisotropic, two-dimensional boron polymorphs. *Science* **2015**, *350*, 1513–1516. [CrossRef]
8. Sachdev, H. Disclosing boron's thinnest side. *Science* **2015**, *350*, 1468–1469. [CrossRef]
9. Federici, G.; Skinner, C.H.; Brooks, J.N.; Coad, J.P.; Grisolia, C.; Haasz, A.A.; Hassanein, A.; Philipps, V.; Pitcher, C.S.; Roth, J.; et al. Plasma–material interactions in current tokamaks and their implications for next step fusion reactors. *Nucl. Fusion* **2001**, *41*, 1967–2137. [CrossRef]
10. Gunduz, G.; Uslu, I.; Durmazucar, H.H. Boron-nitride-coated nuclear fuels. *Nucl. Technol.* **1996**, *116*, 78–90. [CrossRef]
11. Shao, Q.; Voss, L.F.; Conway, A.M.; Nikolic, R.J.; Dar, M.A.; Cheung, C.L. High aspect ratio composite structures with 48.5% thermal neutron detection efficiency. *Appl. Phys. Lett.* **2013**, *102*, 063505. [CrossRef]
12. Fabian Heide, F.; McDougall, M.; Harder-Viddal, C.; Roshko, R.; Davidson, D.; Wu, J.; Aprosoff, C.; Moya-Torres, A.; Lin, F.; Stetefeld, J. Boron rich nanotube drug carrier system is suited for boron neutron capture therapy. *Sci. Rep.* **2021**, *11*, 15520. [CrossRef]
13. Son, S.; Fisch, N.J. Aneutronic fusion in a degenerate plasma. *Phys. Lett. A* **2004**, *329*, 76–82. [CrossRef]
14. Harris, P.C. Chemistry and Rheology of Borate-Crosslinked Fluids at Temperatures to 300°F. *JPT* **1993**, *45*, 264–269. [CrossRef]
15. Matiasovsk, K.; Chrenkova-Paucirova, M.; Fellner, P.; Makyta, M. Electrochemical and Thermochemical Boriding in Molten Salts. *Surf. Coat. Technol.* **1988**, *35*, 133–149. [CrossRef]
16. Kulka, M.; Makuch, N.; Popławski, M. Two Stage Gas Boriding of Nisil in N_2-H_2-BCl_3 Atmosphere. *Surf. Coat. Technol.* **2014**, *244*, 78–86. [CrossRef]
17. Goeuriot, P.; Thevenot, F.; Driver, J.H. Surface Treatment of Steels: Borudif, a New Boriding Process. *Thin Solid Film.* **1981**, *78*, 67–76. [CrossRef]

18. Oks, E.; Anders, A.; Nikolaev, A.; Yushkov, Y.G. Sputtering of pure boron using a magnetron without a radio-frequency supply. *Rev. Sci. Instrum.* **2017**, *88*, 043506. [CrossRef]
19. Oks, E.M.; Tyunkov, A.V.; Yushkov, Y.G.; Zolotukhin, D.B. Synthesis of boron-containing coatings through planar magnetron sputtering of boron targets. *Vacuum* **2018**, *155*, 38–42. [CrossRef]
20. Gushenets, V.; Oks, E.; Bugaev, A. Characteristics of a Pulsed Vacuum Arc Discharge with Pure Boron Cathode. In Proceedings of the 28th International Symposium on Discharges and Electrical Insulation in Vacuum (ISDEIV), Greifswald, Germany, 23–28 September 2018. [CrossRef]
21. Gushenets, V.I.; Oks, E.M.; Bugaev, A.S. A pulsed vacuum arc ion source with a pure boron cathode. *AIP Conf. Proc.* **2011**, 090006. [CrossRef]
22. Bugaev, A.S.; Vizir, A.V.; Gushenets, V.I.; Nikolaev, A.G.; Oks, E.M.; Savkin, K.P.; Yushkov, Y.G.; Tyunkov, A.V.; Frolova, V.P.; Shandrikov, M.V.; et al. Generation of Boron Ions for Beam and Plasma Technologies. *Russ. Phys. J.* **2019**, *62*, 1117–1122. [CrossRef]
23. Yushkov, Y.G.; Zolotukhin, D.B.; Oks, E.M.; Tyunkov, A.V. Different stages of electron-beam evaporation of ceramic target in medium vacuum. *J. Appl. Phys.* **2020**, *127*, 113303. [CrossRef]
24. Yushkov, Y.G.; Tyunkov, A.V.; Oks, E.M.; Zolotukhin, D.B. Electron beam evaporation of boron at forevacuum pressures for plasma-assisted deposition of boron-containing coatings. *J. Appl. Phys.* **2016**, *120*, 233302. [CrossRef]
25. Ivanov, Y.F.; Koval, N.N.; Krysina, O.V.; Baumbach, T.; Doyle, S.; Slobodsky, T.; Timchenko, N.A.; Galimov, R.M.; Shmakov, A.N. Superhard nanocrystalline Ti–Cu–N coatings deposited by vacuum arc evaporation of a sintered cathode. *Surf. Coat. Technol.* **2012**, *207*, 430–434. [CrossRef]
26. Tyunkov, A.V.; Burdovitsin, V.A.; Oks, E.M.; Yushkov, Y.G.; Zolotukhin, D.B. An experimental test-stand for investigation of electron-beam synthesis of dielectric coatings in medium vacuum pressure range. *Vacuum* **2019**, *163*, 31–36. [CrossRef]
27. Greiner, E.S.; Gutowski, J.A. Electrical resistivity of boron. *J. Appl. Phys.* **1957**, *28*, 1364–1365. [CrossRef]
28. Zolotukhin, D.B.; Tyunkov, A.V.; Yushkov, Y.G. Distribution of potential upon the surface of non-conductive boron-containing target during irradiation by an electron beam in the fore-vacuum. *Appl. Phys.* **2017**, *6*, 39–43. Available online: https://applphys.orion-ir.ru/appl-17/17-6/PF-17-6-39.pdf (accessed on 10 September 2022). (In Russian).
29. Kazakov, A.V.; Medovnik, A.V.; Burdovitsin, V.A.; Oks, E.M. Pulsed cathodic arc for forevacuum-pressure plasma-cathode electron sources. *IEEE Trans. Plasma Sci.* **2015**, *43*, 2345–2348. [CrossRef]
30. Zolotukhin, D.B.; Tyunkov, A.V.; Yushkov, Y.G.; Oks, E.M. Modified quadrupole mass analyzer RGA-100 for beam plasma research in forevacuum pressure range. *Rev. Sci. Instrum.* **2015**, *86*, 123301. [CrossRef]
31. Yushkov, Y.G.; Oks, E.M.; Tyunkov, A.V.; Corbella, C.; Zolotukhin, D.B. Deposition of boron-containing coatings by electron-beam evaporation of boron-containing targets. *Ceram. Int.* **2020**, *46*, 4519–4525. [CrossRef]
32. Fitting, H.-J. Six laws of low-energy electron scattering in solids. *J. Electron Spectrosc. Relat. Phenom.* **2004**, *136*, 265–272. [CrossRef]
33. Karataev, V.D. *Lecture course "Experimental Methods of Nuclear Physics"*; Tomsk Polytechnic Institute: Tomsk, Russia, 1986.
34. Meisner, L.L.; Markov, A.B.; Rotshtein, V.P.; Ozur, G.E.; Meisner, S.N.; Yakovlev, E.V.; Gudimova, E.Y. Formation of microcraters and hierarchically-organized surface structures in TiNi shape memory alloy irradiated with a low-energy, high-current electron beam. *AIP Conf. Proc.* **2015**, *1683*, 020145. [CrossRef]

Article

CaO-SiO$_2$-B$_2$O$_3$ Glass as a Sealant for Solid Oxide Fuel Cells

Andrey O. Zhigachev *, Ekaterina A. Agarkova, Danila V. Matveev and Sergey I. Bredikhin

The Institute of Solid State Physics of the Russian Academy of Sciences, 142432 Chernogolovka, Russia
* Correspondence: zhigachev@issp.ac.ru

Abstract: Solid oxide fuel cells (SOFCs) are promising devices for electrical power generation from hydrogen or hydrocarbon fuels. The paper reports our study of CaO-SiO$_2$-B$_2$O$_3$ material with composition 36 mol.% SiO$_2$, 26 mol.% B$_2$O$_3$, and 38 mol.% CaO as a high-temperature sealant for SOFCs with an operating temperature of 850 °C. The material was studied as an alternative to presently existing commercial glass and glass-ceramics sealants for SOFCs with operating temperature of 850 °C. Many of these sealants have limited adhesion to the surface of Crofer 22APU steel, commonly used in these SOFCs. The present study included X-ray diffraction, dilatometric, thermal, and microstructural analysis The study has shown that the softening point of the CaO-SiO$_2$-B$_2$O$_3$ glass is around 900 °C, allowing sealing of the SOFCs with this glass at convenient temperature of 925 °C. The CaO-SiO$_2$-B$_2$O$_3$ glass sealant has shown excellent adhesion to the surface of Crofer 22APU steel; SEM images demonstrated evidences of chemical reaction and formation of strong interface on sealant–steel contact surface. Furthermore, the glass has shown a coefficient of thermal expansion about 8.4×10^{-6} 1/K after sealing, making it thermomechanically compatible with the existing SOFC materials.

Keywords: glass; glass-ceramic; microstructure; sealant; SOFC

1. Introduction

Solid oxide fuel cells (SOFCs) are promising electrochemical devices for conversion of chemical energy of hydrogen and hydrocarbon fuels into electricity. The energy conversion process goes directly through electrochemical reactions, bypassing the fuel-burning stage. Compatibility of SOFCs with conventional hydrocarbon fuels along with high efficiency and nearly noiseless operation make them an attractive solution for industrial and domestic power generation.

SOFCs owe their compatibility with hydrocarbon fuels to their high operation temperatures; most of the commercial SOFC systems work at 800–100 °C. It should be noted that in the last two decades, active research on SOFCs operating at intermediate (600–800 °C) or low temperatures (500–600 °C) has been conducted [1,2], but these systems have not yet reached the commercialization level. High temperatures used in SOFCs not only allow use of hydrocarbon fuels, but also lead to high-quality exhaust heat that can be used for domestic or industrial heating. Unfortunately, high the operating temperature of SOFCs has its drawbacks. It increases duration of startup and stopping of SOFCs, introduces additional requirements to thermomechanical and chemical compatibility of the materials used in SOFCs, and complicates the choice of sealing materials for SOFCs [3].

Primary functions of the sealing materials in SOFCs include dividing fuel and air streams inside the fuel cell, isolating the internal gases flow from the environment, and mechanical consolidation of separate cells into a battery. No organic-based seals can be used at 800 1000 °C, so glasses and glass-ceramics are employed instead. Choice of sealants for SOFCs is a complicated task because they should meet a number of requirements. A sealant should have a coefficient of thermal expansion (CTE) close to that of both ceramic membrane (electrolyte) and metal interconnect (usually high-chromium steel or CFY alloy), typically in the $8\text{–}12 \times 10^{-6}$ 1/K range. In addition, a strong adhesion should exist between a sealant and steel interconnect to ensure both impermeability for gas products and high mechanical strength of the battery. The last requirement for SOFC sealing materials is the stability of their chemical composition and physical properties during the operation of the battery.

A number of glass and glass-ceramic sealant compositions for SOFCs are reported in the literature and implemented in commercial products. Among the most frequently used chemical compositions, one may note the following:

1. Barium aluminosilicate glass-ceramics. Their approximate chemical composition is 45–55 mol.% SiO_2, 5–15 mol.% B_2O_3, 20–30 mol.% BaO, 5–15 mol.% Al_2O_3, minor amounts of ZrO_2, NiO, ZnO, and Cr_2O_3, and some other oxides. These glass-ceramics usually have CTE of $10.5-11 \times 10^{-6}$ 1/K, nicely matching CTE of SDC and ScSZ electrolytes. They show decent adhesion to heat-resistant steel such as Crofer 22 APU [4–10]. In these glass-ceramics, the choice of the appropriate amount of flux is crucial to obtain material with the desired softening and glass-transition temperature.
2. Diopside-based glass-ceramics. These glass-ceramics are also frequently used to seal SOFCs. Approximate chemical composition is 15–20 mol.% CaO, 25–35 mol.% MgO, and 45–50 mol.% Al_2O_3, and minor quantities of Al_2O_3, SiO_2, and B_2O_3 [6,11,12]. These sealants also have CTE of about 10×10^{-6} 1/K and decent adhesion to conventional metal interconnect.
3. Modified soda–lime glass. Soda–lime glass, with the addition of alumina and increased calcia content, is sometimes considered as glass sealant for SOFCs [4]. An example of composition used is 50–60 % mol.% SiO_2, 5–10 mol.% Al_2O_3, 24–28 mol.% CaO, and 10–14 mol.% Na_2O.

The materials mentioned above, especially barium aluminosilicate glass-ceramics, provide good sealing in SOFCs according to the literature. The procedure of SOFC sealing with these materials includes heating of the whole battery well-above its operating temperature. For example, for batteries operating at 850 °C, typical sealing temperature is 950 °C (Schott 394 glass). Dwell at the elevated temperature usually takes a few hours if conventional sealants are used so that reaction between the sealant and metal interconnect takes place and a strong interface is formed. The dwell at elevated temperatures may cause undesirable changes in the structure of electrodes due to sintering processes and may lower power output of the cell.

It is thus desirable to use glass or glass-ceramic sealant that quickly reacts with the material of a bipolar plate forming strong interface such that it remains stable under operating conditions. Indeed, other requirements for sealants still hold true for such sealants. They should be thermomechanically compatible with high-chromium steel or CFY and with the electrolyte used in the cell. Their properties should not deteriorate during the operation of the fuel cell through crystallization or through further reaction with either steel or the gas environment.

In the present work, we have investigated the glass-ceramic sealant of $CaO-SiO_2-B_2O_3$ system with high boron oxide content of 26 mol.% B_2O_3. This system was chosen because calcium borosilicate glasses are known to have CTE in the range $8-10 \times 10^{-6}$ 1/K and glass-transition temperature about 650–700 °C [13–16]. $CaO-SiO_2-B_2O_3$ glasses with boron oxide content above 7–8 wt. % are rarely used as sealants for SOFCs; they primarily find applications in the sealing of microelectronic components working at near-room temperature [17–19]. The main reason for the limited high-temperature use of these glasses is volatility of boron oxide in glasses and its ability to cause poisoning of the cathode materials when volatilized [20,21]. However, in the present research, we hoped to achieve a high degree of crystallization of the sealant material after heat treatment to bind the boron oxide into the crystal lattice and into reaction products with steel and thus, lower boron mobility.

The main aims of the study were the following:

1. Study interface between CaO-SiO$_2$-B$_2$O$_3$ sealant and Crofer 22 metal interconnect after brief heat treatment or after treatment at temperatures lower than are conventionally used;
2. investigate thermomechanical compatibility of the CaO-SiO$_2$-B$_2$O$_3$ sealant and Crofer 22 interconnects;
3. investigate degree of crystallinity and phase composition of the sealant after the sealing procedure both in and out of contact with steel interconnect.

2. Materials and Methods

We prepared a sealant with chemical composition 38 mol.% CaO, 26 mol.% B$_2$O$_3$, and 36 mol.% SiO$_2$. Choice of the composition was based on the previous publications on low-temperature applications of CaO-B$_2$O$_3$-SiO$_2$ seals [13]. As initial materials for preparation of the sealant, we used >99 % pure calcium carbonate, boric acid, and silicon dioxide (Chemcraft, Russia). In order to mix the precursors homogeneously, we added bidistilled water and mixed the slurry with a laboratory mixer UED-20 (UED Group, Russia) for 5 min. The slurry was then left to dry in a laboratory oven at 100 °C for 15 h to remove water.

The dried mixture was then placed into platinum crucible and heated in air to 1500 °C at heating rate of 2.5 °C/min. Heating was conducted in a vertical load furnace LHT 02/17 LB (Nabertherm, Germany). We chose low heating rate to allow full decomposition of precursors to take place before the melting. We held the melt at 1500 °C for 1 h for the components to form homogeneous melt. Then, we quenched the melt into bidistilled water. After cooling of the water, we extracted the pieces of the resulting material and dried them at 100 °C for 2 h.

The dried pieces of the material were ground with a mechanical mortar Pulverisette 2 (Fritsch, Germany). We used zirconia pestle and mortar to minimize contamination of the material during grinding. We estimated the size of the ground powder with laser diffraction method (Analysette 22 Next, Fritsch, Germany) and compared it to size of commercially available powder glass Schott 394. For the purpose of the measurement small quantity of ground powder was dispersed in bidistilled water, ultrasonicated to break any agglomerates, and subjected to static light scattering measurement. It was performed to ensure that there was no significant particle size difference that could influence comparison of properties of the materials.

We performed X-ray diffraction (XRD) analysis of as-prepared sealants after heat treatment both in and out of contact with Crofer 22 APU. XRD analysis was performed on finely ground powder with copper X-ray tube in Bragg–Brentano reflective geometry with Smartlab (Rigaku, Japan) diffractometer. Peaks were identified with the use of PDF database. The latter was used as a reference material of a bipolar plate. XRD data were used to estimate degree of crystallinity of the material and identify crystalline phases if any were present.

We investigated high-temperature behavior of the prepared material. For this purpose, we have put about 0.3 g of the studied material on Crofer 22 APU (ThyssenKruppVDM, Berlin, Germany) plate with dimension 20 × 20 × 2 mm. The powder formed a loose cone with 10-milimeter base and about 10-milimeter height. The plate with the powder was then put to furnace and heated to a maximum temperature of 850–950 °C at 2 °C/min rate, held there for 1 h, and cooled down at the same rate. Low heating and cooling rates were chosen to minimize effects caused by CTE mismatch. Plate with the sealant was then visually assessed to estimate suitable sealing temperature of the prepared material. We ground the sealant after high-temperature treatment to prepare fine powder and performed XRD analysis of its phase compositions. We also cut rectangular samples with approximate dimensions 3 × 1 × 1 mm from the heat-treated sealant and measured its CTE in air with dilatometer L75Vertical (LINSEIS, Berlin, Germany). The samples were measured with heating and cooling rates of 3 °C/min and were held at 850 °C for 2 h. Ends of the samples were covered with thin zirconia pellets to minimize reaction of the sealant with the measurement chamber. Expansion of zirconia pellets was accounted for and subtracted from the final data.

In order to assess strength of the sealant adhesion to the Crofer 22 APU plate, we performed a mechanical test. The steel plate was fixed in a vertical position by a lower part not covered with the sealant in clamps. Then, we applied bending load to the upper part of the plate in a way such that the sealant/steel interface was under tensile stress. We registered the applied load at which delamination of the sealant occurred and calculated the corresponding flexural stress.

We also investigated microstructure of sealant/steel interface. To prepare samples for this study, we deposited a hollow square pattern of sealant on 20 × 20 × 2 mm Crofer 22 APU plate. For the deposition of the sealant, we used F4200N (Fisnar, Copenhagen, Denmark) dispenser with terpineol-glass-PVB paste. Organic components for the paste were supplied by Chemcraft, Russia. The as-deposited sealant was dried and covered on top with similar Crofer plate. This "sandwich" structure was loaded with a force of approximately 1 N and heat-treated in a furnace on the described above routine. A 150-micrometer zirconia delimiter was put in the center of the assembly to avoid leaking of the sealant under the load at elevated temperature. After cooling of the assembly, we prepared and polished cross-sections of the assembly. We prepared SEM images of the cross-sections with Supra 50 VP (Carl Zeiss, Germany) microscope. Both our material and commercial Schott 394 glass were investigated this way. We used energy-dispersive X-ray spectroscopy (EDXS) for elemental analysis of the cross-section and sealant/steel interface.

Thermal cycling tests were conducted on "sandwich" assemblies after heat treatment at 925 °C. For thermal cycling, we put the assemblies with no load applied into a furnace and heated them to 850 °C (typical temperature of SOFCs operation) with 2 °C/min heating rate, held at this temperature for 2 h, and then cooled down to room temperature with 2 °C/min cooling rate.

Schematic diagram of the sequence of the experiments conducted in the present study is given for reference in Figure S1 (Supplementary Materials).

3. Results and Discussion

CaO-SiO_2-B_2O_3 sealant melted at 1500 °C flowed freely from the platinum crucible into water. Its viscosity, as assessed visually, was much lower than that of barium aluminosilicate, diopside, and modified soda–lime glass sealants melted at the same temperature. The as-quenched calcium borosilicate sealant was opaque white contrary to what we observed on other types of sealants, which were transparent.

Mechanical strength of the as-quenched CaO-SiO_2-B_2O_3 sealant was similar to that of other sealants as judged from time required to mill quenched chunks of the material into fine powder.

Particle size distribution of the CaO-SiO_2-B_2O_3 powder is presented in Figure 1 along with the data for commercial powder Schott 394 measured under the same conditions. It is clear from the presented particle size distribution that the prepared sealant was similar to the commercial product, although with slightly narrower distribution especially for larger particles. Peaks of the distribution for our sealant lie at ~1.5 μm and at ~10–12 μm. Commercial powder has a wider distribution slightly shifted toward larger particle size with the rightmost peak at 30–40 μm. It should be noted that this peak corresponding to 30–40 μm particle can also be observed on the powder of our sealant, but there the peak is much less intensive. The difference between the distributions can be illustrated by d_{10}, d_{50}, and d_{90} values of the presented data. The values are shown on the insert in Figure 1. All the values are slightly lower for the CaO-SiO_2-B_2O_3 sealant, indicating that its particle size distribution is shifted toward lower values. The most notable difference is in d_{90} value, which is predictable given the difference in the right part of the distributions.

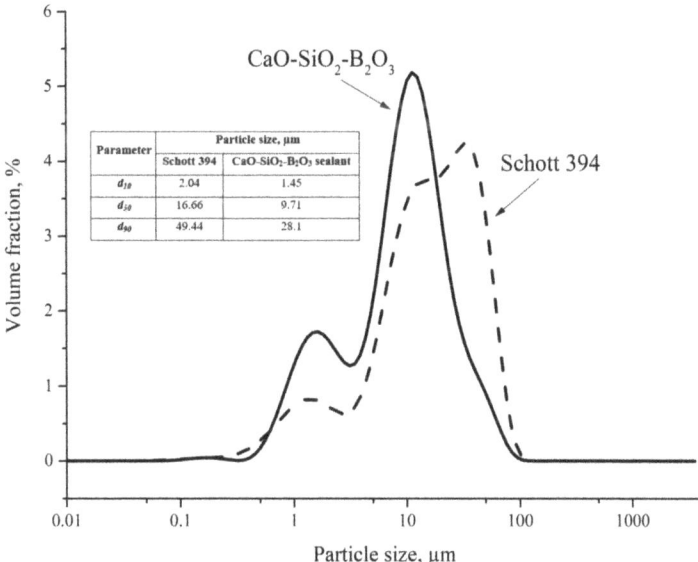

Figure 1. Particle size distribution in a ground powder of CaO-SiO$_2$-B$_2$O$_3$ sealant and in commercial Schott powder.

Figure 2 shows XRD pattern of the ground CaO-SiO$_2$-B$_2$O$_3$ sealant. No crystalline peaks can be seen; three broad amorphous halos are present instead. They correspond from left to right to interatomic distances 9.23 Å, 3.132 Å, and 2.01 Å. Among these distances, only 2.01 Å may be clearly attributed to B–O bond length [22]. Other distance cannot be attributed to either Si–O or Ca–O bond, which have lengths of 1.60–1.65 Å and 2.45–2.54 Å [23,24]. The low-angle peaks may correspond not to distances between adjacent atoms, but to distance between adjacent "strands" of the glass network.

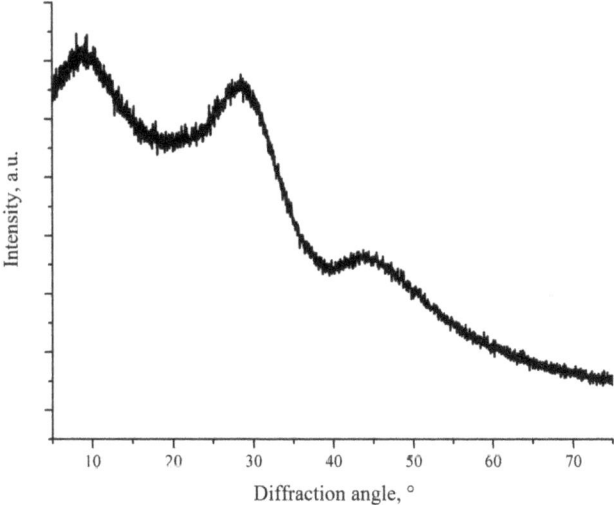

Figure 2. XRD pattern of as-quenched CaO-SiO$_2$-B$_2$O$_3$ sealant.

Figure 3 shows schematic depiction of the sealant-on-plate assembly after heat treatment at different temperatures for 1 h. We did not insert photographs here; because of the relatively small size of the samples, it was difficult to make a high-quality "macrophotograph", showing the necessary details. Colors used in the figure are nearly identical to actual colors of different zones of the sealant after heat treatment. The color of Crofer 22 APU plate is shown only schematically with hatching. Relative dimensions of the sealants and the steel plate are lifelike with factual size of the plate 20 × 20 × 2 mm. Wetting angles are also shown true to life.

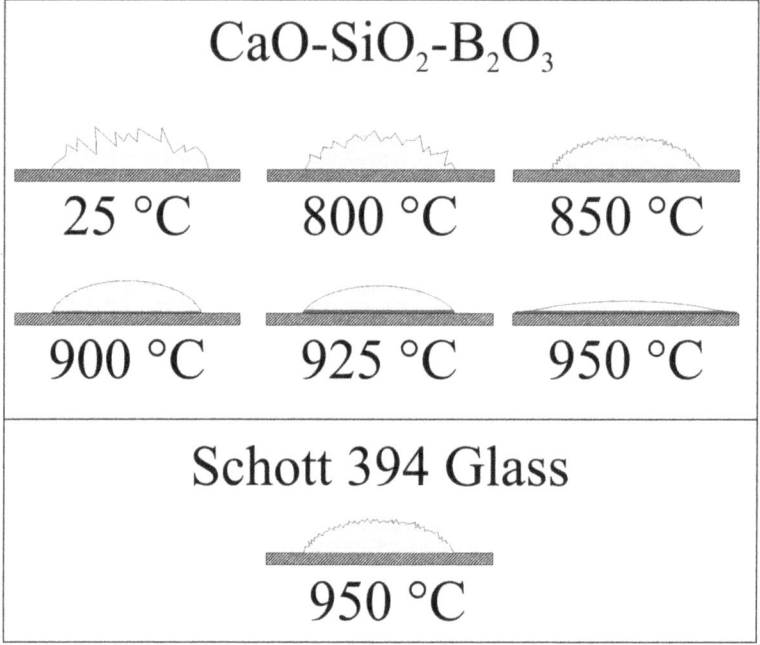

Figure 3. Shape and color of sealants after heat treatment for 1 h.

For CaO-SiO$_2$-B$_2$O$_3$ sealant, heat treatment at 800 °C for 1 h did not cause any notable change in the geometry of the powder pile. Sharp edges of the particles remained and the particles themselves were only loosely bound to each other. We observed no reaction zone on sealant/steel interface. Increase in the temperature to 850 °C led to some softening of the angles of the powder pile and to the formation of a thin (~ 100 um), but notable layer of dark blue color at the interface with steel. Even further increase in temperature led to progressive softening of the relief of the pile and thickening of the blue layer to ~300 μm at 925 °C. At this temperature, surface of the sealant became smooth, forming a drop, slightly wetting the steel plate. It indicates that the glass softening point lies between 850 °C and 925 °C. For samples heated to 950 °C, we observed spreading of the sealant over the surface of the plate, demonstrating excellent wetting and low viscosity of the sealant. The latter is undesirable for sealing of SOFCs because low-viscosity sealant may flow from the designed areas and block air or fuel channels. In the following discussions, part of the sealant that was not in direct contact with the steel plate and did not change color significantly is named "White area". The part that was in contact with the plate and changed its color to saturated blue is named "Blue area". The key difference between these regions is that the former represents change in the sealant itself after the heat treatment, while the latter represents products of interaction of the sealant and Crofer plate. Experiments with Schott 394 glass were conducted only for 950 °C because it is a recommended sealing temperature for this sealant. The degree of the sealant softening at this temperature was similar to that of CaO-SiO_2-B_2O_3 sealant at 850 °C. Significant visual difference lies in the color of the contact layer on the interface with steel. The color is saturated yellow as opposed to the dark blue of CaO-SiO_2-B_2O_3.

Mechanical test of adhesion strength was performed on samples with CaO-SiO_2-B_2O_3 sealant heat-treated at 925 °C and on Schott 394 sealant heat-treated at 950 °C. We registered the bending load was applied when the delamination of the sealant from the steel plate took place. In the case of Schott 394, sealant delamination occurred at a load of 11 N, which corresponded to flexural stress of 0.12 MPa. In the case of CaO-SiO_2-B_2O_3, delamination of the sealant took place at a load of 28 N, corresponding to flexural stress of 0.31 MPa. In the present experiment setup, flexural stress represents tensile stress at steel-sealant interface in the interface plane; therefore, it can be said that the strength of the CaO-SiO_2-B_2O_3 adhesion to steel surface is approximately 2.5 times higher than that of Schott 394 sealant. It should be noted that during exfoliation of the studied sealant, it came of the steel surface with significant starting velocity unlike what was observed on the Schott 394 sealant. Such behavior may follow from both external mechanical stresses introduced during the test and from thermomechanical stresses generated at sealant–steel interface due to possible CTE mismatch. Since it is likely that the appearance of the blue zone on the sealant/steel interface was a result of the reaction between the CaO-SiO_2-B_2O_3 sealant and steel, we performed XRD study of both white and blue areas of the sealant after heat treatment at 925 °C. Figure 4 shows XRD patterns for both of these areas.

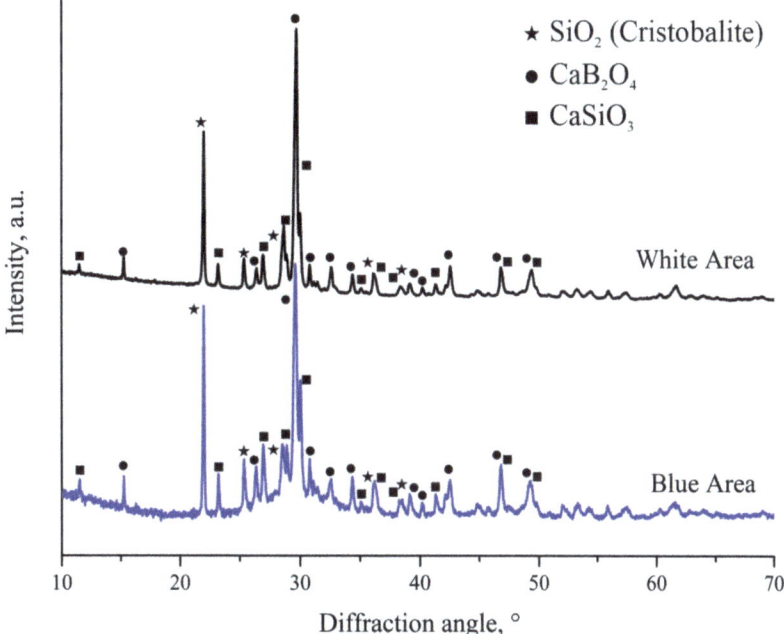

Figure 4. XRD patterns of white and blue areas of the CaO-SiO_2-B_2O_3 after heat treatment at 925 °C. Symbols mark positions of the most intensive peaks of different phases.

Sealant that was not in contact in Crofer plate is fully crystallized (upper curve in Figure 4). The most prominent crystalline phases attributed to the observed peaks are CaB_2O_4 (PDF 00-032-0155), SiO_2 (Cristobalite modification, PDF 00-039-1425), and $CaSiO_3$ (PDF 00-043-1460). The amount of amorphous phase in the white area is negligible. The ratio between $CaSiO_3$ and CaB_2O_4 allowed us to estimate boron content in the heat-treated sealant; it turned out to be close to the nominal value. XRD pattern of the blue area shows the same crystalline phases with similar molar fractions of the phases with a slightly higher $CaSiO_3/CaB_2O_4$ ratio, which may serve as an indication that part of boron oxide took part in reaction with the steel plate. The products of the reaction are likely to be amorphous, as evidenced by the appearance of an amorphous halo on the XRD pattern. We detected no crystalline phases containing iron or chromium (main components of Crofer 22 APU steel) in the blue area. However, the coloration itself hints at the presence of Cr^{3+} ions in the interface area. It may be assumed, thus, that the amorphous phase is formed through reaction of at least boron oxide and components of the steel plate. Of course, other components of the sealant could take part in the formation of the amorphous layers. XRD patterns show that the sealant after heat treatment is glass-ceramic in the area near the bipolar plate. Taking into account that thickness of the seal is usually well-below 500 μm, we may safely assume that the entire seal will be composed of the "blue" zone, meaning that the seal will be in the glass-ceramic state after the sealing procedure.

It was difficult to calculate molar fractions of the crystalline phases in powder prepared from the blue area because of the significant contribution of the amorphous phase. Roughly, ratios of the crystalline phases may be estimated to be 55 mol.% $CaSiO_3$, 30 mol.% CaB_2O_4, and 15 mol.% SiO_2. We managed to calculate the amount of amorphous phase only approximately to be 60–70% of total crystalline phases content. CTEs of the present phases are, according to the literature:

1. $CaSiO_3$ has a CTE of 11.2×10^{-6} 1/K in a wide range of temperatures, as calculated from high-temperature crystallographic data on $CaSiO_3$ reported in the literature [25].
2. We found no results on CTEs of CaB_2O_4 in the literature. The closest match we found in the literature is a paper by Kluev et al. [26], where the authors reported CTE of glass consisting of 40 mol.% CaO and 60 mol.% B_2O_3 to be 7.29×10^{-6} 1/K in 20–300 °C range. Other results mentioned in the paper imply that material with 50 mol.% CaO and 50 mol.% B_2O_3 may have slightly higher CTE.
3. SiO_2 in β-cristobalite has a CTE changing significantly with temperature, having CTE of 10.9×10^{-6} 1/K at 100–500 °C rapidly falling to 1.7×10^{-6} 1/K in 500–1000 °C [27].

Among the present crystalline phases, cristobalite, predictably, has the lowest average CTE in the 100–850 °C range. It means that cristobality may serve as a main source of thermomechanical stresses on the sealant-interconnect interface, which may be somewhat mitigated by the glassy phase presenting dampening CTE mismatch. Crystallization of SiO_2 is an undesirable process, lowering overall CTE of the sealant and causing internal thermomechanical stresses. This process should be suppressed by the introduction of additional components to the sealant. We plan to use them in the process of further development of the reported sealant.

To assess thermomechanical compatibility of the sealant and Crofer 22 APU, we cut a rectangular sample of the sealant after heat treatment at 925 °C. We managed to prepare suitable samples from the white area of the treated sealant because the blue area was too thin (<300 μm). Of course, in the case of SOFCs sealing, the sealing space will be filled with the material with phase composition of the blue area because of typically low thickness of the seal (usually well-below 500 μm). Nonetheless, dilatometric data obtained on the white (Figure 5) area may give a useful insight into properties of the blue area because of the closeness of their phase composition.

It should be noted that the data presented in Figure 5, obtained on the sealant after heat treatment, does not illustrate properties of the original glass. Instead, it shows properties of the sealant after the sealing procedure, which, we believe, are more important when studying thermomechanical compatibility of the sealant and steel. Average CTE in range 400–850 °C is 8.4×10^{-6} 1/K. This value is lower than CTE of Crofer 22 APU ($11-12 \times 10^{-6}$ at 800–900 °C) and than that of zirconia electrolyte ($10.5-11 \times 10^{-6}$ 1/K). CTE of the $CaO-SiO_2-B_2O_3$ sealant is close to that of Schott 394 glass–8.6×10^{-6} 1/K. It allows us to assume that a sealant with CTE equal to 8.4×10^{-6} 1/K is thermomechanically compatible with Crofer 22 APU and zirconia-based electrolytes. Since the data were obtained on the sealant crystallized after heat treatment, it is impossible to determine the glass-transition temperature from the presented dilatometry data.

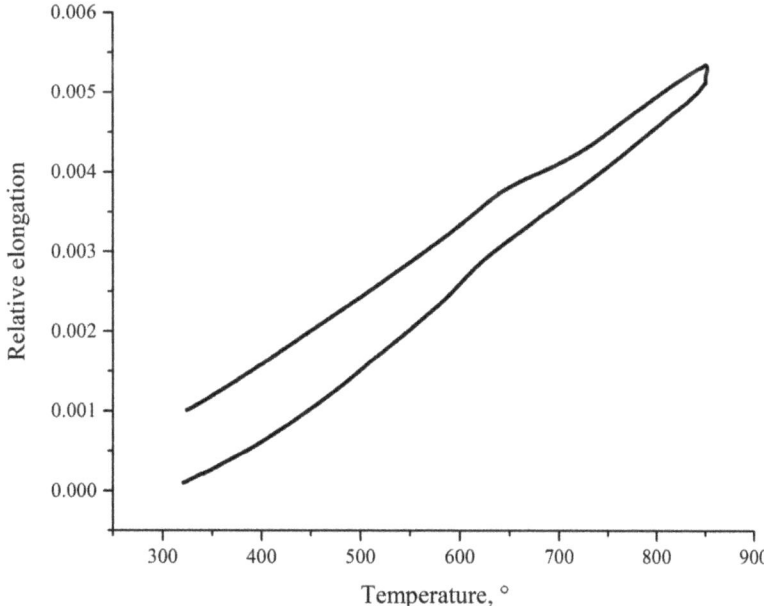

Figure 5. Dilatometric curve of the CaO-SiO$_2$-B$_2$O$_3$ sealant after heat treatment at 925 °C.

SEM images of the steel–sealant–steel assembly cross-sections are presented in Figure 6. Figure 6a,b features an assembly with the CaO-SiO$_2$-B$_2$O$_3$ sealant. It can be seen that a strongly pronounced interaction area is formed; it consists of elongated formations with predominant orientation normally to the interface surface. The interaction area appears consistently along the entire sealant–steel contact surface. The sealant layer all along the studied surface showed a solid structure without cracks or extensive pore networks, which could lead to the loss of impermeability of the material. We have also held some samples at the operation temperature of 850 °C for 100 h to examine if the reaction zone propagates further into the sealant. We found no evidence of such behavior; SEM images of as-heat-treated and aged samples were hard to distinguish from each other.

In the assembly with Schott 394, the interaction area was not pronounced, as can be seen in Figure 6c. SEM images of the assembly with Schott 394 sealant show regions where delamination of the sealant from steel surface is observed (central part of the lower interface in Figure 6c). The delamination might have taken place due to possible factors: (a) poor adhesion combined with thermal stresses at the interface or (b) contamination of the steel plate surface prior to sealant deposition. The latter is unlikely because of the careful preparation of the samples and consistency of the appearance of the areas with delamination, but it cannot be entirely ruled out.

The reported CTE of the sealant is lower than that of steel interconnect or zirconia electrolyte. The CTE mismatch causes thermomechanical stress, which was not high enough to lead to immediate delamination of the sealant, according to conducted mechanical tests. It, however, may lead to the degradation of the interface upon thermal cycling. SOFCs rarely undergo a high number of thermal cycles to room temperature, since they usually operate at a nearly constant temperature for prolonged periods. Nevertheless, a sealant-interconnect interface should withstand a small number of cycles inevitably occurring at startups or stops of the SOFC battery. We have thermally cycled the steel-sealant–steel assembly 20 times and after that obtained an SEM image of the polished cross-section. We have observed no delamination or cracking of the sealant after 20 cycles. It may be supposed that CTE of the reaction zone (blue zone) is different from that of the bulk sealant, perhaps having intermediate value between bulk sealant and steel interconnect, thus avoiding generation of large amounts of thermomechanical stress on the interface. However, it is only an assumption, since we were not able to prepare a sample for dilatometry from the reaction zone because of its small size. Another possible assumption, which is indirectly supported by the SEM images of the interface, is that adhesion on the interface is strong enough to withstand the thermomechanical stress.

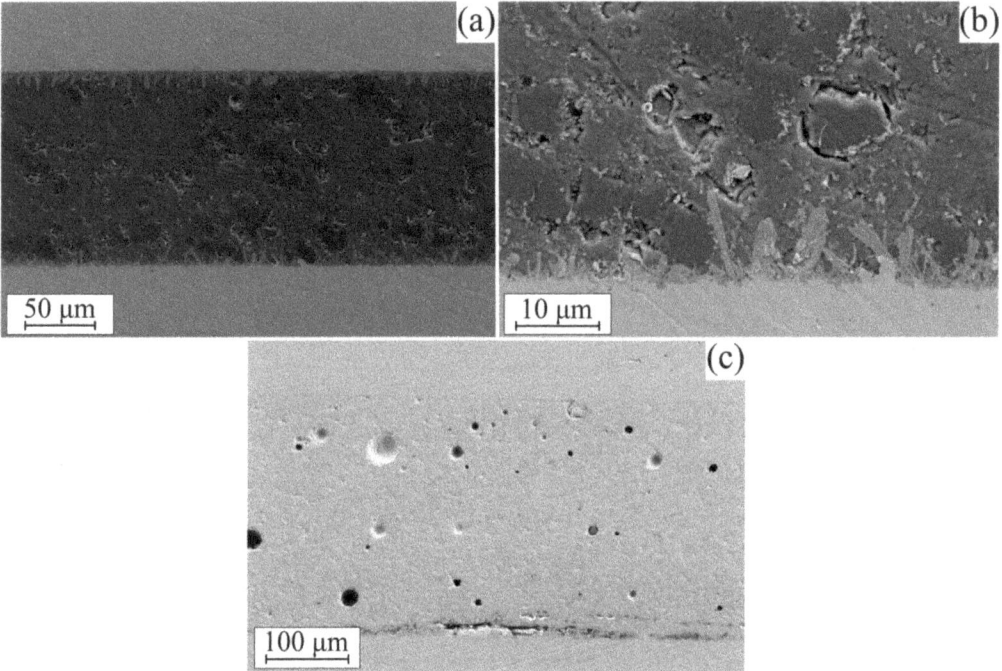

Figure 6. SEM images of the cross-sections of sealant–steel assembly: (a) CaO-SiO$_2$-B$_2$O$_3$ sealant, (b) close-up image of CaO-SiO$_2$-B$_2$O$_3$ sealant–steel interface, and (c) Schott 394 sealant.

For EDXS analysis of the sealant–steel assembly, we intentionally performed heat treatment of the assembly at 925 °C for 6 h instead of 1 h. The purpose of such treatment was to grow the sealant–steel interaction zone to obtain more reliable data on the elemental composition. Prolonged exposure to high temperatures caused intensive growth of the interaction zone with dendrite formations growing deep into the sealant layer. Figure 7 shows points where elemental composition was measured; these points are: (a) large dendrite formations originating at sealant–steel interface, (b) dark contrast regions in the sealant, and (c) gray-contrast areas in the bulk of the sealant. Table 1 summarizes our findings on the compositions in these points. It should be noted that EDXS does not allow detection of boron so boron is not listed in the table, but is indeed present as evidenced by XRD data (Figure 4).

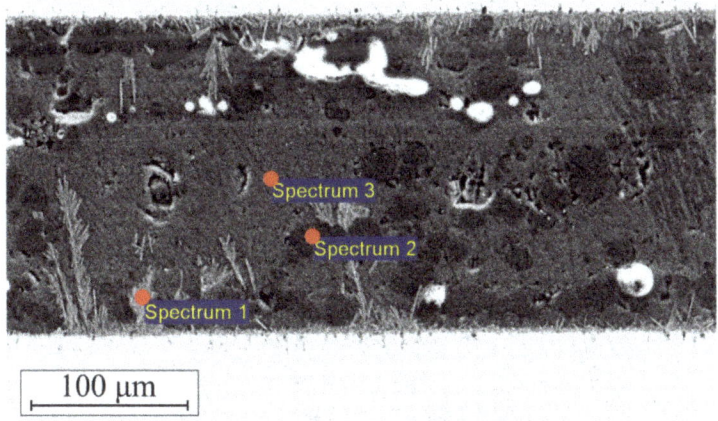

Figure 7. Choice of points for EDXS of heat-treated sealant–steel assembly.

Table 1. Elemental composition in the points shown in Figure 7.

Type of Atom	Content, mol.%		
	Point 1	Point 2	Point 3
Si	11.89	35.26	9.29
Ca	2.24	0.74	23.60
Cr	26.35	-	0.84
Fe	0.40	-	0.32
O	59.12	64.00	65.94

The gray-contrast area (point 3 in Figure 7) contains mostly Ca, Si, and O with minor quantities of Cr and Fe. Presented in Table 1, elemental composition in the point cannot be attributed to either $CaSiO_3$, SiO_2, or their mixture because of Ca/Si and Ca/O ratios. It is evident that the disbalance in the amount of the elements is due to the presence of the CaB_2O_4 phase. Darker contrast regions (point 2) show almost purely SiO_2 with some minor amounts of $CaSiO_3$. Dendrite formations at the interface contain large amounts of Cr and Si with a little Ca. It shows that at the sealant–steel interface, the sealant predominantly reacts with Cr. The ratios Cr/Si and Cr/Ca allow us to safely assume that the reaction product contains significant amounts of B. The latter was expectable because of the high reactivity of boron. Furthermore, it agrees with the observations made from XRD patterns of the reaction zone.

The main observed shortcoming of the studied sealant is its relatively low CTE of 8.4×10^{-6} 1/K, which is lower than CTEs of most materials used in SOFC. For example, metal interconnects and zirconia electrolytes usually have a CTE of about 11×10^{-6} 1/K. CTE of the studied sealant is also low relatively to CTEs usually offered by reported sealants. For example, barium aluminosilicate sealants often have CTEs of $8.5–14 \times 10^{-6}$ [4,9]. Calcium aluminosilicates may have CTEs in a wide range including $7–12 \times 10^{-6}$ 1/K, depending on the composition with higher values of CTE observed at low-silica and high-calcia contents [28]. Despite this difference in CTEs, it is close to that of commercially supplied sealants for SOFC, e.g., Schott 394 glass designed for high-temperature SOFCs has a CTE of 8.6×10^{-6}. Low CTE of the studied sealant is likely caused by the presence of cristobalite and calcium metaborate phases with low CTE. It may be possible to improve thermomechanical properties of the sealant by substituting part of CaO for SrO or by introducing some additional components such as La_2O_3 [4].

It should be noted that we only managed to measure CTE of the heat-treated sealant that was not in contact with the steel interconnect. According to our XRD data, the part that was in contact with the interconnect contained a significant amount of glassy phase with some dissolved Cr^{3+} ions, as evidenced by EDXS data and by coloration of the reaction zone. This glassy phase likely has CTE intermediate between that of heat-treated sealant and steel interconnect, as indirectly evidenced by SEM images showing no cracks at the cross-section of steel–sealant assembly.

The glass-transition temperature of the prepared sealant was not measured, but it can be estimated from the literature data to be around 650–700 °C [17–19]. A high content of glassy phase after sealing along with a relatively low glass-transition temperature and excellent adhesion to the metal interconnect surface suggest that the prepared sealant may be suitable for the sealing of SOFCs with an operation temperature of 800–850 °C.

4. Conclusions

The $CaO\text{-}SiO_2\text{-}B_2O_3$ sealant prepared in the study may be considered a promising glass-ceramic sealant for SOFCs with an operation temperature of about 800–850 °C. It has excellent adhesion to the surface of Crofer 22 APU steel frequently used as a material of bipolar plates in SOFCs. The prepared sealant has an acceptable CTE of 8.4×10^{-6} 1/K comparable to that of commercial sealants, although significantly lower than that of the most SOFC sealants reported in the literature.

SEM images of the sealant-interconnect interface demonstrate that reaction took place between the components' forming layer, providing strong adhesion at the interface. In addition, the sealant formed a dense structure with no cracks or pore clusters, which indicates that this sealant can be used to prepare airtight seals for SOFCs. An excellent adhesive layer was formed after sealing at 925 °C for 1 h. Usually, the sealing of SOFCs operating at 850 °C requires heating to at least 950 °C for a few hours. The ability of the studied sealant to adhere strongly at relatively mild sealing conditions is beneficial for the preservation of the electrodes structure during sealing. Short-term aging at an operation temperature of 850 °C for 100 h caused no visible change in the morphology of the interface, indicating that the formed reaction zone is relatively stable.

We would like to note, however, that the 100-h experiment gives only preliminary information about the stability of the sealant–steel interface and that of the sealant itself. Further experiments with longer exposures should be carried out. A study of the sealant stability in humid environments at operation temperatures will be of particular interest. Furhtermore, experiments studying long-term stability of SOFCs sealed with the proposed sealants should be conducted. These experiments will allow for assessing the effect of high boron oxide content on electrode poisoning and studying stability of dielectric and thermomechanical properties of the sealant.

Supplementary Materials: The following supporting information can be downloaded at: https://www.mdpi.com/article/10.3390/ceramics5040047/s1.

Author Contributions: Conceptualization, S.I.B. and D.V.M.; methodology A.O.Z. and E.A.A.; validation, A.O.Z. and D.V.M.; resources, E.A.A.; writing—original draft preparation, A.O.Z.; supervision, S.I.B.; project administration, S.I.B. and A.O.Z.; funding acquisition—S.I.B. and A.O.Z. All authors have read and agreed to the published version of the manuscript.

Funding: This research was partially funded by the Russian Science Foundation, project number 17-79-30071 П (microstructural and XRD study of the materials) and was partially funded by the Grant of the President of Russian Federation, project number MK-3060.2022.1.2 (preparation of the sealants and thermal tests). The APC was paid by the authors.

Institutional Review Board Statement: Not applicable.

Informed Consent Statement: Not applicable.

Data Availability Statement: Not applicable.

Conflicts of Interest: The authors declare no conflict of interest.

References

1. Kaur, G. *Intermediate Temperature Solid Oxide Fuel Cells: Electrolytes, Electrodes and Interconnects*; Elsevier Science: Amsterdam, The Netherlands, 2019.
2. Shao, Z.; Tadé, M.O. *Intermediate-Temperature Solid Oxide Fuel Cells*; Springer: Luxembourg, 2016.
3. Tietz, F. Thermal expansion of SOFC materials. *Ionics* **1999**, *5*, 129–139. [CrossRef]
4. Singh, K.; Walia, T. Review on silicate and borosilicate-based glass sealants and their interaction with components of solid oxide fuel cell. *Int. J. Energy Res.* **2021**, *45*, 20559–20582. [CrossRef]
5. Sohn, S.-B.; Choi, S.-Y.; Kim, G.-H.; Song, H.-S.; Kim, G.-D. Stable sealing glass for planar solid oxide fuel cell, *J. Non. Cryst. Solids* **2002**, *297*, 103–112. [CrossRef]
6. Fergus, J.W. Sealants for solid oxide fuel cells, *J. Power Sources* **2005**, *147*, 46–57. [CrossRef]
7. Ghosh, S.; Kundu, P.; Das Sharma, A.; Basu, R.N.; Maiti, H.S. Microstructure and property evaluation of barium aluminosilicate glass–ceramic sealant for anode-supported solid oxide fuel cell. *J. Eur. Ceram. Soc.* **2008**, *28*, 69–76. [CrossRef]
8. Meinhardt, K.D.; Kim, D.-S.; Chou, Y.-S.; Weil, K.S. Synthesis and properties of a barium aluminosilicate solid oxide fuel cell glass—Ceramic sealant. *J. Power Sources* **2008**, *182*, 188–196. [CrossRef]
9. Puig, J.; Ansart, F.; Lenormand, P.; Antoine, L.; Dailly, J. Sol-gel synthesis and characterization of barium (magnesium) aluminosilicate glass sealants for solid oxide fuel cells. *J. Non. Cryst. Solids* **2011**, *357*, 3490–3494. [CrossRef]
10. Kermani, P.S.; Ghatee, M.; Yazdani, A. Synthesis and Characterization of Barium Aluminosilicate Glass as the Sealant for Solid Oxide Fuel Cell Application. *Adv. Ceram. Prog.* **2020**, *6*, 25–30. [CrossRef]
11. Reddy, A.A.; Tulyaganov, D.U.; Pascual, M.J.; Kharton, V.V.; Tsipis, E.V.; Kolotygin, V.A.; Ferreira, J.M.F. Diopside–Ba disilicate glass–ceramic sealants for SOFCs: Enhanced adhesion and thermal stability by Sr for Ca substitution. *Int. J. Hydrogen Energy* **2013**, *38*, 3073–3086. [CrossRef]
12. Reddy, A.A.; Tulyaganov, D.U.; Goel, A.; Pascual, M.J.; Kharton, V.V.; Tsipis, E.V.; Ferreira, J.M.F. Diopside–Mg orthosilicate and diopside–Ba disilicate glass–ceramics for sealing applications in SOFC: Sintering and chemical interactions studies. *Int. J. Hydrog. Energy* **2012**, *37*, 12528–12539. [CrossRef]
13. Dai, B.; Zhu, H.; Zhou, H.; Xu, G.; Yue, Z. Sintering, crystallization and dielectric properties of CaO-B_2O_3-SiO_2 system glass ceramics. *J. Cent. South Univ.* **2012**, *19*, 2101–2106. [CrossRef]
14. Veron, E.; Garaga, M.N.; Pelloquin, D.; Cadars, S.; Suchomel, M.; Suard, E.; Massiot, D.; Montouillout, V.; Matzen, G.; Allix, M. Synthesis and structure determination of $CaSi_{1/3}B_{2/3}O_{8/3}$: A new calcium borosilicate. *Inorg. Chem.* **2013**, *52*, 4250–4258. [CrossRef] [PubMed]
15. Yan, T.; Zhang, W.; Mao, H.; Chen, X.; Bai, S. The effect of CaO/SiO_2 and B_2O_3 on the sintering contraction behaviors of CaO-B2O3-SiO2 glass-ceramics. *Int. J. Mod. Phys. B* **2019**, *33*, 1950070. [CrossRef]
16. Chang, C.-R.; Jean, J.-H. Crystallization kinetics and mechanism of low-dielectric, low-temperature, cofirable CaO-B_2O_3-SiO_2 glass-ceramics. *J. Am. Ceram. Soc.* **1999**, *82*, 1725–1732. [CrossRef]
17. Zhu, H.; Liu, M.; Zhou, H.; Li, L.; Lv, A. Study on properties of CaO-SiO_2-B_2O_3 system glass-ceramic. *Mater. Res. Bull.* **2007**, *42*, 1137–1144. [CrossRef]
18. Shao, H.; Wang, T.; Zhang, Q. Preparation and properties of CaO–SiO_2–B_2O_3 glass-ceramic at low temperature. *J. Alloys Compd.* **2009**, *484*, 2–5. [CrossRef]
19. Chiang, C.-C.; Wang, S.-F.; Wang, Y.-R.; Wei, W.-C.J. Densification and microwave dielectric properties of CaO–B_2O_3–SiO_2 system glass–ceramics. *Ceram. Int.* **2008**, *34*, 599–604. [CrossRef]
20. Rodríguez-López, S.; Haanappel, V.A.C.; Durán, A.; Muñoz, F.; Mather, G.C.; Pascual, M.J.; Gross-Barsnick, S.M. Glass–ceramic seals in the system MgO-BaO-B_2O_3-SiO_2 operating under simulated SOFC conditions. *Int. J. Hydrog. Energy* **2016**, *41*, 15335–15345. [CrossRef]
21. Zhang, T.; Fahrenholtz, W.G.; Reis, S.T.; Brow, R.K. Borate volatility from SOFC sealing glasses. *J. Am. Ceram. Soc.* **2008**, *91*, 2564–2569. [CrossRef]
22. Inoue, H.; Aoki, N.; Yasui, I. Molecular dynamics simulation of the structure of borate glasses. *J. Am. Ceram. Soc.* **1987**, *70*, 622–627. [CrossRef]
23. Henderson, G.S. A Si K-edge EXAFS/XANES study of sodium silicate glasses. *J. Non. Cryst. Solids* **1995**, *183*, 43–50. [CrossRef]
24. Mastelaro, V.R.; Zanotto, E.D.; Lequeux, N.; Cortès, R. Relationship between short-range order and ease of nucleation in $Na_2Ca_2Si_3O_9$, $CaSiO_3$ and $PbSiO_3$ glasses. *J. Non. Cryst. Solids* **2000**, *262*, 191–199. [CrossRef]
25. Swamy, V.; Dubrovinksy, L.S.; Tutti, F. High-temperature Raman spectra and thermal expansion of wollastonite. *J. Am. Ceram. Soc.* **1997**, *80*, 2237–2247. [CrossRef]
26. Klyuev, V.P.; Pevzner, B.Z. Thermal expansion and glass transition temperature of calcium borate and calcium aluminoborate glasses. *Glass Phys. Chem.* **2003**, *29*, 127–136. [CrossRef]
27. Aumento, F. Stability, lattice parameters, and thermal expansion of β-cristobalite. *Am. Miner.* **1966**, *51*, 1167–1176.
28. Shelby, J.E. Formation and properties of calcium aluminosilicate glasses. *J. Am. Ceram. Soc.* **1985**, *68*, 155–158. [CrossRef]

MDPI
St. Alban-Anlage 66
4052 Basel
Switzerland
www.mdpi.com

Ceramics Editorial Office
E-mail: ceramics@mdpi.com
www.mdpi.com/journal/ceramics

Disclaimer/Publisher's Note: The statements, opinions and data contained in all publications are solely those of the individual author(s) and contributor(s) and not of MDPI and/or the editor(s). MDPI and/or the editor(s) disclaim responsibility for any injury to people or property resulting from any ideas, methods, instructions or products referred to in the content.

www.ingramcontent.com/pod-product-compliance
Lightning Source LLC
LaVergne TN
LVHW070454100526
838202LV00014B/1721